Lecture Notes in Artificial Intelligence   4755

Edited by J. G. Carbonell and J. Siekmann

Subseries of Lecture Notes in Computer Science

Vincent Corruble   Masayuki Takeda
Einoshin Suzuki (Eds.)

Discovery Science

10th International Conference, DS 2007
Sendai, Japan, October 1-4, 2007
Proceedings

 Springer

Series Editors

Jaime G. Carbonell, Carnegie Mellon University, Pittsburgh, PA, USA
Jörg Siekmann, University of Saarland, Saarbrücken, Germany

Volume Editors

Vincent Corruble
Université Pierre et Marie Curie (Paris 6)
Laboratoire d'Informatique de Paris 6
104 avenue du Président Kennedy, 75016 Paris, France
E-mail: Vincent.Corruble@lip6.fr

Masayuki Takeda
Einoshin Suzuki
Kyushu University
Department of Informatics
744 Motooka, Nishi, Fukuoka 819-0395, Japan
E-mail: {takeda, suzuki}@i.kyushu-u.ac.jp

Library of Congress Control Number: 2007935939

CR Subject Classification (1998): I.2, H.2.8, H.3, J.1, J.2

LNCS Sublibrary: SL 7 – Artificial Intelligence

ISSN	0302-9743
ISBN-10	3-540-75487-3 Springer Berlin Heidelberg New York
ISBN-13	978-3-540-75487-9 Springer Berlin Heidelberg New York

Springer is a part of Springer Science+Business Media

springer.com

© Springer-Verlag Berlin Heidelberg 2007
Printed in Germany

Typesetting: Camera-ready by author, data conversion by Scientific Publishing Services, Chennai, India
Printed on acid-free paper   SPIN: 12170210   06/3180   5 4 3 2 1 0

Preface

This volume contains the papers presented at DS-2007: The Tenth International Conference on Discovery Science held in Sendai, Japan, October 1–4, 2007.

The main objective of the Discovery Science (DS) conference series is to provide an open forum for intensive discussions and the exchange of new ideas and information among researchers working in the area of automating scientific discovery or working on tools for supporting the human process of discovery in science. It has been a successful arrangement in the past to co-locate the DS conference with the International Conference on Algorithmic Learning Theory (ALT). This combination of ALT and DS allows for a comprehensive treatment of the whole range, from theoretical investigations to practical applications. Continuing this tradition, DS 2007 was co-located with the 18th ALT conference (ALT 2007). The proceedings of ALT 2007 were published as a twin volume 4754 of the LNCS series.

The International Steering Committee of the Discovery Science conference series provided important advice on a number of issues during the planning of Discovery Science 2007. The members of the Steering Committee are Einoshin Suzuki (Kyushu University, Chair), Achim G. Hoffmann (University of New South Wales, Vice Chair), Setsuo Arikawa (Kyushu University), Hiroshi Motoda (Osaka University), Masahiko Sato (Kyoto University), Satoru Miyano (University of Tokyo), Thomas Zeugmann (Hokkaido University), Ayumi Shinohara (Tohoku University), Alberto Apostolico (Geogia Institute of Technology and University of Padova), Massimo Melucci (University of Padova), Tobias Scheffer (Max Planck Institute for Computer Science), Ken Satoh (National Institute of Informatics), Nada Lavrac (Jozef Stefan Institute), Ljupco Todorovski (University of Ljubljana), and Hiroki Arimura (Hokkaido University).

In response to the call for papers 55 manuscripts were submitted. The Program Committee selected for publication 17 submissions as long papers and 10 submissions as regular papers. Each submission was reviewed by at least two members of the Program Committee, which consisted of international experts in the field. The selection was made after careful evaluation of each paper based on clarity, significance, technical quality, and originality, as well as relevance to the field of discovery science. This volume consists of three parts. The first part contains the papers/abstracts of the invited talks, the second part contains the accepted long papers, and the third part contains the accepted regular papers.

We are deeply indebted to the Program Committee members as well as their subreferees who played the critically important role of reviewing the submitted papers and contributing to the intense discussions which resulted in the selection of the papers published in this volume. Without their enormous effort, ensuring the high quality of the work presented at Discovery Science 2007 would not have been possible. Furthermore, we would like to thank all individuals and institutions who contributed to the success of the conference: the authors for submitting papers, the invited speakers for their acceptance of the invitation and their stimulating contributions to the conference, the Steering Committee, and the sponsors for their support. In particular, we acknowledge the generous financial support from the Air Force Office of Scientific Research (AFOSR), Asian Office of Aerospace Research and Development (AOARD)[1]; the Graduate School of Information Sciences (GSIS), Tohoku University for providing secretarial assistance and equipment; the Research Institute of Electrical Communication (RIEC), Tohoku University; New Horizons in Computing, MEXT Grant-in-Aid for Scientific Research on Priority Areas; and the Semi-Structured Data Mining Project, MEXT Grant-in-Aid for Specially Promoted Research.

July 2007   Vincent Corruble
                                                            Masayuki Takeda
                                                            Einoshin Suzuki

---

[1] AFOSR/AOARD support is not intended to express or imply endorsement by the U. S. Federal Government.

Conference Organization

Conference Chair

Ayumi Shinohara   Tohoku University, Japan

Program Committee

Vincent Corruble (Co-chair)	Université Pierre et Marie Curie, Paris, France
Masayuki Takeda (Co-chair)	Kyushu University, Japan
Jean-Francois Boulicaut	INSA Lyon, France
Will Bridewell	CSLI, Stanford, USA
Simon Colton	Imperial College London, UK
Antoine Cornuejols	Université Paris-Sud, France
Andreas Dress	Shanghai Institutes for Biological Sciences, China
Saso Dzeroski	Jozef Stefan Institute, Slovenia
Tapio Elomaa	Tampere University of Technology, Finland
Johannes Fuernkranz	Technical University of Darmstadt, Germany
Dragan Gamberger	Rudjer Boskovic Institute, Hungary
Ricard Gavalda	Technical University of Catalonia, Spain
Gunter Grieser	Technical University of Darmstadt, Germany
Fabrice Guillet	Ecole Polytechnique of the University of Nantes, France
Mohand-Said Hacid	Université Lyon 1, France
Udo Hahn	Jena University, Germany
Makoto Haraguchi	Hokkaido University, Japan
Tomoyuki Higuchi	The Institute of Statistical Mathematics, Japan
Kouichi Hirata	Kyushu Institute of Technology, Japan
Tu Bao Ho	Japan Advanced Institute of Science and Technology, Japan
Achim Hoffmann	University of New South Wales, Australia
Tamás Horváth	Fraunhofer Institute for Intelligent Analysis and Information Systems, Germany
Daisuke Ikeda	Kyushu University, Japan
Kentaro Inui	Nara Institute of Science and Technology, Japan
Szymon Jaroszewicz	National Institute of Telecommunications, Poland
Hisashi Kashima	IBM Research, Tokyo Research Laboratory, Japan
Kristian Kersting	Universitaet Freiburg, Germany
Ross King	University of Wales, UK
Andras Kocsor	University of Szeged, Hungary
Kevin Korb	Monash University, Australia
Stefan Kramer	Technical University of Munich, Germany
Nicolas Lachiche	Université de Strasbourg, France

Nada Lavrac	Jozef Stefan Institute, Slovenia
Aleksandar Lazarevic	United Technologies Research Center, USA
Jinyan Li	Institute for Infocomm Research, Singapore
Yuji Matsumoto	Nara Institute of Science and Technology, Japan
Michael May	Fraunhofer Institute for Autonomous Intelligent Systems, Germany
Taneli Mielikäinen	Nokia Research Center, Palo Alto, USA
Tetsuhiro Miyahara	Hiroshima City University, Japan
Dunja Mladenic	Jozef Stefan Institute, Slovenia
Yasuhiko Morimoto	Hiroshima University, Japan
Tsuyoshi Murata	Tokyo Institute of Technology, Japan
Bernhard Pfahringer	University of Waikato, New Zealand
Jan Rauch	University of Economics, Czech Republic
Kazumi Saito	NTT Communication Science Laboratories, Japan
Michèle Sébag	Université Paris-Sud, France
Rudy Setiono	National University of Singapore
Einoshin Suzuki	Kyushu University, Japan
Kai Ming Ting	Monash University, Australia
Ljupco Todorovski	University of Ljubljana, Slovenia
Luis Torgo	University of Porto, Portugal
Kuniaki Uehara	Kobe University, Japan
Takashi Washio	Osaka University, Japan
Gerhard Widmer	Johannes Kepler University, Austria
Akihiro Yamamoto	Kyoto University, Japan
Mohammed Zaki	Rensselaer Polytechnic Institute, USA
Filip Zelezny	Czech Technical University in Prague, Czech Republic
Jean-Daniel Zucker	Université Paris 13, France

Local Arrangements

Akira Ishino	Tohoku University, Japan

Subreferees

Vineet Chaoji	Saori Kawasaki	Saeed Salem
Jure Ferlež	Attila Kertesz-Farkas	Yacine Sam
Blaz Fortuna	Richard Kirkby	Kazuhiro Seki
Gemma Garriga	Jussi Kujala	Ivica Slavkov
Miha Grcar	Thanh-Phuong Nguyen	Jan-Nikolas Sulzmann
Hakim Hacid	Yoshiaki Okubo	Swee Chuan Tan
Corneliu Henegar	Benjarath Phoophakdee	Shyh Wei Teng
Sachio Hirokawa	Pedro Rodrigues	Yoshiaki Yasumura
Frederik Janssen	Ulrich Rückert	Monika Zakova

Table of Contents

Invited Papers

Challenge for Info-plosion 1
  Masaru Kitsuregawa

Machine Learning in Ecosystem Informatics 9
  Thomas G. Dietterich

Simple Algorithmic Principles of Discovery, Subjective Beauty, Selective
Attention, Curiosity & Creativity 26
  Jürgen Schmidhuber

A Theory of Similarity Functions for Learning and Clustering 39
  Avrim Blum

A Hilbert Space Embedding for Distributions 40
  Alex Smola, Arthur Gretton, Le Song, and Bernhard Schölkopf

Long Papers

Time and Space Efficient Discovery of Maximal Geometric Graphs 42
  Hiroki Arimura, Takeaki Uno, and Shinichi Shimozono

Iterative Reordering of Rules for Building Ensembles Without
Relearning 56
  Paulo J. Azevedo and Alípio M. Jorge

On Approximating Minimum Infrequent and Maximum Frequent
Sets 68
  Mario Boley

A Partially Dynamic Clustering Algorithm for Data Insertion and
Removal 78
  Haytham Elghazel, Hamamache Kheddouci,
  Véronique Deslandres, and Alain Dussauchoy

Positivism Against Constructivism: A Network Game to Learn
Epistemology 91
  Hélène Hagège, Christopher Dartnell, and Jean Sallantin

Learning Locally Weighted C4.4 for Class Probability Estimation 104
  Liangxiao Jiang, Harry Zhang, Dianhong Wang, and Zhihua Cai

User Preference Modeling from Positive Contents for Personalized
Recommendation   116
  Heung-Nam Kim, Inay Ha, Jin-Guk Jung, and Geun-Sik Jo

Reducing Trials by Thinning-Out in Skill Discovery   127
  *Hayato Kobayashi, Kohei Hatano, Akira Ishino, and
  Ayumi Shinohara*

A Theoretical Study on Variable Ordering of Zero-Suppressed BDDs
for Representing Frequent Itemsets   139
  Shin-ichi Minato

Fast NML Computation for Naive Bayes Models   151
  Tommi Mononen and Petri Myllymäki

Unsupervised Spam Detection Based on String Alienness Measures   161
  *Kazuyuki Narisawa, Hideo Bannai, Kohei Hatano, and
  Masayuki Takeda*

A Consequence Finding Approach for Full Clausal Abduction   173
  Oliver Ray and Katsumi Inoue

Literature-Based Discovery by an Enhanced Information
Retrieval Model   185
  Kazuhiro Seki and Javed Mostafa

Discovering Mentorship Information from Author Collaboration
Networks   197
  *V. Suresh, Narayanan Raghupathy, B. Shekar, and
  C.E. Veni Madhavan*

Active Contours as Knowledge Discovery Methods   209
  Arkadiusz Tomczyk, Piotr S. Szczepaniak, and Michal Pryczek

An Efficient Polynomial Delay Algorithm for Pseudo Frequent Itemset
Mining   219
  Takeaki Uno and Hiroki Arimura

Discovering Implicit Feedbacks from Search Engine Log Files   231
  Ashok Veilumuthu and Parthasarathy Ramachandran

Regular Papers

Pharmacophore Knowledge Refinement Method in the Chemical
Structure Space   243
  Satoshi Fujishima, Yoshimasa Takahashi, and Takashi Okada

An Attempt to Rebuild C. Bernard's Scientific Steps   248
  Jean-Gabriel Ganascia and Bassel Habib

Semantic Annotation of Data Tables Using a Domain Ontology   253
    Gaëlle Hignette, Patrice Buche, Juliette Dibie-Barthélemy, and
    Ollivier Haemmerlé

Model Selection and Estimation Via Subjective User Preferences   259
    Jaakko Hollmén

Detecting Concept Drift Using Statistical Testing   264
    Kyosuke Nishida and Koichiro Yamauchi

Towards Future Technology Projection: A Method for Extracting
Capability Phrases from Documents   270
    Risa Nishiyama, Hironori Takeuchi, and Hideo Watanabe

Efficient Incremental Mining of Top-K Frequent Closed Itemsets   275
    Andrea Pietracaprina and Fabio Vandin

An Intentional Kernel Function for RNA Classification   281
    Hiroshi Sankoh, Koichiro Doi, and Akihiro Yamamoto

Mining Subtrees with Frequent Occurrence of Similar Subtrees   286
    Hisashi Tosaka, Atsuyoshi Nakamura, and Mineichi Kudo

Semantic Based Real-Time Clustering for PubMed Literatures   291
    Ruey-Ling Yeh, Ching Liu, Ben-Chang Shia, I-Jen Chiang,
    Wen-Wen Yang, and Hsiang-Chun Tsai

Author Index   297

Challenge for Info-plosion

Masaru Kitsuregawa

Institute of Industrial Science, The University of Tokyo
4-6-1 Komaba, Meguro-ku, Tokyo 153-8505, Japan
kitsure@tkl.iis.u-tokyo.ac.jp

Abstract. Information created by people has increased rapidly since the year 2000, and now we are in a time which we could call the "information-explosion era." The project "Cyber Infrastructure for the Information-explosion Era" is a six-year project from 2005 to 2010 supported by Grant-in-Aid for Scientific Research on Priority Areas from the Ministry of Education, Culture, Sports, Science and Technology (MEXT) of Japan. The project aims to establish the following fundamental technologies in this information-explosion era: novel technologies for efficient and trustable information retrieval from explosively growing and heterogeneous information resources; stable, secure, and scalable information systems for managing rapid information growth; and information utilization by harmonized human-system interaction. It also aims to design a social system that cooperates with these technologies. Moreover, it maintains the synergy of cutting-edge technologies in informatics.

1 New IT Infrastructure for the Information Explosion Era

The volume of information generated by mankind has increased exponentially, i.e., "exploded" since 2000. The purpose of our research project, "Cyber Infrastructure for the Info-plosion Era" in the Ministry of Education, Culture, Sports, Science and Technology (MEXT) Grand-in-Aid for Scientific Research on Priority Area, is to build advanced IT infrastructure technologies for this information explosion era [1]. According to the research by the University of California at Berkeley, the volume of information created by human is explosively increasing [2, 3]. Huge volume of data is also created by sensors and machines. We have considered that the most important theme for researchers in the field of computer science is the research on new IT infrastructure for the Information-explosion Era.

The project has three major research components (Research Groups) to achieve this goal: build technologies to search for needed information efficiently, without bias and without being at risk from the rapidly growing volume of information (A01); build new and sustainable technologies that can operate large-scale information systems managing enormous amounts of information safely and securely (A02), and build human-friendly technologies to enable flexible dialogue between men and machines and enable everyone to utilize information (A03).

V. Corruble, M. Takeda, and E. Suzuki (Eds.): DS 2007, LNAI 4755, pp. 1–8, 2007.

Underlying these three components is the research into new social systems that can facilitate the use of advanced, information-based IT services (B01). In addition, Large-scale Info-plosion Platform (LIP) is implemented as the shared platforms used by all Research Groups. The project is interdisciplinary in its structure, bringing together advanced research methods in information-related areas. (Project leader: Masaru Kitsuregawa)

2   Infrastructure for Information Management, Fusion and Utilization in the Information Explosion Era (A01)

This Group focuses on the shortfalls of present internet searches and looks into new search methods, including better ranking systems (where minority opinions are not overlooked), interactive searches, reliability assessments and time-space searches. Currently, for knowledge workers, about 30% of their time on intellectual activities is spent just for retrieval information [4]. The Group will attempt to create a system of search platforms to search a massive volume of web page contents.

Another important issue for information retrieval is the dangers associated with information rankings. Search engines are extensively used in the web world. When a general word is given, it may hit millions of pages, and only about 10

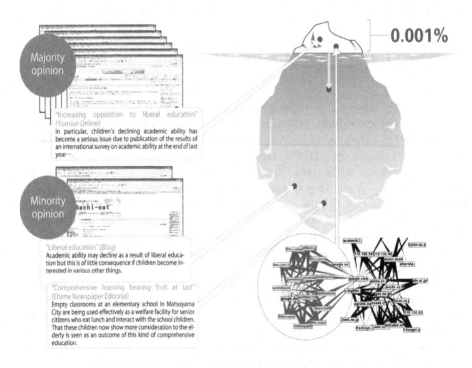

Fig. 1. Dangers Associated with Information Rankings

probable candidates are listed as top rankers on the first page of the search result. Now, is this ranking really reasonable? Who guarantee the correctness of it? Actually, this is controlled by just a private company. There should be a possibility that the ranking is sold and bought. In addition, it is possible to control the ranking deliberately. We have found, as shown in the Fig.1, that the red island of links suddenly appear on February 2004, while only the black island of links is found in 2003. This is an example of a trick to raise their own rank by linking them to a well-known site.

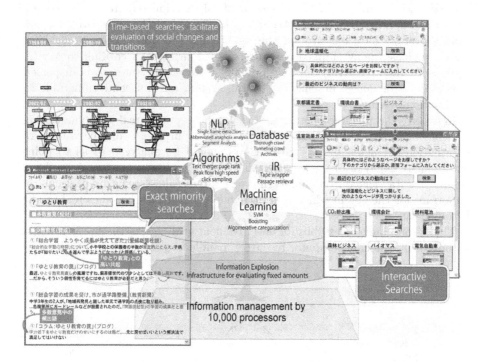

Fig. 2. Next generation searches

Also, in the current ranking system, majority opinions are highlighted while minority opinions are buried. For example, when "Yutori Kyouiku" (liberal education) is searched, majority opinions are found easily like "increasing opposition to liberal education". Although there are minority opinions such as "comprehensive learning bearing fruit at last", they could be completely neglected.

Various technologies, including information searching, natural language processing, machine learning, artificial intelligence (AI) and database technology, are integrated into this system to enable quantitative analysis to be performed (Fig.2). We hope to enable a remarkable level of interdisciplinary synergy among different fields. It is better for human to understand by presenting relational information also than information itself. Several innovative researches are expected in this area, for example, presenting comparative information [5] and information on time sequence [6]. Research on the methods of management, integration and processing of "exploding" real-world information, including cyber information and information obtained from remote sensors, will also be conducted. (Group leader: Masaru Kitsuregawa, University of Tokyo)

3   Infrastructure for Information-Explosion-Proof IT Systems (A02)

The exponential increase in the amount of information requires far larger IT systems to handle the volume. According to [7], Google has now over 450,000 servers over 25 locations around the world. The ratio of computers used for search engines among shipped computers is 5% in total, according to MSRA (Microsoft Research Asia) Summit in 2006. Data on the Internet is dispersed over millions of nodes, making the overall system unstable and vulnerable to information overload.

In order to keep stable operation of such huge systems, real-time monitoring of behavior of software is inevitable. For such a purpose, explosive volume of information extracted by software sensors should be analyzed so as to point out anomaly behavior of the system and stabilize it. Researches on mining huge volume of data yielded from monitoring very-large-scale systems are particularly important for the age in which a nation-wide cyber attack becomes reality like Estonian case.

This Group aims to establish a new "resilient grid" infrastructure which can automatically allocate computer resources, handle large-scale system faults over

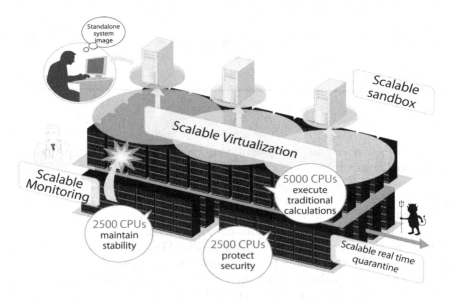

Fig. 3. Infrastructure for large scale system

the network without human intervention and without the modern-day concerns of security breaches and intrusion (Fig.3). This is partially based on Autonomic Computing [8] and interesting research works performed by UC Berkeley and Stanford University [9]. The resilient grid will allow a high-performance virtual computing environment to be configured autonomously, on which applications can be deployed safely and securely. (Group leader: Satoshi Matsuoka, Tokyo Institute of Technology)

4   Infrastructure for Human Communication in the Information Explosion Era (A03)

The information explosion has two aspects: qualitative (volume) and quantitative (complexity). This Group proposes studies of the advancement of human communication to address the issues related to complexity. The underlying concept is a mutually adaptable multi-modal interaction that can fill the communication gaps between people and information systems. This is the key to overcoming the complexity resulting from highly functional and multi-functional information systems, and establishing a secure and user-friendly interactive environment (Fig.4).

Searching information from explosively huge size of information space still requires advanced skills, since existing tools for such a purpose is not necessarily easy to use for naive users. Human-friendly interfaces as well as communications

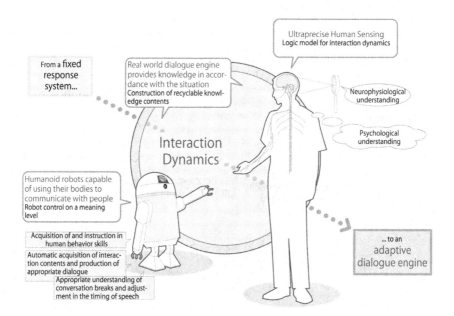

Fig. 4. Infrastructure for human

with a robot in the near future are extremely important, which is covered by this Group. Monitoring human behavior also produces explosive volume of data, which plays essential role to analyze nonverbal communication dynamics.

Recently, new ideas have been demonstrated such as scientific formulation of "knacks", which establishes a firm theoretical ground for connecting highly complex physical dynamics and symbolic information of skillful actions [10]. The effect of the theory is demonstrated by the world's first experiment on highly dynamic whole body motion of a humanoid robot. The ultimate goal of the research is to establish a new framework fusing the robot and information technology that solves the above stated bidirectional connection problem. (Group leader: Takashi Matsuyama, Kyoto University)

5   Governing the Development of a Knowledge-Based Society in the Information Explosion Era (B01)

Advances in technology often race far ahead of consideration of how people will actually use the accompanying new benefits. Engineers focus on the technology rather than the laws and regulations that will be required for the technologies they develop. Support for engineers and keeping them aware of the legal and social implications of technological advances is indispensable. Practical experiments require cooperation and coordination with a number of partners in the

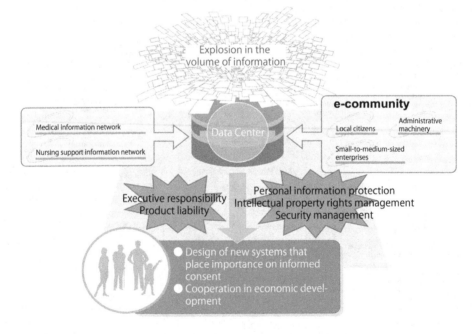

Fig. 5. Promotion of Society-based Research

real world. Advances in IT have both a "bright" side and a "dark" side. Social problems arising from the abuse of the dark side need to be addressed.

The fourth feature of our project is promotion of society-based research. For investigating IT infrastructure for the Information-explosion era, society-based research is inevitable, as shown in Fig.5. In order to realize universal social infrastructure including medical information network and nursing support information network, various kind of problems should be resolved, for example, product liability, personal information protection, intellectual property rights management, and security management.

In such a situation, cooperation in economic development between technology providers and receivers is required. Design of new systems that place importance on informed consent is important. We are making researches into information governance that cover the explosion in the volume of information. This Group focuses on social issues and will study the governance of information technology development, its relationship with the legislative process, and its implications for new legal and social systems [11]. (Group leader: Osamu Sudoh, University of Tokyo)

6   The Aims and Roles of the Large-Scale Info-plosion Platform in the Research Project

One of the most remarkable distinctions of this research project from other national scientific projects is that we will inject a quarter of the total research fund into the Large-scale Info-plosion Platform (LIP). This fund will be spent to implement shared platforms which informatics researchers intriguingly and collaboratively construct for innovative researches. These platforms enable each research team to conduct novel and original research activities that cannot be realized only within a Group. The resources integrated in the platforms will be expected to be also used by researchers outside the research project.

7   Concluding Remarks

In this article, an overview of our Info-plosion project is introduced. The relation of all Research Groups is shown in Fig.6. This research project is approved in 2005 and started full part since April 2006. The budget of the project is about 600 million Yen/yr and over 200 researchers from various fields are participating currently. They are consisting of Planning Researches and Proposal Researches. Over 300 research themes are applied, and more than 60 Proposal Researches are accepted and started their innovative research works. Info-plosion project is in coalition with the Consortium for New Project on "Intellectual Access to Information" (jouhou-daikoukai) sponsored by the Ministry of Economy, Trade and Industry (METI) started in 2006 [12].

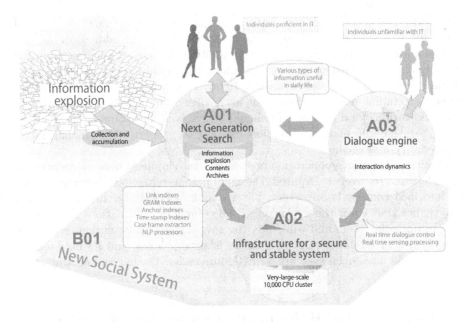

Fig. 6. Overall image of research project

References

[1] http://itkaken.ex.nii.ac.jp/i-explosion/
[2] http://www2.sims.berkeley.edu/research/projects/how-much-info-2003/
[3] http://www.horison.com/
[4] http://www.delphiweb.com/knowledgebase/documents/upload/pdf/1802.pdf
[5] Nadamoto, A., Tanaka, K.: A Comparative Web Browser (CWB) for Browsing and Comparing Web Pages. In: Proc of the 12th Int. World Wide Web Conf. (WWW2003) (2003)
[6] Toyoda, M., Kitsuregawa, M.: What's Really New on the Web? Identifying New Pages from a Series of Unstable Web Snapshots. In: Proc of the 15th Int. World Wide Web Conf. (WWW2006), pp. 233–241 (2006)
[7] http://www.nytimes.com/2006/06/14/technology/14search.html
?pagewanted=2&ei=5090&en=d96a72b3c5f91c47&ex=1307937600
[8] http://www.research.ibm.com/autonomic/
[9] Bodk, P., Friedman, G., et al.: Combining Visualization and Statistical Analysis to Improve Operator Confidence and Efficiency for Failure Detection and Localization. In: Proc. 2nd International Confidence on Autonomic Computing (ICAC'05) (2005)
[10] Kuniyoshi, Y., Ohmura, Y., Terada, K., Nagakubo, A., Eitoku, S., Yamamoto, T.: Embodied Basis of Invariant Features in Execution and Perception of Whole Body Dynamic Actions — Knacks and Focuses of Roll-and-Rise Motion. Robotics and Autonomous Systems 48(4), 189–201 (2004)
[11] Sudoh, O. (ed.): Digital Economy and Social Design, pp. 1–236. Springer, Heidelberg (2005)
[12] http://www.jyouhoudaikoukai-consortium.jp/

Machine Learning in Ecosystem Informatics

Thomas G. Dietterich

Oregon State University, Corvallis, Oregon, USA
tgd@eecs.oregonstate.edu,
http://web.engr.oregonstate.edu/~tgd

Abstract. The emerging field of Ecosystem Informatics applies methods from computer science and mathematics to address fundamental and applied problems in the ecosystem sciences. The ecosystem sciences are in the midst of a revolution driven by a combination of emerging technologies for improved sensing and the critical need for better science to help manage global climate change. This paper describes several initiatives at Oregon State University in ecosystem informatics. At the level of sensor technologies, this paper describes two projects: (a) wireless, battery-free sensor networks for forests and (b) rapid throughput automated arthropod population counting. At the level of data preparation and data cleaning, this paper describes the application of linear gaussian dynamic Bayesian networks to automated anomaly detection in temperature data streams. Finally, the paper describes two educational activities: (a) a summer institute in ecosystem informatics and (b) an interdisciplinary Ph.D. program in Ecosystem Informatics for mathematics, computer science, and the ecosystem sciences.

1 Introduction

The late Jim Gray (Gray & Szalay, 2003) describes four general approaches to scientific research:

- Observational science, in which scientists make direct observations,
- Analytical science, in which scientists develop analytical models capable of making predictions,
- Computational science, in which scientists employ massive computing power to study the behavior of analytical models and to make predictions at much wider scales of time and space, and
- Data exploration science, in which massive amounts of data are automatically collected from sensors, and scientists employ data mining and statistical learning methods to build models and test hypotheses.

The ecosystem sciences currently employ analytical and computational methods as illustrated, for example, by the extensive work on coupled ocean-atmosphere climate models. However, with the exception of data collected via remote sensing, the ecosystem sciences do not yet have large networks of sensors that automatically collect massive data sets.

Three steps are required to enable ecological research to become a data exploration science. First, sensors that can measure ecologically-important quantities

V. Corruble, M. Takeda, and E. Suzuki (Eds.): DS 2007, LNAI 4755, pp. 9–25, 2007.

must be developed and deployed in sensor networks. Second, methods for automatically managing and cleaning the resulting data must be developed. Third, data mining and machine learning algorithms must be applied to generate, refine, and test ecological hypotheses.

This paper briefly reviews work at Oregon State University on each of these three steps. Oregon State University has a long history of excellence in the ecosystem sciences. It includes world-leading research groups in forestry, oceanography, and atmospheric sciences, as well as strong teams in machine learning, data mining, and ecological engineering. The campus leadership has made a significant investment in new faculty positions in mathematics, computer science, and forestry with the goal of developing strong interdisciplinary education and research programs in ecosystem informatics.

This paper is organized as follows. The paper begins with a discussion of two sensor development projects, one in wireless sensor networks for plant physiology and the other on computer vision for automated population counting. Then the paper discusses work on automated data cleaning. Finally, the paper briefly describes two educational initiatives aimed at preparing computer scientists, mathematicians, and ecologists to work together in interdisciplinary teams to address the important scientific problems confronting the ecosystem sciences.

2   New Sensor Technologies for Ecology

The study of complex ecosystems is limited by the kinds of data that can be reliably and feasibly collected. Two recent US National Science Board studies (NSB, 2000; NSB, 2002) emphasize the importance of developing new instrumentation technologies for ecological research. At Oregon State, we are pursuing several projects include the following two: (a) wireless, battery-free temperature sensors for forest physiology and (b) computer vision for rapid throughput arthropod population counting.

2.1   Battery-Free Forest Sensors

Forests play an important role in absorbing carbon dioxide and producing oxygen. A central challenge in the study of forest physiology is to understand the exchange of these gasses between the forest and the atmosphere. Existing models of this exchange only capture vertical interactions, under the simplifying assumption that the forest can be modeled as a planar array of trees. But real forests are often on mountain slopes where breezes tend to move up the slope during the day and down the slope at night. Hence, to obtain a more realistic understanding of forest-atmosphere gas exchange, we need to measure and model these lateral winds as well.

Many research groups around the world have developed wireless sensor networks that rely on on-board batteries to provide electric power (Kahn et al., 1999; Elson & Estrin, 2004). Unfortunately, these batteries typically contain toxic chemicals, which means that these sensors must be retrieved after the batteries have run down. This can be impractical in ecologically-sensitive and

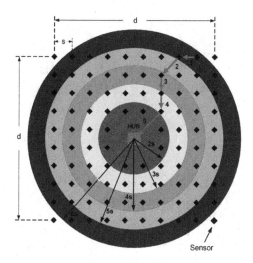

Fig. 1. Spatial layout of battery-free sensor network with powered base station at center

inaccessible locations, and it also limits the period of time that the sensor network can be collecting data.

This was the motivation for a team consisting of Barbara Bond (Forest Science), and Terri Fiez, Karti Mayaram, Huaping Liu, and Thinh Nguyen (Electrical Engineering), and Mike Unsworth (Atmospheric Sciences) to develop battery-free sensors for use in the forests of the Pacific Northwest.

The basic design concept is to have a base station that is connected to standard electric power. This base station broadcasts radio frequency energy across the RF spectrum. This energy is harvested by ultra-low power sensor units. They store the energy in a capacitor and use it to make sensor readings and to receive data from and transmit data to other sensors. The data is relayed from the peripheral sensors to the central base station in a series of hops (see Figure 1).

The development of such passively-powered sensor nodes requires that all components of the sensor employ ultra-low power methods. The initial design includes a temperature sensor, an RF energy harvesting circuit, a binary frequency shift keying (BFSK) receiver, and a BFSK transmitter. The receiver and transmitter share a single antenna. Figure 2 shows the layout of the current prototype sensor.

Note that this prototype contains only a temperature sensor. While it will be easy to add other sensors to the chip, it turns out that by measuring temperatures, it is possible to infer the lateral winds. So this initial sensor chip will be sufficient to address the forest physiology question that motivated the project.

The ultra-low power temperature sensor measures the outside temperature from -10 to 40 degrees Celsius with an accuracy of ±0.5 degrees. It is able to achieve this accuracy while consuming only 1nJ per measurement, which is a factor of 85 less energy than is required by state-of-the-art sensors.

The energy harvesting circuit employs a 36-stage "floating gate" design (Le et al., 2006). It is able to harvest energy up to a distance of 15 meters, which is

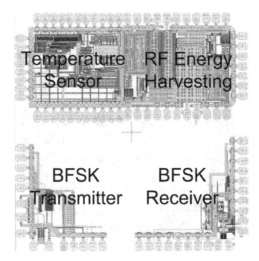

Fig. 2. Layout of prototype battery-free temperature sensor chip

substantially better than the best previously-reported method which only works out to 4.5 meters. Hence, the maximum size of the sensor network region will be approximately 30 meters in diameter.

The transceiver consumes the largest amount of power in the sensor. A low power super-regenerative design based on binary frequency shift keying is employed in the prototype. Experiments in the Oregon coastal mountains with a separate test platform show that even when the sensors are only 10cm above the ground, this design should be able to transmit 10 meters with a raw bit error rate of 10^{-4} (see Figure 3). By applying error-correcting coding, the effective bit error rate will be much lower.

The first version of the chip will be fabricated in summer 2007, which will make it possible to test the complete sensor network design, including energy harvesting and communications protocols.

2.2   Rapid-Throughput Arthropod Population Counting

Two central questions in ecology are (a) to explain the observed distribution of species around the world and (b) to understand the role of biodiversity in maintaining the health and stability of ecosystems. The key data necessary to study these questions consists of counting the number of individuals belonging to each species at many different sites.

There are many thousands of species of arthropods. They populate many different habitats including freshwater streams, lakes, soils, and the oceans. They are also generally easy to collect. Despite all of these advantages, the great drawback of using arthropod population data is the tedious and time-consuming process of manually classifying each specimen to the genus and species level. At

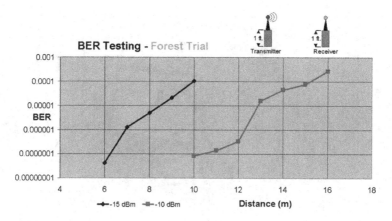

Fig. 3. Bit error rate experiments at two different transmission power levels

Oregon State, a team consisting of Tom Dietterich, Eric Mortensen (Computer Science), Robert Paasch (Mechanical Engineering), Andrew Moldenke (Botany and Plant Pathology), David Lytle (Zoology) along with Linda Shapiro (Computer Science) from the University of Washington is developing a rapid-throughput system that combines robotic manipulation with computer vision to automatically classify and count arthropod specimens.

The first application project has been to classify stonefly larvae that live in the substrate of freshwater streams. Stoneflies are an excellent indicator of stream health. They are highly sensitive to pollution, and, because they live in the stream, they provide a more reliable measurement than a single-point-in-time chemical assay. Figure 4 shows the mechanical apparatus that we have developed. In the left image, each individual stonefly specimen is dropped into the plastic reservoir in the lower right part of the image. This reservoir (and the rest of the apparatus) contains alcohol, and the specimen is manipulated via pumps and alcohol jets. The blue part of the apparatus contains a diamond-shaped channel that is covered with transparent plastic. The specimen is pumped into this tube. Infrared detectors (not shown, but located at the two vertical posts and the circular mirror) detect the specimen, cut off the main pump, and turn on a side jet (see the small metal tube emerging from the left side of the blue base). This side jet "captures" the specimen within the field of the microscope (see image (b)). When the side jet is turned off, the specimen falls to the bottom of the channel and a photo is taken. Then the side jet is turned on, which causes the specimen to rotate rapidly. The jet is again turned off, and another picture taken. This continues until a good image of the back (dorsal) side of the specimen is obtained. The pictures are taken through a mirror apparatus (upper right of (a)), which allows us to capture two views of the specimen with each photo of the camera. This increases the likelihood of capturing a good dorsal view.

Figure 5 shows example images captured by the apparatus for four different taxa. Notice the large variation in size, pose, and coloration.

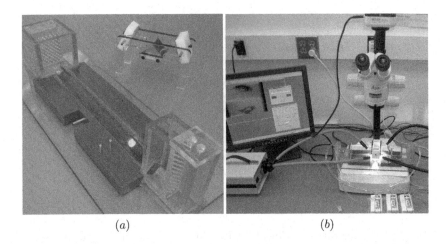

(a)   (b)

Fig. 4. (a) Prototype mirror and transportation apparatus. (b) Entire stonefly transportation and imaging setup (with microscope and attached digital camera, light boxes, and computer controlled pumps for transporting and rotating the specimen.

(a)   (b)   (c)   (d)

Fig. 5. Example images of different stonefly larvae species. (a) *Calineuria*, (b) *Doroneuria*, (c) *Hesperoperla* and (d) *Yoraperla*.

The next step in the process is to apply a learned visual classifier to classify the dorsal views into the class. To do this, we employ a variation on the bag-of-interest-points approach to generic object recognition. This approach consists of the following steps:

1. Apply region detectors to the image to find "interesting" regions. We apply three different detectors: The Hessian Affine detector (Mikolajczyk & Schmid, 2004), the Kadir Entropy detector (Kadir & Brady, 2001), and our own PCBR detector (Deng et al., 2007). Figure 6 shows examples of the detected regions.
2. Represent each detected region as a 128-element SIFT vector (Lowe, 2004). The SIFT descriptor vector is a set of histograms of the local intensity

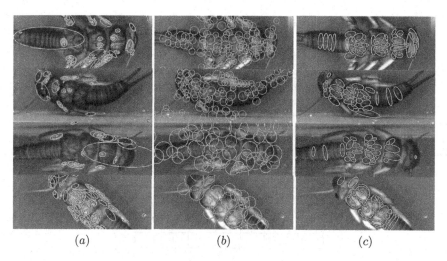

$$(a) \qquad\qquad (b) \qquad\qquad (c)$$

Fig. 6. Visual Comparison of the regions output by the three detectors on three *Calineuria* specimens. (a) Hessian-affine, (b) Kadir Entropy, (c) PCBR.

gradient direction. Although SIFT was originally developed for object tracking, it has been found to work well for object recognition.

3. Compute a feature vector from the set of detected SIFT vectors. Let $D : R^{128} \mapsto \{1, \dots, N_D\}$ be a visual dictionary that maps each SIFT vector into an integer between 1 and N_D (N_D varied from 65 to 90 in our experiments). The visual dictionary is constructed by fitting a gaussian mixture model with N_D components to the SIFT vectors observed on a separate "clustering" data set. The function D takes a SIFT vector and maps it to the gaussian mixture component most likely to have generated that vector.

   Given the visual dictionary, the set of SIFT vectors computed from the image is converted into a feature vector $\mathbf{x}$ such that $\mathbf{x}[i]$ is the number of SIFT vectors $\mathbf{v}$ in the image such that $D(\mathbf{v}) = i$. In effect, $\mathbf{x}$ is a histogram where the ith element counts the number of SIFT vectors that matched the ith dictionary entry.

4. Apply a learned classifier to map $\mathbf{x}$ to one of the K possible taxa.

In our work, we learn a separate dictionary $D_{s,d}$ for each species s and each detector d. Consequently, we compute a separate histogram vector $\mathbf{x}_{s,d}$ for each dictionary. In our case, we have 3 detectors and 4 species, so we compute 12 dictionaries and 12 histograms. We then concatenate all of these feature vectors to obtain one very long feature vector which is processed by the learned classifier.

To train the system, our entomology collaborators (Lytle and Moldenke) collected and independently classified 263 stonefly specimens. These were then photographed resulting in the data summarized in Table 1. These data were then randomly partitioned into 3 folds (stratifying by specimen and by class), and a 3-fold cross-validation was performed. In each iteration, one fold of the data was employed to learn the visual dictionaries, one fold to train the classifier, and one fold to evaluate the results.

Table 1. Specimens and images employed in the study

Taxon	Specimens	Images
Calineuria	85	400
Doroneuria	91	463
Hesperoperla	58	253
Yoraperla	29	124

Table 2. Confusion matrix of the combined Kadir, Hessian-affine and PCBR detectors

predicted as ⇒	Cal.	Dor.	Hes.	Yor.
Calineuria	315	79	6	0
Doroneuria	80	381	2	0
Hesperoperla	24	22	203	4
Yoraperla	1	0	0	123

We employed bagged logistic model trees as implemented in the WEKA system (Landwehr et al., 2005) as the classifier (with 20 iterations of bagging). Table 2 shows the results. Overall, the classifier correctly classifies 82.4% of the images (with a 95% confidence interval of $\pm 2.1\%$). The distinction between *Calineuria* and *Doroneuria* is the most challenging. Separate experiments have shown that our accuracy on this 2-class problem is statistically indistinguishable from human performance, when humans are given the same whole-specimen images that our program observes.

We have recently extended this work to apply to 9 stonefly taxa, with an overall accuracy of 85%. This level of accuracy is more than sufficient for use in routine biomonitoring tasks. Consequently, we are planning a trial with standard field samples later this year. More details on this work can be found in Larios et al. (Larios et al., In Press).

We have now begun working on a new apparatus and algorithms for recognizing and classifying soil mesofauna and freshwater zooplankton. We anticipate that this apparatus will have a broader range of applications in ecological studies of biodiversity.

3   Automated Data Cleaning for Sensor Networks

As sensors collect data, various things can go wrong. First, the sensors can fail. Second, the data recording process (e.g., the network connection) can fail. Third, the semantic connection between the sensor and the environment can be broken. For example, a thermometer measuring stream water temperature will change to measuring air temperature if the water level falls too low.

To catch these errors, we need methods for automated data cleaning. These methods can be applied to automatically flag data values so that scientists using this data can take appropriate steps to avoid propagating errors into their model building and testing.

Ethan Dereszynski, a doctoral student at Oregon State, has developed an automated data cleaning system for identifying anomalies in temperature data collected at the H. J. Andrews Experimental Forest, which is one of the NSF-funded Long Term Ecological Research (LTER) sites. In this forest, there are three major meteorological stations at three different altitudes. At each station, there is a tower with four temperature sensors which measure and report temperature every 15 minutes. Hence, for this simple sensor network, there are 12 parallel data streams, one for each thermometer.

This data is collected and posted on a web site in raw form. At regular intervals, the LTER staff manually inspect the data to find and remove errors. They then post a clean version of the data, which is the version intended for use by scientists around the world. Our goal is to replace this human data cleaning with an automated process. But a nice side effect of the existing practice is that we have several years of supervised training data for constructing and testing data cleaning methods.

We have adopted a density estimation approach to anomaly detection. Our goal is to develop a model that can evaluate the probability of a new sensor reading given past sensor readings. If the new reading is highly unlikely, it is marked as an anomaly, and it is not used in making subsequent probability estimates. In our work to date, we have focused only on anomaly detection for a single sensor data stream. In future work, we will study simultaneous anomaly detection over the 12 parallel data streams.

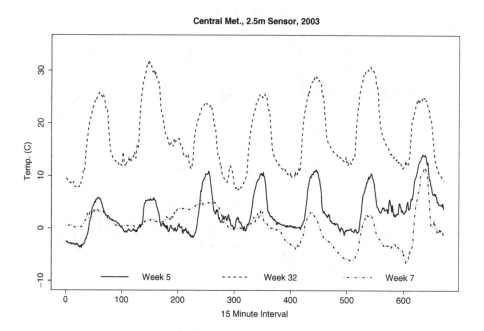

Fig. 7. Seasonal, Diurnal, and Weather effects

Figure 7 shows typical temperature readings as a function of time for the 2.5m sensor at the Central Meteorological station. Observe that there are seasonal effects (it is colder in the winter and warmer in the summer), diurnal (daily) effects (colder at night; warmer in the day), and weather system effects. The weather system effects are the hardest to model. They generally cause the temperature to be systematically warmer or colder than normal over a period of 3-10 consecutive days.

Anomalies can be divided into easy, medium, and hard cases. The easy cases are things such as the failure of the connection between the sensor and the data logger. If the data logger loses contact with the sensor, it records a fixed value of −53.3. Similarly, if the data logger receives an input voltage outside the legal bounds, it records a fixed value of −6999. Obviously, these anomalous values are easy to detect.

Medium anomalies can be detected from a single sensor, but they require more subtle analysis. Figure 8 (top) shows a case in which the heat shield on a sensor has been damaged. This causes the sensor to warm up too quickly, measure incorrectly high readings in the hottest part of the day, and then cool down too quickly in the evening. Figure 8(bottom) shows what happens when snow buries the 1.5m and 2.5m sensors. The 1.5m sensor records a steady value of zero (the freezing point), while the 2.5m sensor's readings are damped toward zero. As the snow melts, first the 2.5m sensor recovers and then the 1.5m sensor recovers.

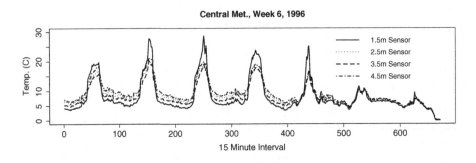

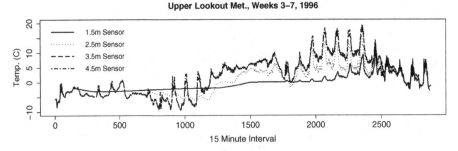

Fig. 8. Top: Broken Sun Shield, Bottom: 1.5m Sensor buried under snowpack, 2.5m Sensor dampened

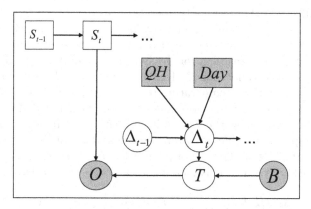

Fig. 9. Dynamic Bayesian network for anomaly detection. Square nodes denote discrete variables; circular nodes denote normally-distributed variables. Grey nodes are observed in the data.

Hard anomalies require the analysis of multiple data streams. One of the most interesting anomalies arose when the cables for two of the sensors were interchanged during maintenance. Normally, the 1.5m, 2.5m, 3.5m, and 4.5m sensors exhibit a monotonic temperature ordering. At night, the 1.5m sensor is warmest, because it is closest to the warm soil. In the day time, the 4.5m sensor is warmest and the 1.5m sensor is coldest. To detect the cable-swap anomaly, we need to model the joint distribution of the four sensors and detect that this monotonic relationship is violated. As indicated above, this will be a topic of our future work.

Figure 9 shows our dynamic Bayesian network for anomaly detection. The heart of the model consists of three variables O (the observed temperature), T (the predicted temperature), and S_t (the state of the sensor). The state of the sensor is quantized into four levels ("very good", "good", "bad", and "very bad"). If the sensor is "very good", then O should be equal to T with some slight variation. This is captured by asserting that

$$P(O|T) = Norm(T, 1.0).$$

That is, the mean value of O is T with a standard deviation of 1.0. If S_t is "good", then the standard deviation is 5.0. If S_t is "bad", the standard deviation is 10.0, and if S_t is "very bad", the standard deviation is 100,000 (i.e., effectively infinite).

In practice, we observe O and, based on previously-observed values, compute the probability distribution of T. Then the most likely value of S_t is determined by how different O and T are.

The key to good anomaly detection in this model is therefore to make good predictions for T. To do this, we need to capture the seasonal, diurnal, and weather system variation in temperature. We capture the first two via a "baseline" temperature B. The weather system variation is captured by a first-order Markov variable Δ.

Conceptually, B is the average temperature reading that would be expected for this particular quarter hour and day of the year ignoring short-term changes due to weather systems. However, we have only four years of training data, so if we average only the four readings for the specific time of day and day of year, we will get a very poor estimate for B. To overcome this problem, we combine the observed values from the 5 temperature readings before and after the particular quarter hour and the 3 days before and after the target day. The local trend within each day and across the 7 days is computed and removed and then the de-trended temperature values are averaged across the years in the training data.

The Δ variable attempts to capture the local departure from the baseline caused by weather systems. It is modeled as a first-order Markov process:

$$P(\Delta_t | QH, D, \Delta_{t-1}) = Norm(\mu_{QH,D} + \Delta_{t-1}, \sigma^2_{QH,D}).$$

QH denotes the quarter hour of each measurement $(1, \ldots, 96)$; Day (or D) denotes the day of the year $(1, \ldots, 365)$. The main idea is that Δ_t is approximately equal to Δ_{t-1} but with a slight offset $\mu_{QH,D}$ that depends on the time of day and the day of the year and a variance that similarly depends on the time of day and the day of the year. A warm spell is represented by $\Delta_t > 0$, and a cold period by $\Delta_t < 0$. If $\Delta_t > 0$, then it will tend to stay > 0 for a while, and similarly if $\Delta_t < 0$, it will tend to stay < 0 for a while.

Figure 10 illustrates the relationship between the baseline B, the Δ process, and the observed and predicted temperatures. The fact that Δ varies somewhat

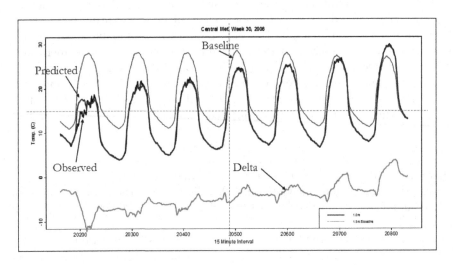

Fig. 10. Relationship between the baseline, Δ, and the observed and predicted temperatures. Note that the baseline curve captures the diurnal variation. It is also slowly dropping, which captures the gradual seasonal change. The Δ curve starts out negative and then gradually increases so that the sum of the baseline plus Δ, which gives the predicted temperature T almost exactly matches the observed temperature O. Where these two curves differ, the model will declare anomalies.

erratically reveals that the model still has room for improvement, since ideally, it would be a fairly smooth curve.

The model is applied one temperature reading at a time. First the observed temperature O, and the QH and D are asserted as evidence. Then probabilistic reasoning is performed to compute updated probability distributions for Δ_t and T and the most likely value of S_t. The data point is tagged with this most likely value. If the most likely value is "very bad", then the observed temperature is removed as evidence, and the value of Δ_t is recomputed. Also, the variance $\sigma^2_{QH,D}$ is set to a small value, so that the distribution of Δ_t remains concentrated near the value of Δ_{t-1}. Then the next data point is processed and tagged.

The model was trained using four years of data and then evaluated on the remaining three years. The model correctly detects all of the easy anomalies. Quantitative evaluation of the medium anomalies is more difficult, because the domain expert tended to mark long contiguous intervals of time as anomalous when there was a problem, whereas the model is more selective. For example, when a sun shield was missing, the expert would label whole days as incorrect, whereas the model only marks the afternoon temperatures as bad, because the sensor is still measuring the correct temperature at night. Figure 11 shows the performance of the model in this case. Notice that it not only detects that the peak temperatures are too high but also that the temperature rises and falls too quickly.

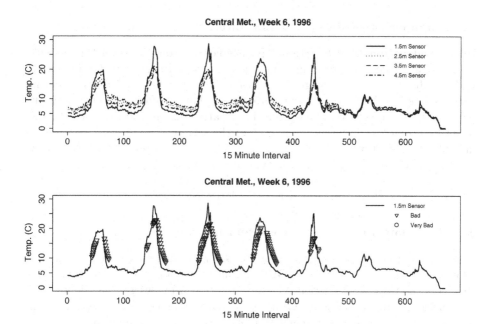

Fig. 11. Top: Lost sun shield in 1.5m sensor. Bottom: Data cleaning applied to 1.5m sensor. Triangles and circles are plotted at points declared to be anomalous. They mark the mean of the predicted temperature distribution.

Our overall assessment is that we are achieving near-100% recall for anomalies but with a false positive rate of roughly 5.3%. This means that we are reducing by over 94% the amount of data that the domain expert must review manually without missing any anomalies. More details are available in Dereszynski and Dietterich (Dereszynski & Dietterich, 2007).

This work shows that carefully-designed dynamic Bayesian networks can do an excellent job of anomaly detection for challenging single-sensor data streams. As more sensor networks are deployed, the need for data cleaning will become much greater, because it will be impossible for human experts to manually inspect and clean the data. We hope that the methods described here will be able to help address this challenge.

4   Education and Training

Ecosystem informatics is inherently an interdisciplinary research area that addresses the scientific problems that arise in various ecological sciences (botany, zoology, population genetics, forest science, natural resource management, earth sciences, etc.) with the modeling and computational methods of mathematics, computer science, and statistics. At Oregon State University, we have developed two educational programs to prepare students for research careers in ecosystem informatics.

4.1   Summer Institute in Ecoinformatics

Under funding from the US National Science Foundation, Professor Desiree Tullos leads a 10-week summer institute in ecosystem informatics for advanced undergraduate and first-year graduate students. Students spend the summer in residence at the Andrews Experimental forest. For the first 3 weeks, they attend an intensive course in ecosystem informatics that introduces them to the scientific problems, research methods, and the terminology of ecosystem informatics. The next 6 weeks involves working on a research project supervised by faculty and doctoral students. This typically involves a mix of field work, data analysis, and mathematical modeling. The final week consists of a series of oral presentations of the results of their research projects.

4.2   Graduate Program in Ecosystem Informatics

The second educational program is a Ph.D. minor in Ecosystem Informatics. This was initiated by a five-year IGERT grant (Julia Jones, Principal Investigator) from the US National Science Foundation that provides graduate fellowship support for students in the program. This was complemented by the hiring of four new faculty members to teach and lead research in this program.

One of the challenges of interdisciplinary education is to prepare people to work together across disciplinary lines without requiring them to become experts in multiple fields. To address this challenge, we decided to structure the program so that students must have a "home" Ph.D. department, and they receive a doctoral degree in their home department. In addition, they receive a Ph.D. minor in Ecosystem Informatics. The minor involves the following:

- Participation in the Ecosystem Informatics "Boot Camp", which is a one week residential course held at the Andrews Experimental Forest prior to the start of classes in the fall.
- Participation in a year-long Introduction to Ecosystem Informatics class. In this class, students are introduced to the problems and terminology of ecosystem informatics, and they work in cross-disciplinary student teams to study emerging problems in ecosystem informatics.
- Participation in a 6-month internship, preferably at an institution outside the US. The goal of this is to expose students to research questions motivated by ecological problems outside the US and to give them a more global perspective. Often, this results in a published paper or an idea that can form the basis of their doctoral research.
- Inclusion of an ecosystem informatics chapter in the doctoral dissertation. This chapter is devoted to interdisciplinary work, sometimes with another student in the program. The research topic for this chapter sometimes grows out of the year-long class or the internship. In addition, to help students develop these topics, we organize cross-disciplinary brainstorming sessions for each student. The student presents a proposed problem, and faculty members and other students brainstorm ideas for how to formulate and study the problem.

We are now entering the fourth year of this graduate program. One of the biggest benefits so far has been the development of interesting mathematical models for analyzing disturbance in forests and habitats in streams. In addition, the program has served as a nexus for fostering new interdisciplinary projects including the battery-free sensor network program described in this paper.

5   Concluding Remarks

Many of the most important scientific and policy questions facing humanity require major advances in the ecological sciences. Ecology has traditionally been a difficult area to study because of the difficulty of measuring the primary data: the fluxes of chemicals and nutrients and the distribution and interaction of living organisms. Fortunately, we are in the midst of a revolution in sensor technology that is going to make it possible to measure this primary data continuously with dense networks of sensors. This will enable the ecosystem sciences to apply the methods of data exploration science including data mining, machine learning, and statistical model building to make rapid progress.

This paper has briefly described some of the activities in sensors and ecosystem informatics at Oregon State University. At the level of sensor development, we have discussed the development of ultra-low power temperature sensor nodes that can operate by harvesting power from spread-spectrum RF broadcast from a central powered base station. We have also described our work on applying computer vision and robotics to automatically manipulate and classify arthropod specimens. At the level of data analysis, we have described work on automated data cleaning for temperature data streams collected over a 7-year period at the Andrews Experimental Forest. Finally, we have discussed two new educational programs that seek to train researchers to work in interdisciplinary teams.

Much more research is required in all of these areas. Furthermore, there is a great need for new kinds of data analysis and data management tools. In particular, machine learning and data mining methods must be developed that can deal with spatially explicit models and that can model interactions among hundreds or thousands of species in time and space. I hope this paper will motivate the reader to consider contributing new ideas to this exciting and important research area.

Acknowledgements

The research described in this paper is funded by several grants from the US National Science Foundation. The battery-free sensor network research is funded by NSF grant BDI-0529223 (Barbara Bond, PI). The arthropod classification project is funded by NSF grant IIS-0326052 (Tom Dietterich, PI). The data cleaning project and the graduate fellowship program are funded by an IGERT grant DGE-0333257 (Julia Jones, PI). And the Summer Institute in Ecoinformatics is funded by grant EEC-0609356 (Desiree Tullos, PI). The author gratefully acknowledges the assistance of Barbara Bond, Adam Kennedy, Ethan Dereszynski, Huaping Liu, and Karti Mayaram in preparing this paper.

References

Deng, H., Zhang, W., Mortensen, E., Dietterich, T., Shapiro, L.: Principal curvature-based region detector for object recognition. In: Proceedings of the IEEE Conference on Computer Vision and Pattern Recognition (CVPR-2007) (2007)

Dereszynski, E., Dietterich, T.: Probabilistic models for anomaly detection in remote sensor data streams. In: 23rd Conference on Uncertainty in Artificial Intelligence (UAI-2007) (2007)

Elson, J., Estrin, D.: Wireless sensor networks: A bridge to the physical world. In: Raghavendra, Sivalingam, Znat (eds.) Wireless sensor networks, Kluwer, Dordrecht (2004)

Gray, J., Szalay, A.: Online science: The world-wide telescope as a prototype for the new computational science (Technical Report Powerpoint Presentation). Microsoft Research (2003)

Kadir, T., Brady, M.: Saliency, scale and image description. Int. J. Computer Vision 45, 83–105 (2001)

Kahn, J.M., Katz, R.H., Pister, K.S.J.: Next century challenges: Mobile networking for Smart Dust. In: Proceedings of the Fifth Annual ACM/IEEE international Conference on Mobile Computing and Networking, pp. 271–278. ACM, New York (1999)

Landwehr, N., Hall, M., Frank, E.: Logistic model trees. Machine Learning 59, 161–205 (2005)

Larios, N., Deng, H., Zhang, W.: Automated insect identification through concatenated histograms of local appearance features. Machine Vision and Applications (in Press)

Le, T., Mayaram, K., Fiez, T.S.: Efficient far-field radio frequency power conversion system for passively powered sensor networks. In: IEEE 2006 Custom Integrated Circuits Conference (CICC 2006), pp. 293–296. IEEE, Los Alamitos (2006)

Lowe, D.G.: Distinctive image features from scale-invariant keypoints. Int. J. Comput. Vision 60, 91–110 (2004)

Mikolajczyk, K., Schmid, C.: Scale and affine invariant interest point detectors. International J. Computer Vision, 63–83 (2004)

NSB, Environmental science and engineering for the 21st century (Technical Report NSB-00-22). National Science Foundation (2000)

NSB, Science and engineering infrastructure for the 21st century: The role of the national science foundation (Technical Report NSF-02-190). National Science Foundation (2002)

Simple Algorithmic Principles of Discovery, Subjective Beauty, Selective Attention, Curiosity & Creativity

Jürgen Schmidhuber

TU Munich, Boltzmannstr. 3, 85748 Garching bei München, Germany & IDSIA,
Galleria 2, 6928 Manno (Lugano), Switzerland
juergen@idsia.ch,
http://www.idsia.ch/~juergen

Abstract. I postulate that human or other intelligent agents function or should function as follows. They store all sensory observations as they come—the data is 'holy.' At any time, given some agent's current coding capabilities, part of the data is compressible by a short and hopefully fast program / description / explanation / world model. In the agent's subjective eyes, such data is more regular and more *beautiful* than other data. It is well-known that knowledge of regularity and repeatability may improve the agent's ability to plan actions leading to external rewards. In absence of such rewards, however, *known* beauty is boring. Then *interestingness* becomes the *first derivative* of subjective beauty: as the learning agent improves its compression algorithm, formerly apparently random data parts become subjectively more regular and beautiful. Such progress in data compression is measured and maximized by the *curiosity* drive: create action sequences that extend the observation history and yield previously unknown / unpredictable but quickly learnable algorithmic regularity. I discuss how all of the above can be naturally implemented on computers, through an extension of passive unsupervised learning to the case of active data selection: we reward a general reinforcement learner (with access to the adaptive compressor) for actions that improve the subjective compressibility of the growing data. An unusually large compression breakthrough deserves the name *discovery*. The *creativity* of artists, dancers, musicians, pure mathematicians can be viewed as a by-product of this principle. Several qualitative examples support this hypothesis.

1   Introduction

A human lifetime lasts about 3×10^9 seconds. The human brain has roughly 10^{10} neurons, each with 10^4 synapses on average. Assuming each synapse can store not more than 3 bits, there is still enough capacity to store the lifelong sensory input stream with a rate of roughly 10^5 bits/s, comparable to the demands of a movie with reasonable resolution. The storage capacity of affordable technical systems will soon exceed this value.

Hence, it is not unrealistic to consider a mortal agent that interacts with an environment and has the means to store the entire history of sensory inputs, which partly depends on its actions. This data anchors all it will ever know about itself and its role in the world. In this sense, the data is 'holy.'

What should the agent do with the data? How should it learn from it? Which actions should it execute to influence future data?

V. Corruble, M. Takeda, and E. Suzuki (Eds.): DS 2007, LNAI 4755, pp. 26–38, 2007.

Some of the sensory inputs reflect external rewards. At any given time, the agent's goal is to maximize the remaining reward or reinforcement to be received before it dies. In realistic settings external rewards are rare though. In absence of such rewards through teachers etc., what should be the agent's motivation? Answer: It should spend some time on *unsupervised learning*, figuring out how the world works, hoping this knowledge will later be useful to gain external rewards.

Traditional unsupervised learning is about finding regularities, by clustering the data, or encoding it through a factorial code [2,14] with statistically independent components, or predicting parts of it from other parts. All of this may be viewed as special cases of data compression. For example, where there are clusters, a data point can be efficiently encoded by its cluster center plus relatively few bits for the deviation from the center. Where there is data redundancy, a non-redundant factorial code [14] will be more compact than the raw data. Where there is predictability, compression can be achieved by assigning short codes to events that are predictable with high probability [3]. Generally speaking we may say that a major goal of traditional unsupervised learning is to improve the compression of the observed data, by discovering a program that computes and thus explains the history (and hopefully does so quickly) but is clearly shorter than the shortest previously known program of this kind.

According to our complexity-based theory of beauty [15,17,26], the agent's currently achieved compression performance corresponds to subjectively perceived beauty: among several sub-patterns classified as 'comparable' by a given observer, the subjectively most beautiful is the one with the simplest (shortest) description, given the observer's particular method for encoding and memorizing it. For example, mathematicians find beauty in a simple proof with a short description in the formal language they are using. Others like geometrically simple, aesthetically pleasing, low-complexity drawings of various objects [15,17].

Traditional unsupervised learning is not enough though—it just analyzes and encodes the data but does not choose it. We have to extend it along the dimension of active action selection, since our unsupervised learner must also choose the actions that influence the observed data, just like a scientist chooses his experiments, a baby its toys, an artist his colors, a dancer his moves, or any attentive system its next sensory input.

Which data should the agent select by executing appropriate actions? Which are the *interesting* sensory inputs that deserve to be targets of its curiosity? I postulate [26] that in the absence of external rewards or punishment the answer is: Those that yield *progress* in data compression. What does this mean? New data observed by the learning agent may initially look rather random and incompressible and hard to explain. A good learner, however, will *improve* its compression algorithm over time, using some application-dependent learning algorithm, making parts of the data history subjectively more compressible, more explainable, more regular and more 'beautiful.' A beautiful thing is interesting only as long as it is new, that is, as long as the algorithmic regularity that makes it simple has not yet been fully assimilated by the adaptive observer who is still learning to compress the data better. So the agent's goal should be: create action sequences that extend the observation history and yield previously unknown / unpredictable but quickly learnable algorithmic regularity or compressibility. To rephrase this principle in an informal way: maximize the *first derivative* of subjective beauty.

An unusually large compression breakthrough deserves the name *discovery*. How can we motivate a reinforcement learning agent to make discoveries? Clearly, we cannot simply reward it for executing actions that just yield a compressible but boring history. For example, a vision-based agent that always stays in the dark will experience an extremely compressible and uninteresting history of unchanging sensory inputs. Neither can we reward it for executing actions that yield highly informative but uncompressible data. For example, our agent sitting in front of a screen full of white noise will experience highly unpredictable and fundamentally uncompressible and uninteresting data conveying a lot of information in the traditional sense of Boltzmann and Shannon [30]. Instead, the agent should receive reward for creating / observing data that allows for *improvements* of the data's subjective compressibility.

The appendix will describe formal details of how to implement this principle on computers. The next section will provide examples of subjective beauty tailored to human observers, and illustrate the learning process leading from less to more subjective beauty. Then I will argue that the *creativity* of artists, dancers, musicians, pure mathematicians as well as unsupervised *attention* in general is just a by-product of our principle, using qualitative examples to support this hypothesis.

2   Visual Examples of Subjective Beauty and Its 'First Derivative' Interestingness

Figure 1 depicts the drawing of a female face considered *'beautiful'* by some human observers. It also shows that the essential features of this face follow a very simple geometrical pattern [17] to be specified by very few bits of information. That is, the data stream generated by observing the image (say, through a sequence of eye saccades) is more compressible than it would be in the absence of such regularities. Although few people are able to immediately see how the drawing was made without studying its grid-based explanation (right-hand side of Figure 1), most do notice that the facial features somehow fit together and exhibit some sort of regularity. According to our postulate, the observer's reward is generated by the conscious or subconscious discovery of this compressibility. The face remains interesting until its observation does not reveal any additional previously unknown regularities. Then it becomes boring even in the eyes of those who think it is beautiful—beauty and interestingness are two different things.

Figure 2 provides another example: a butterfly and a vase with a flower. The image to the left can be specified by very few bits of information; it can be constructed through a very simple procedure or algorithm based on fractal circle patterns [15]. People who understand this algorithm tend to appreciate the drawing more than those who do not. They realize how simple it is. This is not an immediate, all-or-nothing, binary process though. Since the typical human visual system has a lot of experience with circles, most people quickly notice that the curves somehow fit together in a regular way. But few are able to immediately state the precise geometric principles underlying the drawing. This pattern, however, is learnable from the right-hand side of Figure 2. The conscious or subconscious discovery process leading from a longer to a shorter description of the data, or from less to more compression, or from less to more subjectively perceived beauty, yields reward depending on the first derivative of subjective beauty.

Fig. 1. Left: *Drawing of a female face based on a previously published construction plan [17] (1998). Some human observers report they feel this face is 'beautiful.' Although the drawing has lots of noisy details (texture etc) without an obvious short description, positions and shapes of the basic facial features are compactly encodable through a very simple geometrical scheme. Hence the image contains a highly compressible algorithmic regularity or pattern describable by few bits of information. An observer can perceive it through a sequence of attentive eye movements or saccades, and consciously or subconsciously discover the compressibility of the incoming data stream.* **Right:** *Explanation of how the essential facial features were constructed [17]. First the sides of a square were partitioned into 2^4 equal intervals. Certain interval boundaries were connected to obtain three rotated, superimposed grids based on lines with slopes ± 1 or $\pm 1/2^3$ or $\pm 2^3/1$. Higher-resolution details of the grids were obtained by iteratively selecting two previously generated, neighbouring, parallel lines and inserting a new one equidistant to both. Finally the grids were vertically compressed by a factor of $1 - 2^{-4}$. The resulting lines and their intersections define essential boundaries and shapes of eyebrows, eyes, lid shades, mouth, nose, and facial frame in a simple way that is obvious from the construction plan. Although this plan is simple in hindsight, it was hard to find: hundreds of my previous attempts at discovering such precise matches between simple geometries and pretty faces failed.*

3 Compressibility-Based Rewards of Art and Music

The examples above indicate that works of art and music may have important purposes beyond their social aspects [1] despite of those who classify art as superfluous [10]. Good observer-dependent art deepens the observer's insights about this world or possible worlds, unveiling previously unknown regularities in compressible data, connecting previously disconnected patterns in an initially surprising way that makes the combination of these patterns subjectively more compressible, and eventually becomes known and less interesting. I postulate that the active creation and attentive perception of all kinds of artwork are just by-products of my curiosity principle yielding reward for compressor improvements.

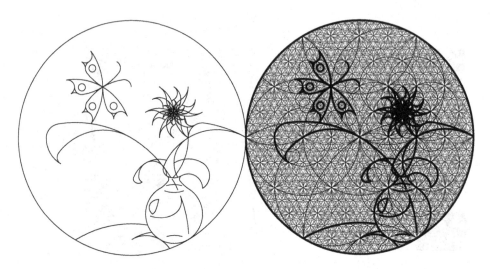

Fig. 2. Left: *Image of a butterfly and a vase with a flower, reprinted from* Leonardo *[15,26].*
Right: *Explanation of how the image was constructed through a very simple algorithm exploiting fractal circles [15]. The frame is a circle; its leftmost point is the center of another circle of the same size. Wherever two circles of equal size touch or intersect are centers of two more circles with equal and half size, respectively. Each line of the drawing is a segment of some circle, its endpoints are where circles touch or intersect. There are few big circles and many small ones. In general, the smaller a circle, the more bits are needed to specify it. The drawing to the left is simple (compressible) as it is based on few, rather large circles. Many human observers report that they derive a certain amount of pleasure from discovering this simplicity. The observer's learning process causes a reduction of the subjective complexity of the data, yielding a temporarily high derivative of subjective beauty. (Again I needed a long time to discover a satisfactory way of using fractal circles to create a reasonable drawing.)*

Let us elaborate on this idea in more detail, following the discussion in [26]. Artificial or human observers must perceive art sequentially, and typically also actively, e.g., through a sequence of attention-shifting eye saccades or camera movements scanning a sculpture, or internal shifts of attention that filter and emphasize sounds made by a pianist, while surpressing background noise. Undoubtedly many derive pleasure and rewards from perceiving works of art, such as certain paintings, or songs. But different subjective observers with different sensory apparati and compressor improvement algorithms will prefer different input sequences. Hence any objective theory of what is good art must take the subjective observer as a parameter, to answer questions such as: Which action sequences should he select to maximize his pleasure? According to our principle he should select one that maximizes the quickly learnable compressibility that is new, relative to his current knowledge and his (usually limited) way of incorporating or learning new data.

For example, which song should some human observer select next? Not the one he just heard ten times in a row. It became too predictable in the process. But also not the new weird one with the completely unfamiliar rhythm and tonality. It seems too irregular and contain too much arbitrariness and subjective noise. He should try a song that is unfamiliar enough to contain somewhat unexpected harmonies or melodies or beats etc., but familiar enough to allow for quickly recognizing the presence of a new learnable regularity or compressibility in the sound stream. Sure, this song will get boring over time, but not yet.

The observer dependence is illustrated by the fact that Schönberg's twelve tone music is less popular than certain pop music tunes, presumably because its algorithmic structure is less obvious to many human observers as it is based on more complicated harmonies. For example, frequency ratios of successive notes in twelve tone music often cannot be expressed as fractions of very small integers. Those with a prior education about the basic concepts and objectives and constraints of twelve tone music, however, tend to appreciate Schönberg more than those without such an education.

All of this perfectly fits our principle: The current compressor of a given subjective observer tries to compress his history of acoustic and other inputs where possible. The action selector tries to find history-influencing actions that improve the compressor's performance on the history so far. The interesting musical and other subsequences are those with previously unknown yet learnable types of regularities, because they lead to compressor improvements. The boring patterns are those that seem arbitrary or random, or whose structure seems too hard to understand.

Similar statements not only hold for other dynamic art including film and dance (taking into account the compressibility of controller actions), but also for painting and sculpture, which cause dynamic pattern sequences due to attention-shifting actions [29] of the observer.

Just as observers get intrinsic rewards from sequentially focusing attention on artwork that exhibits new, previously unknown regularities, the 'creative' artists get reward for making it. For example, I found it extremely rewarding to discover (after hundreds of frustrating failed attempts) the simple geometric regularities that permitted the construction of the drawings in Figures 1 and 2. The distinction between artists and observers is not clear though. Artists can be observers and vice versa. Both artists and observers execute action sequences. The intrinsic motivations of both are fully compatible with our simple principle. Some artists, however, crave *external* reward from other observers, in form of praise, money, or both, in addition to the *internal* reward that comes from creating a new work of art. Our principle, however, conceptually separates these two types of reward.

From our perspective, scientists are very much like artists. They actively select experiments in search for simple laws compressing the observation history. For example, different apples tend to fall off their trees in similar ways. The discovery of a law underlying the acceleration of all falling apples helps to greatly compress the recorded data.

The framework in the appendix is sufficiently formal to allow for implementation of our principle on computers. The resulting artificial observers will vary in terms of the computational power of their history compressors and learning algorithms. This will influence what is good art / science to them, and what they find interesting.

References

1. Balter, M.: Seeking the key to music. Science 306, 1120–1122 (2004)
2. Barlow, H.B., Kaushal, T.P., Mitchison, G.J.: Finding minimum entropy codes. Neural Computation 1(3), 412–423 (1989)
3. Huffman, D.A.: A method for construction of minimum-redundancy codes. In: Proceedings IRE, vol. 40, pp. 1098–1101 (1952)
4. Hutter, M.: Universal Artificial Intelligence: Sequential Decisions based on Algorithmic Probability (On J. Schmidhuber's SNF grant 20-61847), Springer, Heidelberg (2004)
5. Hutter, M.: On universal prediction and Bayesian confirmation. Theoretical Computer Science (2007)
6. Kaelbling, L.P., Littman, M.L., Moore, A.W.: Reinforcement learning: a survey. Journal of AI research 4, 237–285 (1996)
7. Kolmogorov, A.N.: Three approaches to the quantitative definition of information. Problems of Information Transmission 1, 1–11 (1965)
8. Levin, L.A.: Universal sequential search problems. Problems of Information Transmission 9(3), 265–266 (1973)
9. Li, M., Vitányi, P.M.B.: An Introduction to Kolmogorov Complexity and its Applications, 2nd edn. Springer, Heidelberg (1997)
10. Pinker, S.: How the mind works (1997)
11. Schmidhuber, J.: Adaptive curiosity and adaptive confidence. Technical Report FKI-149-91, Institut für Informatik, Technische Universität München (April 1991) See also [12]
12. Schmidhuber, J.: Curious model-building control systems. In: Proceedings of the International Joint Conference on Neural Networks, vol. 2, pp. 1458–1463. IEEE, Los Alamitos (1991)
13. Schmidhuber, J.: Learning complex, extended sequences using the principle of history compression. Neural Computation 4(2), 234–242 (1992)
14. Schmidhuber, J.: Learning factorial codes by predictability minimization. Neural Computation 4(6), 863–879 (1992)
15. Schmidhuber, J.: Low-complexity art. Leonardo, Journal of the International Society for the Arts, Sciences, and Technology 30(2), 97–103 (1997)
16. Schmidhuber, J.: What's interesting? Technical Report IDSIA-35-97, IDSIA, (1997) ftp://ftp.idsia.ch/pub/juergen/interest.ps.gz (extended abstract in Proc. Snowbird'98, Utah (1998) see also [18])
17. Schmidhuber, J.: Facial beauty and fractal geometry. Technical Report TR IDSIA-28-98, IDSIA,(1998) Published in the Cogprint Archive, http://cogprints.soton.ac.uk
18. Schmidhuber, J.: Exploring the predictable. In: Ghosh, A., Tsutsui, S. (eds.) Advances in Evolutionary Computing, pp. 579–612. Springer, Heidelberg (2002)
19. Schmidhuber, J.: Hierarchies of generalized Kolmogorov complexities and nonenumerable universal measures computable in the limit. International Journal of Foundations of Computer Science 13(4), 587–612 (2002)
20. Schmidhuber, J.: The Speed Prior: a new simplicity measure yielding near-optimal computable predictions. In: Kivinen, J., Sloan, R.H. (eds.) COLT 2002. LNCS (LNAI), vol. 2375, pp. 216–228. Springer, Heidelberg (2002)
21. Schmidhuber, J.: Gödel machines: self-referential universal problem solvers making provably optimal self-improvements. Technical Report IDSIA-19-03, arXiv:cs.LO/0309048, IDSIA, Manno-Lugano, Switzerland (2003)
22. Schmidhuber, J.: Optimal ordered problem solver. Machine Learning 54, 211–254 (2004)

23. Schmidhuber, J.: Overview of artificial curiosity and active exploration, with links to publications since, 1990 (2004), http://www.idsia.ch/~juergen/interest.html
24. Schmidhuber, J.: Completely self-referential optimal reinforcement learners. In: Duch, W., Kacprzyk, J., Oja, E., Zadrożny, S. (eds.) ICANN 2005. LNCS, vol. 3697, pp. 223–233. Springer, Heidelberg (2005)
25. Schmidhuber, J.: Gödel machines: Towards a technical justification of consciousness. In: Kudenko, D., Kazakov, D., Alonso, E. (eds.) Adaptive Agents and Multi-Agent Systems III. LNCS (LNAI), vol. 3394, pp. 1–23. Springer, Heidelberg (2005)
26. Schmidhuber, J.: Developmental robotics, optimal artificial curiosity, creativity, music, and the fine arts. Connection Science 18(2), 173–187 (2006)
27. Schmidhuber, J.: Gödel machines: fully self-referential optimal universal problem solvers. In: Goertzel, B., Pennachin, C. (eds.) Artificial General Intelligence, pp. 199–226. Springer, Heidelberg (2006)
28. Schmidhuber, J., Heil, S.: Sequential neural text compression. IEEE Transactions on Neural Networks 7(1), 142–146 (1996)
29. Schmidhuber, J., Huber, R.: Learning to generate artificial fovea trajectories for target detection. International Journal of Neural Systems 2(1& 2), 135–141 (1991)
30. Shannon, C.E.: A mathematical theory of communication (parts I and II. Bell System Technical Journal XXVII, 379–423 (1948)
31. Solomonoff, R.J.: A formal theory of inductive inference. Part I. Information and Control 7, 1–22 (1964)
32. Solomonoff, R.J.: Complexity-based induction systems. IEEE Transactions on Information Theory IT-24(5), 422–432 (1978)
33. Storck, J., Hochreiter, S., Schmidhuber, J.: Reinforcement driven information acquisition in non-deterministic environments. In: Proceedings of the International Conference on Artificial Neural Networks, Paris, vol. 2, pp. 159–164. EC2 & Cie (1995)

A Appendix

This appendix is a compactified, compressibility-oriented variant of parts of [26].

The world can be explained to a degree by compressing it. The compressed version of the data can be viewed as its explanation. Discoveries correspond to large data compression improvements (found by the given, application-dependent compressor improvement algorithm). How to build an adaptive agent that not only tries to achieve externally given rewards but also to discover, in an unsupervised and experiment-based fashion, explainable and compressible data? (The explanations gained through explorative behavior may eventually help to solve teacher-given tasks.)

Let us formally consider a learning agent whose single life consists of discrete cycles or time steps $t = 1, 2, \ldots, T$. Its complete lifetime T may or may not be known in advance. In what follows, the value of any time-varying variable Q at time t ($1 \leq t \leq T$) will be denoted by $Q(t)$, the ordered sequence of values $Q(1), \ldots, Q(t)$ by $Q(\leq t)$, and the (possibly empty) sequence $Q(1), \ldots, Q(t-1)$ by $Q(< t)$. At any given t the agent receives a real-valued input $x(t)$ from the environment and executes a real-valued action $y(t)$ which may affect future inputs. At times $t < T$ its goal is to maximize future success or *utility*

$$u(t) = E_\mu \left[\sum_{\tau=t+1}^{T} r(\tau) \ \middle| \ h(\leq t) \right], \tag{1}$$

where $r(t)$ is an additional real-valued reward input at time t, $h(t)$ the ordered triple $[x(t), y(t), r(t)]$ (hence $h(\leq t)$ is the known history up to t), and $E_\mu(\cdot \mid \cdot)$ denotes the conditional expectation operator with respect to some possibly unknown distribution μ from a set $\mathcal{M}$ of possible distributions. Here $\mathcal{M}$ reflects whatever is known about the possibly probabilistic reactions of the environment. For example, $\mathcal{M}$ may contain all computable distributions [31,32,9,4]. There is just one life, no need for predefined repeatable trials, no restriction to Markovian interfaces between sensors and environment, and the utility function implicitly takes into account the expected remaining lifespan $E_\mu(T \mid h(\leq t))$ and thus the possibility to extend it through appropriate actions [21,24,27,25].

Recent work has led to the first learning machines that are universal and optimal in various very general senses [4,24,27]. Such machines can in principle find out by themselves whether curiosity and world model construction are useful or useless in a given environment, and learn to behave accordingly. The present appendix, however, will assume *a priori* that compression / explanation of the history is good and should be done; here we shall not worry about the possibility that 'curiosity may kill the cat.' Towards this end, in the spirit of our previous work [12,11,33,16,18], we split the reward signal $r(t)$ into two scalar real-valued components: $r(t) = g(r_{ext}(t), r_{int}(t))$, where g maps pairs of real values to real values, e.g., $g(a, b) = a + b$. Here $r_{ext}(t)$ denotes traditional *external* reward provided by the environment, such as negative reward in response to bumping against a wall, or positive reward in response to reaching some teacher-given goal state. But I am especially interested in $r_{int}(t)$, the internal or intrinsic or *curiosity* reward, which is provided whenever the data compressor / internal world model of the agent improves in some sense. Our initial focus will be on the case $r_{ext}(t) = 0$ for all valid t. The basic principle is essentially the one we published before in various variants [11,12,33,16,18,23,26]:

Principle 1. *Generate curiosity reward for the controller in response to improvements of the history compressor.*

So we conceptually separate the goal (explaining / compressing the history) from the means of achieving the goal. Once the goal is formally specified in terms of an algorithm for computing curiosity rewards, let the controller's reinforcement learning (RL) mechanism figure out how to translate such rewards into action sequences that allow the given compressor improvement algorithm to find and exploit previously unknown types of compressibility.

A.1   Predictors vs Compressors

Most of our previous work on artificial curiosity was prediction-oriented, e. g., [11,12,33,16,18,23,26]. Prediction and compression are closely related though. A predictor that correctly predicts many $x(\tau)$, given history $h(< \tau)$, for $1 \leq \tau \leq t$, can be used to encode $h(\leq t)$ compactly: Given the predictor, only the wrongly predicted $x(\tau)$ plus information about the corresponding time steps τ are necessary to reconstruct history $h(\leq t)$, e.g., [13]. Similarly, a predictor that learns a probability distribution of the possible next events, given previous events, can be used to efficiently encode observations with high (respectively low) predicted probability by few (respectively many) bits [3,28], thus achieving a compressed history representation. Generally speaking, we may view the predictor as the essential part of a program p that recomputes $h(\leq t)$. If this program is short in comparison to the rad data $h(\leq t)$, then $h(\leq t)$ is regular or non-random [31,7,9,19], presumably reflecting essential environmental laws. Then p may also be highly useful for predicting future, yet unseen $x(\tau)$ for $\tau > t$.

A.2 Compressor Performance Measures

At any time t $(1 \leq t < T)$, given some compressor program p able to compress history $h(\leq t)$, let $C(p, h(\leq t))$ denote p's compression performance on $h(\leq t)$. An appropriate performance measure would be

$$C_l(p, h(\leq t)) = l(p),\qquad(2)$$

where $l(p)$ denotes the length of p, measured in number of bits: the shorter p, the more algorithimic regularity and compressibility and predictability and lawfulness in the observations so far. The ultimate limit for $C_l(p, h(\leq t))$ would be $K^*(h(\leq t))$, a variant of the Kolmogorov complexity of $h(\leq t)$, namely, the length of the shortest program (for the given hardware) that computes an output starting with $h(\leq t)$ [31,7,9,19].

$C_l(p, h(\leq t))$ does not take into account the time $\tau(p, h(\leq t))$ spent by p on computing $h(\leq t)$. An alternative performance measure inspired by concepts of optimal universal search [8,22] is

$$C_{l\tau}(p, h(\leq t)) = l(p) + \log \tau(p, h(\leq t)).\qquad(3)$$

Here compression by one bit is worth as much as runtime reduction by a factor of $\frac{1}{2}$.

A.3 Compressor Improvement Measures

The previous Section A.2 only discussed measures of compressor performance, but not of performance *improvement*, which is the essential issue in our curiosity-oriented context. To repeat the point made above: *The important thing are the improvements of the compressor, not its compression performance per se.* Our curiosity reward in response to the compressor's progress (due to some application-dependent compressor improvement algorithm) between times t and $t + 1$ should be

$$r_{int}(t + 1) = f[C(p(t + 1), h(\leq t + 1)), C(p(t), h(\leq t + 1))],\qquad(4)$$

where f maps pairs of real values to real values. Various alternative progress measures are possible; most obvious is $f(a, b) = a - b$.

Note that both the old and the new compressor have to be tested on the same data, namely, the complete history so far.

A.4 Asynchronous Framework for Creating Curiosity Reward

Let $p(t)$ denote the agent's current compressor program at time t, $s(t)$ its current controller, and do:

Controller: At any time t $(1 \le t < T)$ do:

1. Let $s(t)$ use (parts of) history $h(\le t)$ to select and execute $y(t+1)$.
2. Observe $x(t+1)$.
3. Check if there is non-zero curiosity reward $r_{int}(t+1)$ provided by the separate, asynchronously running compressor improvement algorithm (see below). If not, set $r_{int}(t+1) = 0$.
4. Let the controller's reinforcement learning (RL) algorithm use $h(\le t+1)$ including $r_{int}(t+1)$ (and possibly also the latest available compressed version of the observed data—see below) to obtain a new controller $s(t+1)$, in line with objective (1).

Compressor: Set p_{new} equal to the initial data compressor. Starting at time 1, repeat forever until interrupted by death T:

1. Set $p_{old} = p_{new}$; get current time step t and set $h_{old} = h(\le t)$.
2. Evaluate p_{old} on h_{old}, to obtain $C(p_{old}, h_{old})$ (Section A.2). This may take many time steps.
3. Let some (application-dependent) compressor improvement algorithm (such as a learning algorithm for an adaptive neural network predictor) use h_{old} to obtain a hopefully better compressor p_{new} (such as a neural net with the same size but improved prediction capability and therefore improved compression performance). Although this may take many time steps, p_{new} may not be optimal, due to limitations of the learning algorithm, e.g., local maxima.
4. Evaluate p_{new} on h_{old}, to obtain $C(p_{new}, h_{old})$. This may take many time steps.
5. Get current time step τ and generate curiosity reward

$$r_{int}(\tau) = f[C(p_{old}, h_{old}), C(p_{new}, h_{old})], \tag{5}$$

e.g., $f(a, b) = a - b$; see Section A.3.

Obviously this asynchronuous scheme may cause long temporal delays between controller actions and corresponding curiosity rewards. This may impose a heavy burden on the controller's RL algorithm whose task is to assign credit to past actions (to inform the controller about beginnings of compressor evaluation processes etc., we may augment its input by unique representations of such events). Nevertheless, there are RL algorithms for this purpose which are theoretically optimal in various senses, to be discussed next.

A.5 Optimal Curiosity & Creativity & Focus of Attention

Our chosen compressor class typically will have certain computational limitations. In the absence of any external rewards, we may define *optimal pure curiosity behavior* relative to these limitations: At time t this behavior would select the action that maximizes

$$u(t) = E_\mu \left[\sum_{\tau=t+1}^{T} r_{int}(\tau) \;\middle|\; h(\le t) \right]. \tag{6}$$

Since the true, world-governing probability distribution μ is unknown, the resulting task of the controller's RL algorithm may be a formidable one. As the system is revisiting previously uncompressible parts of the environment, some of those will tend to become more compressible, that is, the corresponding curiosity rewards will decrease over time. A good RL algorithm must somehow detect and then *predict* this decrease, and act accordingly. Traditional RL algorithms [6], however, do not provide any theoretical guarantee of optimality for such situations. (This is not to say though that sub-optimal RL methods may not lead to success in certain applications; experimental studies might lead to interesting insights.)

Let us first make the natural assumption that the compressor is not super-complex such as Kolmogorov's, that is, its output and $r_{int}(t)$ are computable for all t. Is there a best possible RL algorithm that comes as close as any other to maximizing objective (6)? Indeed, there is. Its drawback, however, is that it is not computable in finite time. Nevertheless, it serves as a reference point for defining what is achievable at best.

A.6   Optimal But Incomputable Action Selector

There is an optimal way of selecting actions which makes use of Solomonoff's theoretically optimal universal predictors and their Bayesian learning algorithms [31,32,9,4,5]. The latter only assume that the reactions of the environment are sampled from an unknown probability distribution μ contained in a set $\mathcal{M}$ of all enumerable distributions— compare text after equation (1). More precisely, given an observation sequence $q(\leq t)$, we only assume there exists a computer program that can compute the probability of the next possible $q(t+1)$, given $q(\leq t)$. In general we do not know this program, hence we predict using a mixture distribution

$$\xi(q(t+1) \mid q(\leq t)) = \sum_i w_i \mu_i(q(t+1) \mid q(\leq t)), \qquad (7)$$

a weighted sum of *all* distributions $\mu_i \in \mathcal{M}$, $i = 1, 2, \ldots$, where the sum of the constant weights satisfies $\sum_i w_i \leq 1$. This is indeed the best one can possibly do, in a very general sense [32,4]. The drawback of the scheme is its incomputability, since $\mathcal{M}$ contains infinitely many distributions. We may increase the theoretical power of the scheme by augmenting $\mathcal{M}$ by certain non-enumerable but limit-computable distributions [19], or restrict it such that it becomes computable, e.g., by assuming the world is computed by some unknown but deterministic computer program sampled from the Speed Prior [20] which assigns low probability to environments that are hard to compute by any method.

Once we have such an optimal predictor, we can extend it by formally including the effects of executed actions to define an optimal action selector maximizing future expected reward. At any time t, Hutter's theoretically optimal (yet uncomputable) RL algorithm AIXI [4] uses an extended version of Solomonoff's prediction scheme to select those action sequences that promise maximal future reward up to some horizon T, given the current data $h(\leq t)$. That is, in cycle $t + 1$, AIXI selects as its next action the first action of an action sequence maximizing ξ-predicted reward up to the given horizon, appropriately generalizing eq. (7). AIXI uses observations optimally [4]: the Bayes-optimal policy p^ξ based on the mixture ξ is self-optimizing in the sense that its average utility value converges asymptotically for all $\mu \in \mathcal{M}$ to the optimal value achieved by the Bayes-optimal policy p^μ which knows μ in advance. The necessary and sufficient condition is that $\mathcal{M}$ admits self-optimizing policies. The policy p^ξ is also

Pareto-optimal in the sense that there is no other policy yielding higher or equal value in *all* environments $\nu \in \mathcal{M}$ and a strictly higher value in at least one [4].

A.7   Computable Selector of Provably Optimal Actions, Given Current System

AIXI above needs unlimited computation time. Its computable variant AIXI*(t,l)* [4] has asymptotically optimal runtime but may suffer from a huge constant slowdown. To take the consumed computation time into account in a general, optimal way, we may use the recent Gödel machines [21,24,27,25] instead. They represent the first class of mathematically rigorous, fully self-referential, self-improving, general, optimally efficient problem solvers. They are also applicable to the problem embodied by objective (6).

The initial software $\mathcal{S}$ of such a Gödel machine contains an initial problem solver, e.g., some typically sub-optimal method [6]. It also contains an asymptotically optimal initial proof searcher based on an online variant of Levin's *Universal Search* [8], which is used to run and test *proof techniques*. Proof techniques are programs written in a universal language implemented on the Gödel machine within $\mathcal{S}$. They are in principle able to compute proofs concerning the system's own future performance, based on an axiomatic system $\mathcal{A}$ encoded in $\mathcal{S}$. $\mathcal{A}$ describes the formal *utility* function, in our case eq. (6), the hardware properties, axioms of arithmetic and probability theory and data manipulation etc, and $\mathcal{S}$ itself, which is possible without introducing circularity [21].

Inspired by Kurt Gödel's celebrated self-referential formulas (1931), the Gödel machine rewrites any part of its own code (including the proof searcher) through a self-generated executable program as soon as its *Universal Search* variant has found a proof that the rewrite is *useful* according to objective (6). According to the Global Optimality Theorem [21,24,27,25], such a self-rewrite is globally optimal—no local maxima possible!—since the self-referential code first had to prove that it is not useful to continue the search for alternative self-rewrites.

If there is no provably useful optimal way of rewriting $\mathcal{S}$ at all, then humans will not find one either. But if there is one, then $\mathcal{S}$ itself can find and exploit it. Unlike the previous *non*-self-referential methods based on hardwired proof searchers [4], Gödel machines not only boast an optimal *order* of complexity but can optimally reduce (through self-changes) any slowdowns hidden by the $O()$-notation, provided the utility of such speed-ups is provable.

A.8   Consequences of Optimal Action Selecton

Now let us apply any optimal RL algorithm to curiosity rewards as defined above. The expected consequences are: at time t the controller will do the best to select an action $y(t)$ that starts an action sequence expected to create observations yielding maximal expected compression *progress* up to the expected death T, taking into accunt the limitations of both the compressor and the compressor improvement algorithm. In particular, ignoring issues of computation time, it will focus in the best possible way on things that are currently still uncompressible but will soon become compressible through additional learning. It will get bored by things that already are compressible. It will also get bored by things that are currently uncompressible but will apparently remain so, given the experience so far, or where the costs of making them compressible exceed those of making other things compressible, etc.

A Theory of Similarity Functions for Learning and Clustering

Avrim Blum

Department of Computer Science
Carnegie Mellon University
Pittsburgh, PA 15213
avrim@cs.cmu.edu

Abstract. Kernel methods have proven to be powerful tools in machine learning. They perform well in many applications, and there is also a well-developed theory of sufficient conditions for a kernel to be useful for a given learning problem. However, while a kernel can be thought of as just a pairwise similarity function that satisfies additional mathematical properties, this theory requires viewing kernels as implicit (and often difficult to characterize) maps into high-dimensional spaces. In this talk I will describe work on developing a theory that applies to more general similarity functions (not just legal kernels) and furthermore describes the usefulness of a given similarity function in terms of more intuitive, direct properties, without need to refer to any implicit spaces.

An interesting feature of the proposed framework is that it can also be applied to learning from purely unlabeled data, i.e., clustering. In particular, one can ask how much stronger the properties of a similarity function should be (in terms of its relation to the unknown desired clustering) so that it can be used to *cluster* well: to learn well without any label information at all. We find that if we are willing to relax the objective a bit (for example, allow the algorithm to produce a hierarchical clustering that we will call successful if some pruning is close to the correct answer), then this question leads to a number of interesting graph-theoretic and game-theoretic properties that are sufficient to cluster well. This work can be viewed as an approach to defining a PAC model for clustering.

This talk is based on work joint with Maria-Florina Balcan and Santosh Vempala.

V. Corruble, M. Takeda, and E. Suzuki (Eds.): DS 2007, LNAI 4755, p. 39, 2007.
© Springer-Verlag Berlin Heidelberg 2007

A Hilbert Space Embedding for Distributions*

Alex Smola[1], Arthur Gretton[2], Le Song[1], and Bernhard Schölkopf[2]

[1] NICTA and ANU, Northbourne Avenue 218, Canberra 0200 ACT, Australia
{alex.smola, le.song}@nicta.com.au
[2] MPI for Biological Cybernetics, Spemannstr. 38, 72076 Tübingen, Germany
{arthur,bernhard.schoelkopf}@tuebingen.mpg.de

While kernel methods are the basis of many popular techniques in supervised learning, they are less commonly used in testing, estimation, and analysis of probability distributions, where information theoretic approaches rule the roost. However it becomes difficult to estimate mutual information or entropy if the data are high dimensional.

We present a method which allows us to compute distances between distributions *without* the need for intermediate density estimation. Our approach allows algorithm designers to specify which properties of a distribution are most relevant to their problems. Our method works by studying the convergence properties of the expectation operator when restricted to a chosen class of functions. In a nutshell our method works as follows: denote by $\mathcal{X}$ a compact domain and let $\mathcal{H}$ be a Reproducing Kernel Hilbert Space on $\mathcal{X}$ with kernel k. Note that in an RKHS we have $f(x) = \langle f, k(x, \cdot) \rangle$ for all functions $f \in \mathcal{H}$. This allows us to denote the expectation operator of a distribution p via

$$\mu[p] := \mathbf{E}_{x \sim p}[k(x, \cdot)] \qquad \text{and hence } \mathbf{E}_{x \sim p}[f(x)] = \langle \mu[p], f \rangle \text{ for } f \in \mathcal{H}.$$

Moreover, for a sample $X = \{x_1, \ldots, x_m\}$ drawn from some distribution p we may denote the empirical counterparts via

$$\mu[X] := \frac{1}{m} \sum_{i=1}^{m} k(x_i, \cdot) \qquad \text{and hence } \frac{1}{m} \sum_{i=1}^{m} f(x_i) = \langle \mu[X], f \rangle \text{ for } f \in \mathcal{H}.$$

This allows us to compute distances between distributions p, q via $D(p, q) := \|\mu[p] - \mu[q]\|$ and empirical samples X, X' via $D(X, X') := \|\mu[X] - \mu[X']\|$ alike. One can show that under rather benign regularity conditions $\mu[X] \to \mu[p]$ at rate $O(m^{-\frac{1}{2}})$. Such a distance is useful in a number of estimation problems:

- Two-sample tests whether X and X' are drawn from the same distribution.
- Density estimation, where we try to find p so as to minimize the distance between $\mu[p]$ and $\mu[X]$, either by mixture models or by exponential families.
- Independence measures where we compute the distance between the joint distribution and the product of the marginals via $D(p(x, y), p(x) \cdot p(y))$.

---

* The full version of this paper is published in the Proceedings of the 18th International Conference on Algorithmic Learning Theory, ALT 2007, Lecture Notes in Artificial Intelligence Vol. 4754.

V. Corruble, M. Takeda, and E. Suzuki (Eds.): DS 2007, LNAI 4755, pp. 40–41, 2007.
© Springer-Verlag Berlin Heidelberg 2007

– Feature selection algorithms which try to find a subset of covariates x maximally dependent on the target random variables y.

Our framework allows us to unify a large number of existing feature extraction and estimation methods, and provides new algorithms for high dimensional nonparametric statistical tests of distribution properties.

Time and Space Efficient Discovery of Maximal Geometric Graphs

Hiroki Arimura[1], Takeaki Uno[2], and Shinichi Shimozono[3]

[1] Hokkaido University, Kita 14-jo, Nishi 9-chome, Sapporo 060-0814, Japan
arim@ist.hokudai.ac.jp
[2] National Institute of Informatics, Tokyo 101–8430, Japan
uno@nii.jp
[3] Kyushu Institute of Technology, Kawazu 680-4, Iizuka 820-8502, Japan
sin@ai.kyutech.ac.jp

Abstract. A *geometric graph* is a labeled graph whose vertices are points in the 2D plane with an isomorphism invariant under geometric transformations such as translation, rotation, and scaling. While Kuramochi and Karypis (ICDM2002) extensively studied the frequent pattern mining problem for geometric subgraphs, the maximal graph mining has not been considered so far. In this paper, we study the maximal (or closed) graph mining problem for the general class of geometric graphs in the 2D plane by extending the framework of Kuramochi and Karypis. Combining techniques of canonical encoding and a depth-first search tree for the class of maximal patterns, we present *a polynomial delay and polynomial space algorithm*, MaxGeo, *that enumerates all maximal subgraphs* in a given input geometric graph without duplicates. This is the first result establishing the output-sensitive complexity of closed graph mining for geometric graphs. We also show that the frequent graph mining problem is also solvable in polynomial delay and polynomial time.

Keywords: geometric graphs, closed graph mining, depth-first search, rightmost expansion, polynomial delay polynomial space enumeration algorithms.

1 Introduction

Background. There has been increasing demands for efficient methods of extracting useful patterns and rules from weakly structured datasets due to rapid growth of both the amount and the varieties of nonstandard datasets in scientific, spatial, and relational domains. *Graph mining* is one of the most promising approaches to knowledge discovery from such weakly structured datasets. The following topics have been extensively studied for the last few years: frequent subgraph mining [6,12,17,27], maximal (closed) subgraph mining [3,9,20,25] and combination with machine learning [21,28]. See surveys, e.g. [8,24], for the overviews.

The Class of Geometric Graphs. In this paper, we address a graph mining problem for the class $\mathcal{G}$ of geometric graphs. *Geometric graphs* (*geographs*, for short) [15] are a special kind of vertex- and edge-labeled graphs whose vertices have coordinates in the 2D plane $\mathbb{R}^2$, while labels represent geometric features and their relationships. The

V. Corruble, M. Takeda, and E. Suzuki (Eds.): DS 2007, LNAI 4755, pp. 42–55, 2007.

matching relation for geographs is defined through the invariance under a class of geo-metric transformations, such as translation, rotation, and scaling in the plane, in addition to the usual constraint for graph isomorphism. We do not consider the mirror projection, but the extension is simple (consider the mirror projection when we compute the canonical form). Geographs are useful in applications concerned with geometric configurations, e.g., the analysis of chemical compounds, geographic information systems, and knowledge discovery from vision and image data.

Maximal Pattern Discovery Problem. For the class of geometric graphs, Kuramochi and Karypis presented an efficient mining algorithm gFSG for the frequent geometric subgraph mining, based on Apriori-like breadth-first search [15]. However, the frequent pattern mining poses a problem in that it can easily produce an extremely large number of solutions, which degrades the performance and the comprehensivity of data mining to a large extent. The *maximal subgraph mining problem*, on the other hand, asks to find only all *maximal patterns* (closed patterns) appearing in a given input geometric graph D, where a *maximal pattern* is a geometric graph which is not included in any properly larger subgraph having the same set of occurrences in D. Since the set $\mathcal{M}$ of all maximal patterns is expected to be much smaller than the set $\mathcal{F}$ of all frequent patterns and still contains the complete information of D, maximal subgraph mining has some advantages as a compact representation to frequent subgraph mining.

Difficulties of Maximal Pattern Mining. However, there are a number of difficulties in maximal subgraph mining for geometric graphs. In general, maximal pattern mining has a large computational complexity [4,26]. So far, a number of efficient maximal pattern algorithms have been proposed for *sets, sequences*, and *graphs* [3,9,20,22,25]. Some algorithms use explicit duplicate detection and maximality test by maintaining a collection of already discovered patterns. This requires a large amount of memory and delay time, and introduces difficulties in the use of efficient search techniques, e.g., depth-first search. For these reasons, output-polynomial time computation for the maximal pattern problem is still a challenge in maximal geometric graphs. Moreover, the invariance under geometric transformation for geometric graphs adds another difficulty to geometric graph mining. In fact, no depth-first algorithm has been known to date even for frequent pattern mining.

Main Result. The goal of this paper is to develop a time and space efficient algorithm that can work well in theory and practice for maximal geometric graphs. As our main result, we present an efficient depth-first search algorithm MaxGeo that, given an input geometric graph, enumerates all frequent maximal pattern P in $\mathcal{M}$ without duplicates in $O(m(m+n)||D||^2 \log ||D||) = O(n^8 \log n)$ time per pattern and in $O(m) = O(n^2)$ space, with the maximum number m of occurrences of a pattern other than trivial patterns, the number n of vertices in the input graph, and the number $||D||$ of vertices and edges in the input graph. This is a polynomial delay and polynomial time algorithm for the maximal pattern discovery problem for geometric graphs. This is the first result establishing the output-sensitive complexity of maximal graph mining for geometric graphs.

Other Contributions of This Paper. To cope with the difficulties mentioned above, we devise some new techniques for geometric graph mining.

(1) We define a polynomial time computable *canonical code* for all geometric graphs in $\mathcal{G}$, which is invariant under geometric transformations. We give the first polynomial delay and polynomial space algorithm FreqGeo for the frequent geometric subgraph mining problem as a bi-product.
(2) We introduce the *intersection* and the *closure operation* for $\mathcal{G}$. Using these tools, we define the *tree-shaped search route $\mathcal{T}$ for all maximal patterns* in $\mathcal{G}$. We propose a new pattern growth technique arising from reverse search and *closure extension* [18] for traversing the search route $\mathcal{R}$ by depth-first search.

Related Works. There have been closely related researches on 1D and 2D point set matching algorithms, e.g. [2], where point sets are the simplest kind of geometric graphs. However, since they have mainly studied exact and approximate matching of point sets, the purpose is different from this work.

A number of efficient maximal pattern mining algorithms have been presented for subclasses of graph, trees, and sequences, e.g., general graphs [25], ordered and unordered trees [9], attribute trees [3,20], and sequences [4,5,23]. Some of them have output-sensitive time complexity as follows. The first group deal with the mining of "elastic" or "flexible" patterns, where the closure is not defined. CMTreeMiner [9], BIDE [23], and MaxFlex [5] are essentially output-polynomial time algorithms for location-based maximal patterns though it is implicit. They are originally used as pruning for document-based maximal patterns [5].

The second group deal with the mining of "rigid" patterns which have *closure*-like operations. LCM [22] proposes ppc-extension for maximal sets, and then CloATT [3] and MaxMotif [4] generalize it for trees and sequences. They together with this paper are polynomial delay and polynomial space algorithms.

Some of the other maximal pattern miners for complex graph classes, e.g.,CloseGraph [25], adopt frequent pattern discovery augmented with,e.g., maximality test and duplicate detection although output-polynomial time computability seems difficult to achieve with this approach.

Organization of this paper. Section 2 introduces the maximal pattern mining for geometric graphs. Section 3 gives the canonical code and the frequent pattern mining. In Section 4, we present polynomial delay and polynomial space algorithm MaxGeo for maximal pattern mining, and in Section 5, we conclude.

2   Preliminaries

We prepare basic definitions and notations for maximal geometric graph mining. We denote by $\mathbb{N}$ and $\mathbb{R}$ the set of all natural numbers and the set of all real numbers, resp.

2.1   Geometric Transformation and Congruence

We briefly prepare basic of plane geometry [11,13]. In this paper, we consider geometric objects, such as points, lines, point sets, and polygons, on the *two-dimensional*

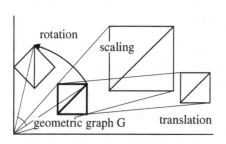

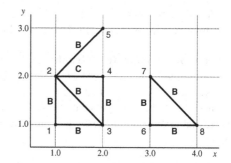

Fig. 1. Three basic types of geometric transformations

Fig. 2. A geometric database D with $V = \{1, \ldots, 8\}$, $\Sigma_V = \emptyset$, and $\Sigma_E = \{B, C\}$

Euclidean space $\mathbb{E} = \mathbb{R}^2$, also called the *2D plane*. A geometric transformation T is any mapping $T : \mathbb{R}^2 \to \mathbb{R}^2$, which transforms geometric objects into other geometric objects in the 2D plane $\mathbb{R}^2$. In this paper, we consider the class $\mathcal{T}_{\text{rgeo}}$ called *rigid transformations* of geometric transformations consisting of three basic types of geometric transformations: *rotation*, *scaling*, and their combinations. In general, any geometric transformation $T \in \mathcal{T}_{\text{rgeo}}$ can be represented as a *2D affine transformation* $T : x \mapsto Ax + t$, where A is a 2×2 nonsingular matrix with $det(A) \neq 0$, and t is a 2-vector. Such T is one-to-one and onto. In addition, if $T \in \mathcal{T}_{\text{rgeo}}$ then T preserves the angle between two lines. It is well-known that any affine transformation can be determined by a set of three non-collinear points and their images. For $\mathcal{T}_{\text{rgeo}}$, we have the following lemma.

Lemma 1 (Determination of Unknown Transformation). *Given two distinct points in the plane* x_1, x_2 *and the two corresponding points* x'_1, x'_2, *there exists a unique rigid transformation* T *in* $\mathcal{T}_{\text{rgeo}}$, *denoted by* $\mathbf{T}(\overline{x_1 x_2}; \overline{x'_1 x'_2})$, *such that* $T(x_i) = x'_i$ *for every* $i = 1, 2$.

$\mathbf{T}(\overline{x_1 x_2}; \overline{x'_1 x'_2})$ is computable in $O(1)$ time. The above lemma is crucial in the following discussion. For any geometric object O and $T \in \mathcal{T}_{\text{rgeo}}$, we denote the image of O via T by $T(O)$. The *inverse image* of O via T is $T^{-1}(O)$.

2.2 Geometric Graphs

We introduce the class of geometric graphs according to [15] as follows. Let Σ_V and Σ_E be mutually disjoint sets of *vertex labels* and *edge labels* associated with total orders $<_\Sigma$ on $\Sigma_V \cup \Sigma_E$. In what follows, a vertex is always an element of $\mathbb{N}$. A *graph* is a vertex and edge-labeled graph $G = (V, E, \lambda, \mu)$ with a set V of *vertices* and a set $E \subseteq V^2$ of *edges*. Each $x \in V$ has a vertex label $\lambda(x) \in \Sigma_V$, and each $e = xy \in E \subseteq V^2$ represents an unordered edge $\{x, y\}$ with an edge label $\mu(e) \in \Sigma_E$. Two graphs $G_i = (V_i, E_i, \lambda_i, \mu_i)$ $(i = 1, 2)$ are *isomorphic* if they are topologically identical to each other, i.e., there is a bijection $\phi : V_1 \to V_2$ such that (i) $\lambda_1(x) = \lambda_2(\phi(x))$, (ii) for every $xy \in (V_1)^2$, $xy \in E_1$ iff $\phi(x)\phi(y) \in E_2$, and (iii) for any $xy \in E_1$, $\mu_1(xy) = \mu_2(\phi(x)\phi(y))$. The mapping ϕ is called an *isomorphism* of G_1 and G_2.

A geometric graph is a representation of some geometric object by a set of features and their relationships on a collection of 2D points.

Definition 1 (Geometric Graph). Formally, a *geometric graph* (or *geograph*, for short) is a structure $G = (V, E, c, \lambda, \mu)$, where (V, E, λ, μ) is an underlying labeled graph and $c : V \to \mathbb{R}^2$ is a one-to-one mapping called the coordinate function. Each vertex $v \in V$ has the associated coordinate $c(v) \in \mathbb{R}^2$ in the 2D plane as well as its vertex label $\lambda(v)$. We refer to the components V, E, c, λ and μ of G as V_G, E_G, c_G, λ_G and μ_G.

We here assume that no two vertices or edges have the same coordinates, i.e., for any two vertices v and u, $c(v) \neq c(v)$. We note that even if there are vertices mapped on the same points, we can shrink them into a vertex. This for all such vertices takes $O(|V_G| \log |V_G|)$ time. We denote by $\mathcal{G}$ the *class of all geometric graphs* over Σ_V and Σ_E.

Alternative Representation for Geographs. Alternatively, a geometric graph can be simply represented as a collection of labeled objects $\underline{G} = \underline{V} \cup \underline{E}$, where $\underline{V} = \{ \langle \boldsymbol{x}_i, \lambda_i \rangle \,|\, i = 1, \ldots, n \} \subseteq \mathbb{R}^2 \times \Sigma_V$, and $\underline{E} = \{ \langle e_i, \mu_i \rangle \,|\, i = 1, \ldots, m \} \subseteq \mathbb{R}^2 \times \mathbb{R}^2 \times \Sigma_E$. Each $\langle \boldsymbol{x}, \lambda \rangle$ is a *labeled vertex* for a vertex v with $c(v) = \boldsymbol{x}$ and $\lambda(v) = \lambda$, and each $\langle c(v), c(u), \mu \rangle$ is a *labeled edge* for an edge $e = vu$ with label $\mu(e) = \mu$. A *labeled object* refers to either a labeled vertex or a labeled edge. Let $OL = (\mathbb{R}^2 \times \Sigma_V) \cup (\mathbb{R}^2 \times \mathbb{R}^2 \times \Sigma_E)$ be a domain of labeled objects. We assume the lexicographic order $<_{OL}$ over OL by extending those over $\mathbb{N}, \mathbb{R}^2, \Sigma_V$ and Σ_E. Since the correspondence between $\underline{G}$ and G is obvious, we will often use both representations interchangeably. For instance, we may write $G \cup \{\langle v, \boldsymbol{x}, \lambda \rangle\}$ or $G \setminus \{\langle e, \mu \rangle\}$. Since c is one-to-one, we may also write $\boldsymbol{x} \in G$ instead of $\boldsymbol{x} \in c(V_G)$.

2.3 Geometric Isomorphism and Matching

Now, let us extend the notions of isomorphisms and matchings for geographs as in [15]. Let $G_1, G_2 \in \mathcal{G}$ be any geographs. Then, G_1 and G_2 are *geometrically isomorphic*, denoted by $G_1 \equiv G_2$, if there are an isomorphism ϕ of G_1 and G_2 and a transformation $T \in \mathcal{T}_{\mathrm{rgeo}}$ such that $T(c(x)) = c(\phi(x))$ for every vertex x of G_1. The pair $\langle \phi, T \rangle$ is a *geometric isomorphism* of G_1 and G_2.

Let $G = (V, E, c, \lambda, \mu)$ be a geograph. A geograph H is a *geometric subgraph* of G, denoted by $H \subseteq G$, if H is a substructure of G, that is, (i) $V_H \subseteq V$ and $E_H \subseteq E$ hold, and (ii) mappings λ_H, μ_H, and c_H are the restrictions of λ, μ, and c, respectively, on V_H. Now, we define the matching of geographs in terms of geometric subgraph isomorphism.

Definition 2 (Geometric Matching). A geograph P *geometrically matches* a geograph G (or, P *matches* G) if there exists some geometric subgraph H of G that is geographically isomorphic to P with a geometric isomorphism $\langle \phi, T \rangle$. Then, we call the rigid transformation T a *geometric matching function* from P to G or an *occurrence* of P in G.

We denote by $\mathcal{M}(P, G) \subseteq \mathcal{T}_{\mathrm{rgeo}}$ *the set of all geometric matching functions* from P to G. We omit ϕ from $\langle \phi, T \rangle$ above because if P matches G then, there is at most one vertex $v = \phi(u) \in V_G$ of G such that $c(v) = T(c(u))$ for each $u \in V_P$ of P. Clearly, P matches G iff $\mathcal{M}(P, G) \neq \emptyset$. If P matches G then we write $P \sqsubseteq G$ and say P *occurs in* G or P *appears in* G. If $P \sqsubseteq Q$ and $Q \not\sqsubseteq P$ then we define $P \sqsubset Q$. We can observe that if both $P \sqsubseteq Q$ and $Q \sqsubseteq P$ hold then $P \equiv Q$, that is, P and Q are geometrically isomorphic. If we take the set $\overline{\mathcal{G}}$ of the equivalence classes of geographs modulo geometric isomorphisms, then $\sqsubseteq$ is a partial order over $\overline{\mathcal{G}}$.

2.4 Patterns, Occurrences, and Frequencies

Let $k \geq 0$ be a nonnegative integer. A k-*pattern* (or k-*geograph*) is any geograph $P \in \mathcal{G}$ with k vertices. From the invariance under $\mathcal{T}_{\mathrm{rgeo}}$, we assume without any loss of generality that if P is a k-pattern then $V_P = \{1, \dots, k\}$, and if $k \geq 2$ then P has the fixed coordinates $c(1) = (0, 0)$ and $c(2) = (0, 1) \in \mathbb{R}^2$ for its first two vertices in the local Cartesian coordinate. An input *geometric database* of size $n \geq 0$ is a single geograph $D = (V, E, c, \lambda, \mu) \in \mathcal{G}$ with $|V| = n$. We denote $|V| + |E|$, which is the total size of D, by $\|D\|$. D is also called an *input geograph*. Fig. 2 shows an example of an input geometric database D with $V = \{1, \dots, 8\}$ over $\Sigma_V = \emptyset$, and $\Sigma_E = \{\mathsf{B}, \mathsf{C}\}$.

Let $P \in \mathcal{G}$ be any k-pattern. Then, the *location list* of pattern P in D is defined by the set $L(P)$ of all rigid transformations that matches P to the input geograph D, i.e., $L(P) = \mathcal{M}(P, D)$. The *frequency* of P is $|L(P)| \in \mathbb{N}$. For an integer $0 \leq \sigma \leq n$, called a *minimum support* (or *minsup*), P is σ-*frequent* in D if its frequency is no less than σ.

Unlike ordinary graphs, the number of distinct matching functions in $L(P)$ is bounded by polynomial in the input size.

Lemma 2. *For any geograph P, $|L(P)|$ is no greater than n^2 under $\mathcal{T}_{\mathrm{rgeo}}$.*

Proof. From Lemma 1, the images $\boldsymbol{x}_1' \boldsymbol{x}_2'$ of just two points $\boldsymbol{x}_1 \boldsymbol{x}_2$ in the plane are sufficient to determine $\mathbf{T}(\boldsymbol{x}_1 \boldsymbol{x}_2; \boldsymbol{x}_1' \boldsymbol{x}_2')$ in $\mathcal{T}_{\mathrm{rgeo}}$. Thus, the result follows.   □

Lemma 3 (Monotonicity). *Let P, Q be any geographs. (i) If $P \equiv Q$ then $L(P) = L(Q)$. (ii) If $P \sqsubseteq Q$ then $L(P) \supseteq L(Q)$. (iii) If $P \sqsubseteq Q$ then $|L(P)| \geq |L(Q)|$.*

2.5 Maximal Pattern Discovery

From the monotonicity of the location list and the frequency in Lemma 3, it is natural to consider maximal subgraphs in terms of $\sqsubseteq$ preserving their location lists as follows.

Definition 3 (Maximal Geometric Patterns). A geometric pattern $P \in \mathcal{G}$ is said to be *maximal* in an input geograph T if there is no other geometric pattern $Q \in \mathcal{G}$ such that (i) $P \sqsubset Q$ and (ii) $L(P) = L(Q)$ hold.

In other words, P is maximal in D if there is no pattern strictly larger than P that has the same location list as P's. Equivalently, P is maximal iff any addition of a labeled object to P makes $L(P)$ strictly smaller than before. We denote by $\mathcal{F}^\sigma \subseteq \mathcal{G}$ be the set of all σ-frequent geometric patterns in D, and by $\mathcal{M} \subseteq \mathcal{G}$ be the set of all maximal geometric patterns in D under $\mathcal{T}$. The set of all σ-frequent maximal patterns is $\mathcal{M}^\sigma = \mathcal{M} \cap \mathcal{F}^\sigma$.

Now, we state our data mining problem as follows.

Definition 4 (Maximal Pattern Enumeration Problem). The *maximal geometric pattern enumeration problem* is, given an input geograph $D \in \mathcal{G}$ of size n and a minimum support $1 \leq \sigma \leq n$, to enumerate every frequent maximal geometric pattern $P \in \mathcal{M}^\sigma$ appearing in D without outputting no isomorphic two.

Our goal is to devise a light-weight and high-throughput mining algorithm for enumerating all maximal patterns appearing in a given input geograph. This is paraphrased in terms of output-sensitive enumeration algorithms in Section 2.6 as a polynomial delay and polynomial space algorithm for solving this problem. This goal has been an open question for $\mathcal{M}$ and even for $\mathcal{F}^\sigma$ so far.

We can define a different notion of location list $D(P)$, called the document list, defined as the set of input graphs in which a pattern appears, and maximality based on $D(P)$ in a similar way. Actually, location-based maximality is a necessary condition for document-based maximality. However, we do not go further in this direction.

2.6 Model of Computation

We make the following standard assumptions in computational geometry [19]: For every point $p = (x, y) \in \mathbb{E}$, we assume that its coordinates x and y have infinite precision. Our model of computation is the *random access machine* (RAM) model with $O(1)$ unit time arithmetic operations over real numbers as well as the standard functions of analysis $((\cdot)^{\frac{1}{2}}, \sin, \cos,$ etc) [1,19].

An enumeration algorithm $\mathcal{A}$ is an *output-polynomial time* algorithm if $\mathcal{A}$ finds all solutions $S \in \mathcal{S}$ without duplicates on a given input I in total polynomial time both in the input size and the output size. $\mathcal{A}$ is *polynomial delay* if the *delay*, which is the maximum computation time between two consecutive outputs, is bounded by polynomials in the input size. If $\mathcal{A}$ is polynomial delay, then $\mathcal{A}$ is also output-polynomial time. $\mathcal{A}$ is a *polynomial space* algorithm if the maximum space $\mathcal{A}$ uses is bounded by a polynomial in the input size.

3 Algorithm for Frequent Pattern Discovery

3.1 Canonical Encoding for Geographs

In this subsection, to properly handle the geometric isomorphism among the isomorphic patterns, we introduce the canonical code for geometric patterns, which is invariant under transformations in $\mathcal{T}_{\mathrm{rgeo}}$. Let P be any k-pattern with $V_P = \{1, \ldots, k\}$. Recall that the first two vertices of P have the fixed coordinates $c(1) = (0,0), c(2) = (0,1) \in \mathbb{R}^2$ in their local 2D plane.

Defining a Code. Suppose that the vertex set V_P of P has at least two vertices. Let $o = (\sum_{v \in V_P} c(v))/|V_P|$ be the centroid (the *center*) of the vertices in P, which is the averages of x-coordinates and y-coordinates of all vertices in P. We choose a point $x \in P, x \neq o$ having the minimum Euclidean distance to o called the *base point*. Denote by Q the pattern obtained by transforming P in a polar coordinate system such that o is mapped to the origin and x is mapped to $(0, 1)$, where the first element of the coordinates gives the angle. We define the coordinate of the origin by $(0, 0)$. Let

---

Elimination Ordering($\underline{P}$)

1: $i = 1; j = 1; P_1 = \underline{P}$;
2: **while** $P_i \neq \emptyset$ **do**
3:    $\langle o, l \rangle = \texttt{tail}(P_i)$ based on the canonical code $\texttt{Code}^*(P_i)$;
4:    $P_{i+1} = P_i - \{\langle o, l \rangle\}; \xi_j = \langle o, l \rangle$ and $j = j + 1$;
5: **end while**
6: **return** $\texttt{elimseq}(P) = (\xi_k, \ldots, \xi_1)$;

Fig. 3. Procedure for computing perfect elimination $\texttt{elimseq}(P)$ for geometric graph P

$O = \underline{V_Q} \cup \{\langle c(v), c(u) - c(v), \mu_{uv} \rangle, \langle c(u), c(v) - c(u), \mu_{uv} \rangle \mid uv \in E_Q\}$. Then, the code $\texttt{Code}(P, x)$ of P is defined by the elements of O sorted in lexicographic order.

Clearly, there are at most k distinct $\texttt{Code}(P, x)$ depending on the choice of the base point x. Then, the *canonical code* $\texttt{Code}^*(P)$ for pattern P is defined by the lexicographically minimum code among the codes of P. A pattern P is said to be *canonical* if (i) it has no vertex, (ii) it has one vertex at $(0, 0)$, or (iii) its vertices are indexed in the order of its canonical code.

Theorem 1 (Characterization of Canonical Code). *For any* $P, Q \in \mathcal{G}$ *of size* $k \geq 0$, $\texttt{Code}^*(P) = \texttt{Code}^*(Q)$ *iff* $P \equiv Q$ *under* $\mathcal{T}_{rgeo}$.

A code can be computed in $O(k^2 \log k)$ time for any k-pattern P and base point x, then the code for another base point is obtained by shifting it. Hence, we can compute the canonical code of P in $O(k^2 \log k)$ time. The purpose of the canonical code and the canonical pattern is to define a representative pattern among the geometric isomorphic patterns. Thus, our task is to enumerate all σ-frequent canonical patterns.

3.2 Perfect Elimination Sequences

Before studying enumeration or generation of each pattern, we consider the reverse process of enumeration, the decomposition of a given geograph. Let $P \in \mathcal{G}$ be any k-geograph. We define *perfect elimination sequence* by the sequence $\texttt{elimseq}(P) = (\xi_k, \ldots, \xi_1) \in OL^*$ obtained by the procedure *Elimination Ordering* in Fig. 3. Note that the elimination sequence $(\xi_k, \ldots, \xi_1)$ for P is not identical to the reverse of the canonical code $\texttt{Code}^*(P)$ since the i-th element ξ_i is selected based on the canonical code of the current geograph P_i not with the order defined on the initial graph $P = P_k$.

3.3 Algorithm for Frequent Pattern Discovery

Fig. 4 shows the algorithm FreqGeo for the frequent geometric subgraph discovery. Starting from the empty graph $\emptyset$, FreqGeo searches $\mathcal{F}^\sigma$ from smaller to larger by growing P with adding new labeled objects one by one. To avoid duplicates, FreqGeo adds a labeled object ξ to the current pattern P only when ξ is the last object in the canonical code $\texttt{Code}^*(P \cup \xi)$ of $P \cup \xi$. It corresponds to that any pattern P is generated in the reverse order of the elimination sequence. Thereby any pattern $Q = P \cup \xi$ is generatedexactly once only from the pattern $Q \setminus \xi$ where ξ is the last object in $\texttt{Code}^*(Q)$.

FreqGeo(σ : minsup, D : input database)
  1: **call** Expand_FG($\emptyset, \sigma, D$);

Expand_FG(P, σ, D)
  1: **if** $|L(P)| < \sigma$ **then return else output** P as a frequent subgraph;

  2: **for** each missing object ξ **do**
  3:    $Q = P \cup \{\xi\}$;
  4:    **if** ξ is the last of Code$^*(Q)$ **then call** Expand_FG(Q, σ, D);
  5: **end for**

Fig. 4. Polynomial delay and polynomial space algorithm for the frequent geometric subgraph enumeration problem

This ensures that each σ frequent pattern is output exactly once.

There are infinitely many candidates for the possible labeled object in Line 2 of Fig. 4. From the next lemma, we can avoid such a blind search by only focusing on *missing objects for P*, which is either labeled vertex or edge ξ such that $L(P) \supseteq L(P \cup \{\xi\}) \neq \emptyset$ holds. From Lemmas 1 and 3, we have the next lemma.

Lemma 4 (Missing Labeled Objects). *Let P be a pattern with nonempty $L(P)$ in D. Any missing object $\xi = \langle o, l \rangle$ for P is the inverse image of some labeled vertex or labeled edge π via T for some matching $T \in L(P)$, i.e., $\xi = T^{-1}(\pi)$ for some $\pi \in \underline{D}$.*

From Lemma 4 above, we know that there are at most $O(|L(P)| \cdot ||D||) = O(|V|^2(|V| + |E|))$ missing objects. Thus, Line 2 can be done in polynomial time. By using the technique called occurrence deliver described in [3,4,5,22], we can compute the frequencies of $P \cup \{\xi\}$ for all missing objects for P in $O(|V|^2(|V| + |E|) \log |V|)$ time. Therefore, the average computation time for each output pattern is $O(|V|^2(|V| + |E|)k^2 \log |V|)$, where k is the maximum size of σ-frequent pattern. Combining the above, we have the following theorem.

Theorem 2 (Frequent Geograph Enumeration). *The algorithm FreqGeo in Fig. 4 enumerates all σ-frequent geometric graphs in a given input database $D \in \mathcal{G}$ in polynomial delay and polynomial space in the total input size.*

4   Algorithm for Maximal Pattern Discovery

In this section, we present an efficient algorithm MaxGeo for the maximal pattern enumeration problem for the class of geographs that runs in polynomial delay and polynomial space in the input size.

4.1   Outline of the Algorithm

Fig. 5 shows our algorithm MaxGeo for enumerating all σ-frequent maximal geometric patterns in $\mathcal{M}^\sigma$ using backtracking. The key to the algorithm is a tree-like search

---

Algorithm MaxGeo: (D : input geograph, σ : $minsup$)

  1: $\bot = \mathrm{Clo}(\emptyset)$;   //*The bottom maximal geograph*
  2: **call** Expand_MaxGeo($\bot, \sigma, D$);

Algorithm Expand_MaxGeo(P, σ, D)

  1: **if** P is not σ-frequent **then return**;   //*Frequency test*
  2: **else output** P as a σ-frequent maximal geograph;
  3: **for** each missing labeled object $\xi = \langle o, \ell \rangle$ **do**   //*Lemma 4*
  4:     $Q = \mathrm{Clo}(P \cup \{\xi\})$;
  5:     **if** ($\mathcal{P}(Q) \equiv P$) **then**
  6:         **call** Expand_MaxGeo(Q, σ, D);   //*Recursive call for children*
  7: **endfor**

Fig. 5. A polynomial delay and polynomial space algorithm MaxGeo for the maximal geometric subgraph enumeration problem

route $\mathcal{R} = \mathcal{R}(\mathcal{M}^\sigma)$ implicitly defined over $\mathcal{M}^\sigma$. Then, starting at the root of the search route $\mathcal{R}$, MaxGeo searches $\mathcal{R}$ by jumping from a smaller maximal pattern to a larger one in a depth-first manner. Each jump is done by expanding each maximal pattern in polynomial time, thus the algorithm is polynomial delay.

4.2 Intersection and Closure Operations for Geographs

Let G_1 and G_2 be two geographs with $V_{G_1} \cap V_{G_2} \neq \emptyset$. The *maximally common geometric subgraph* (MCGS) of G_1 and G_2 is a geograph which is represented by labeled objects common to both G_1 and G_2. MCGS is unique for geographs, while they are not unique for ordinary graphs.

The intersection operation $\cap$ is reflexive, commutative, and associative over $\mathcal{G}$. For a set $\mathbf{G} = \{G_1, \ldots, G_m\}$ of geographs, we define $\cap \mathbf{G} = G_1 \cap G_2 \cap \cdots \cap G_m$. We can see that the computation time for $\cap \mathbf{G}$ are bounded by $O(\|\mathbf{G}\| \log \|\mathbf{G}\|)$. Some literatures [14] give an intersection of labeled graphs or first-order models in a different way which is based on the *cross product* of two structures. However, their iterative applications causes exponentially large intersections unlike $\cap \mathbf{G}$ above. Gariiga *et al.*[10] discussed related issues.

Now, let us define the closure operation for $\mathcal{G}$.

Definition 5 (Closure Operator for Geographs). Let $P \in \mathcal{G}$ be a geograph of size ≥ 2. Then, the *closure* of P in D is defined by the geograph $\mathrm{Clo}(P)$:

$$\mathrm{Clo}(P) = \bigcap \{ T^{-1}(D) \mid T \in L(P) \}.$$

Theorem 3 (Correctness of Closure Operation). *Let P be a geograph of size ≥ 2 and D be an input database. Then, $\mathrm{Clo}(P)$ is the unique, maximal geograph w.r.t. $\sqsubseteq$ satisfying $L(\mathrm{Clo}(P)) = L(P)$.*

Proof. We give a sketch of the proof. Let $T \in \mathcal{T}_{\mathrm{rgeo}}$ be any rigid transformation. Then, we can see that P matches D via T iff P is a geometric subgraph of the inverse image of D via T, i.e., $P \subseteq T^{-1}(D)$. Thus, taking the intersection of the inverse image $T^{-1}(D)$ for all matching T of P, we obtain the unique maximal subgraph having $L(P)$. □

Lemma 5. *For any geographs $P, Q \in \mathcal{G}$, the following properties hold:*

(i) $P \sqsubseteq \mathrm{Clo}(P)$. *(ii)* $L(\mathrm{Clo}(P)) \equiv L(P)$. *(iii)* $\mathrm{Clo}(P) \equiv \mathrm{Clo}(\mathrm{Clo}(P))$.

(iv) $P \sqsubseteq Q$ *iff* $L(P) \supseteq L(Q)$ *for any maximal $P, Q \in \mathcal{M}$.*

(v) $\mathrm{Clo}(P)$ *is the unique, smallest maximal geograph containing P.*

(vi) *For the empty graph $\emptyset$, $\bot = \mathrm{Clo}(\emptyset)$ is the smallest element of $\mathcal{M}$.*

Theorem 4 (Characterization of Maximal Geographs). *Let D be an input geograph and $P \in \mathcal{G}$ be any geograph. Then, P is maximal in D iff $\mathrm{Clo}(P) \equiv P$.*

4.3   Defining the Tree-Shaped Search Route

In this subsection, we define a tree-like search route $\mathcal{R} = (\mathcal{M}^{\sigma}, \mathcal{P}, \bot)$ for the depth-first search of all maximal geographs based on a so-called parent function.

Let $Q \in \mathcal{M}$ be a maximal pattern of vertices at least two such that $Q \neq \bot$. For any labeled object $\xi \in Q$, define the ξ-prefix of Q as the pattern $Q[\xi]$ which is the collection of the labeled objects prior to ξ in $\mathrm{Code}^*(Q)$. Then, the *core index* $\mathrm{core\_i}(Q)$ of Q is the labeled object ξ such that $L(Q[\xi']) \neq L(Q)$ holds for any ξ' prior to ξ in $\mathrm{Code}^*(Q)$. We can show that if $Q \neq \bot$ then $\mathrm{core\_i}(Q)$ is always defined.

$Q[\mathrm{core\_i}(Q)] \subseteq Q$ is the shortest prefix of Q satisfying $L(Q[\xi]) = L(Q)$. Moreover, if we remove $\mathrm{core\_i}(Q)$ from the prefix $Q[\mathrm{core\_i}(Q)]$, then we have a properly shorter prefix, and then the location list changes. Now, we define the parent function $\mathcal{P}$ that gives the predecessor of Q.

Definition 6 (Parent Function $\mathcal{P}$). The *parent* of any maximal pattern $Q \in \mathcal{M}$ ($Q \neq \bot$) is defined by $\mathcal{P}(Q) = \mathrm{Clo}(Q[\xi] \setminus \{\xi\})$, where $\xi = \mathrm{core\_i}(Q)$ is the core index of Q.

Lemma 6. $\mathcal{P}(Q)$ *is (i) always defined, (ii) unique, and (iii) a maximal pattern in $\mathcal{M}$. Moreover, $\mathcal{P}$ satisfies that (iv) $\mathcal{P}(Q) \subset Q$, (v) $|\mathcal{P}(Q)| < |Q|$, and (vi) $L(\mathcal{P}(Q)) \supset L(Q)$.*

Now, we define the *search route* for $\mathcal{M}^{\sigma}$ as a rooted directed graph $\mathcal{R}(\mathcal{M}^{\sigma}) = (\mathcal{M}^{\sigma}, \mathcal{P}, \bot)$, where $\mathcal{M}^{\sigma}$ is the vertex set, $\mathcal{P}$ is the set of reverse edges, and $\bot$ is the root. For the search route, we have the following theorem.

Theorem 5 (Reverse Search Property). *For every σ, the search route $\mathcal{R}(\mathcal{M}^{\sigma})$ is a spanning tree with the root $\bot$ over all the maximal patterns in $\mathcal{M}^{\sigma}$.*

4.4   A Polynomial Space Polynomial Delay Algorithm

The remaining thing is to show how we can efficiently traverse the search route $\mathcal{R}(\mathcal{M}^{\sigma})$ starting from $\bot$. However, this is not a straightforward task since $\mathcal{R}(\mathcal{M}^{\sigma})$ only has the reverse edges. To cope with this difficulty, we introduce the technique so called reverse search [7] and the closure extension [18].

Lemma 7. *For maximal patterns Q and P, P is the parent of Q only if $Q \equiv clo(P \cup \xi)$ holds for a missing object ξ for P.*

Proof. Suppose that P is the parent of Q, and ξ' is the labeled object preceding and next to $\texttt{core\_i}(Q)$ in the canonical code of Q. ξ' is included in $\underline{P}$, since $P = Clo(Q[\xi'])$. Since $L(Q)$ is a collection of $T \in L(Q[\xi'])$ satisfying that $\underline{T^{-1}(D)}$ includes $\texttt{core\_i}(Q)$, together with $L(Q[\xi']) = L(P)$, $L(P \cup \{\xi\}) = L(Q[\underline{\xi'}] \cup \{\xi\}) = L(Q)$. Thus the statement holds.   □

The operation of adding a labeled object and taking its closure is called *closure extension*. Lemma 7 states that any maximal geometric pattern can be obtained by applying to $\perp$ closure extensions repeatedly.

   From Lemma 7, we can see that to find all children of a pattern P, we have to examine the closure extension for all missing objects for P. Clearly, a closure extension $Q = Clo(P \cup \xi)$ of P is a child of P if its parent is P. Since the parent of Q can be obtained by computing its canonical code, we can check whether a closure extension is a child or not in $O(k^2 log k)$ time where k is the number of labeled objects in Q. Since the computation of $clo(Q)$ takes $O(|L(Q)| \times ||D|| \log ||D||)$ time, we obtain the following theorem.

Theorem 6 (correctness and complexity of MaxGeo). *Given an input geograph D with vertex set V and a minimum support threshold $\sigma > 0$, the algorithm MAX-GEO in Fig. 5 enumerates all σ-frequent maximal geographs in $O((m||D||) \times ((m + n)||D|| \log ||D||)) = O(m(m+n)||D||^2 \log |D|)$ per maximal geograph with $O(||D||)$ space, where $m = O(n^2)$ is the maximum size of the location lists.*

If σ is not too small, then the number of missing objects to examine will consequently be small, such as $O(n)$, decreasing the computation time will be short. This is expected in practical computation. Moreover, in practice, usually almost all (maximal) patterns to be output have small frequencies close to σ, thus the computation time for the closure operation is rather short. According to the computational experiments in [4,22], practical computation time is very short in such cases.

Corollary 1. *The maximal geograph enumeration problem is solvable in polynomial delay and polynomial space.*

5 Conclusion

We presented a polynomial delay and polynomial space algorithm that discovers all maximal geographs in a given geometric configuration without duplicates. As future works, we intend to implement and evaluate the experimental performance of the algorithm. Dealing with the input of many geographs and document occurrence is a straightforward work. Dealing with polygons is also straightforward, by using sophisticated labels to identify edges of polygons as a group. Extensions with approximation and constraints, with applications to image processing and geographic information systems, are other future problems.

References

1. Aho, A.V., Hopcroft, J.E., Ullman, J.D.: Data Structures and Algorithms (1983)
2. Akutsu, T., Tamaki, H., Tokuyama, T.: Distribution of distances and triangles in a point set and algorithms for computing the largest common point sets. Discr. & Comp. Geom. 20(3), 307–331 (1998)
3. Arimura, H., Uno, T.: An output-polynomial time algorithm for mining frequent closed attribute trees. In: Kramer, S., Pfahringer, B. (eds.) ILP 2005. LNCS (LNAI), vol. 3625, pp. 1–19. Springer, Heidelberg (2005)
4. Arimura, H., Uno, T.: A polynomial space and polynomial delay algorithm for enumeration of maximal motifs in a sequence. In: Deng, X., Du, D.-Z. (eds.) ISAAC 2005. LNCS, vol. 3827, Springer, Heidelberg (2005)
5. Arimura, H., Uno, T.: Effcient algorithms for mining maximal flexible patterns in texts and sequences, TCS-TR-A-06-20, DCS, Hokkaido Univeristy 2006 (submitting)
6. Asai, T., Abe, K., Kawasoe, S., Arimura, H., Sakamoto, H., Arikawa, S.: Efficient substructure discovery from large semi-structured data. In: Proc. SDM'02 (2002)
7. Avis, D., Fukuda, K.: Reverse search for enumeration. Discrete App. Math. 65, 21–46 (1996)
8. Chi, Y., Muntz, R.R., Nijssen, S., Kok, J.N.: Frequent subtree mining – An overview. Fundam. Inform. 66(1-2), 161–198 (2005)
9. Chi, Y., Yang, Y., Xia, Y., Muntz, R.R.: CMTreeMiner: mining both closed and maximal frequent subtrees. In: Dai, H., Srikant, R., Zhang, C. (eds.) PAKDD 2004. LNCS (LNAI), vol. 3056, Springer, Heidelberg (2004)
10. Garriga, G.C., Khardon, R., De Raedt, L.: On mining closed sets in multi-relational data. In: Proc. IJCAI 2007, pp. 804–809 (2007)
11. Guerra, C.: Vision and image processing algorithms. In: Algorithms and Theory of Computation Handbook. ch. 22, vol. f 22-1–22-23, CRC Press (1999)
12. Inokuchi, A., Washio, T., Motoda, H.: An apriori-based algorithm for mining frequent substructures from graph data. In: Zighed, A.D.A., Komorowski, J., Żytkow, J.M. (eds.) PKDD 2000. LNCS (LNAI), vol. 1910, pp. 13–23. Springer, Heidelberg (2000)
13. Jain, A.: Fundamentals of Digital Image Processing. Prentice-Hall, Englewood Cliffs (1986)
14. Khardon, R.: Learning function-free horn expressions. Machine Learning 37(3), 241–275 (1999)
15. Kuramochi, M., Karypis, G.: Discovering frequent geometric subgraphs. In: Proc. IEEE ICDM'02, pp. 258–265 (2002)
16. Nakano, S.: Efficient generation of plane trees. Information Processing Letters 84, 167–172 (2002)
17. Nijssen, S., Kok, J.N.: Effcient discovery of frequent unordered trees. In: Proc. MGTS'03 (2003)
18. Pasquier, N., Bastide, Y., Taouil, R., Lakhal, L.: Discovering frequent closed itemsets for association rules. In: Beeri, C., Bruneman, P. (eds.) ICDT 1999. LNCS, vol. 1540, pp. 398–416. Springer, Heidelberg (1999)
19. Preparata, F.P., Shamos, M.I.: Computational Geometry: An Introduction. Springer, Heidelberg (1985)
20. Termier, A., Rousset, M.-C., Sebag, M.: DRYADE: a new approach for discovering closed frequent trees in heterogeneous tree databases. In: Proc. ICMD'04 (2004)
21. Tsuda, K., Kudo, T.: Clustering graphs by weighted substructure mining. In: Proc. ICML 2006, pp. 953–960 (2006)
22. Uno, T., Asai, T., Uchida, Y., Arimura, H.: An efficient algorithm for enumerating closed patterns in transaction databases. In: Suzuki, E., Arikawa, S. (eds.) DS 2004. LNCS (LNAI), vol. 3245, pp. 16–30. Springer, Heidelberg (2004)

23. Wang, J., Han, J.: BIDE: Efficient Mining of Frequent Closed Sequences. In: Proc. IEEE ICDE'04, pp. 79–90 (2004)
24. Washio, T., Motoda, H.: State of the art of graph-based data mining. SIGKDD Explor. 5(1), 59–68 (2003)
25. Yan, X., Han, J.: CloseGraph: mining closed frequent graph patterns. In: Proc. KDD'03 (2003)
26. Yang, G.: The complexity of mining maximal frequent itemsets and maximal frequent patterns. In: Proc. KDD'04, pp. 344–353 (2004)
27. Zaki, M.J.: Efficiently mining frequent trees in a forest. In: Proc. KDD'02, pp. 71–80 (2002)
28. Zaki, M.J., Aggarwal, C.C.: XRules: an effective structural classifier for XML data. In: Proc. KDD'03, pp. 316–325 (2003)

Iterative Reordering of Rules for Building Ensembles Without Relearning[*]

Paulo J. Azevedo[1] and Alípio M. Jorge[2,3]

[1] CCTC, Departamento de Informática, Universidade do Minho, Portugal
pja@di.uminho.pt
[2] Fac. de Economia, Universidade do Porto, Portugal
amjorge@fep.up.pt
[3] LIAAD, INESC Porto L.A.

Abstract. We study a new method for improving the classification accuracy of a model composed of classification association rules (CAR). The method consists in reordering the original set of rules according to the error rates obtained on a set of training examples. This is done iteratively, starting from the original set of rules. After obtaining N models these are used as an ensemble for classifying new cases. The net effect of this approach is that the original rule model is clearly improved. This improvement is due to the ensembling of the obtained models, which are, individually, slightly better than the original one. This ensembling approach has the advantage of running a single learning process, since the models in the ensemble are obtained by self replicating the original one.

1 Introduction

The use of association rules for classification has proved to be a promising path in terms of improving predictive performance by enabling a wider search in the set of patterns supported by the data [12,13,15]. Given a set of association rules, using them in the best possible way to perform classification is a challenge proportional to the enormous number of rules that can be produced with reasonable computational resources. Recent work has exploited the use of low-cost ensemble learning (with a single learning process) to further improve the results of association rule classifiers [9]. The idea is to generate a first set of rules and then to obtain replications of this set by sampling it in a manner similar to bootstrap. The replications are then used as an ensemble.

In this paper we study another approach for generating ensembles using a single rule generation step. The main idea is to obtain the models by iteratively reweighting/reordering the rules of the original rule set. The initial rule model M_0 is obtained using a learning algorithm. In this initial model, each rule has an associated predictive value, which can be used to sort the rules for classification.

[*] Supported by Fundação Ciência e Tecnologia, Project Site-o-matic, FEDER e Programa de Financiamento Plurianual de Unidades de I & D.

V. Corruble, M. Takeda, and E. Suzuki (Eds.): DS 2007, LNAI 4755, pp. 56–67, 2007.

Model M_1 is obtained by reweighting the rules on the training set. This reweighting can lead to a different rule ordering, if a decision list approach is used for model evaluation. Each model in the sequence M_i is obtained from the previous one in the same way, until we obtain N models. The ensemble $\{M_i, i = 1..N\}$ is used to classify new cases. The intended effect is that rule ordering is recomputed taking into account global effects on accuracy, instead of local ones. We call this approach Iterative Reordering Ensembling (IRE).

As referred above, one particular feature of this ensemble approach is that the learning process that generates the rules runs only once. The sequence of models is obtained by finding close alternatives to the initial rule ordering. This process has some similarities to boosting [7,19], where a sequence of models is generated from iteratively reweighted sets of examples. In boosting, the weights of the examples are changed, so that misclassified examples get higher weights.

In Iterative Reordering Ensembling, a new model is generated by changing the order of the rules, where rules with more errors go down. Thus, misclassified examples improve their chance of being well classified. The main advantage w.r.t. boosting is the fact that one single learning step is used, whereas in boosting there as many learning steps as models in the ensemble.

In the remaining of the paper we revisit the research done on classification with association rules and also on ensemble learning. We describe in detail this new approach and present an empirical evaluation. The results obtained indicate that IRE improves the predictive accuracy of classification with association rules mainly by reducing the bias component of the classification error.

2   Classification with AR

Association rules have been proposed for the first time as complete and competitive classification models by Liu *et al.* in 1998 [13]. In simple terms, the produced classifier was a decision list, and each new case was classified by the best rule that applied to it, i.e., the rule with highest confidence. Later, Li et al. [12] proposed the use of multiple rules, instead of just one, to classify each new case. The subset of rules that apply to the new case are grouped by anwered class, and each of these groups is assessed with a weighted χ^2 heuristic that tried to identify the strongest group. Meretakis and Wüthrich [15] suggested a well founded procedure to combine multiple rules by using the confidence of the rules to determine the most likely class for each case, in a kind of naïve Bayes approach with less independence assumptions. Jovanoski and Lavrac [10] have studied the effect of simple voting and other simple strategies to improve the prediction ability of a set of association rules. Jorge and Azevedo [9] have proposed an ensemble strategy based on multiple sets of association rules. The work presented here is a follow-up of that general approach.

2.1   Obtaining Classifiers from Association Rules

We can regard classification from association rules as a particular case of the general problem of model combination. Either because we see each rule as a separate model or because we consider subsets of the rules for combination. We first build a set of rules R. Then we select a subset M of rules that will be used in classification, and finally we choose a prediction strategy π that obtains a decision for a given unknown case x. To optimize predictive performance we can fine tune one or more of these three steps.

Strategy for the generation of rules: A standard approach is to employ a sort of coverage strategy [13]. All association rules are derived. Then, one chooses the best rule, removes the covered cases and repeat the selection of rules until all cases are covered. In [12] this standard coverage strategy is generalised to allow more redundancy between rules. A case is only removed from the training data when it is covered by a pre-defined number of rules. In our work, we build the set of rules separately using the *CAREN* system [9]. *CAREN* is specialized in generating association rules for classification and employs a bitwise depth-first frequent patterns mining algorithm.

Choice of the rule subset: We can use the whole set of rules for prediction, and count on the predictive strategy to dynamically select the most relevant ones. Selection of rules is based on some measure of quality, or combination of measures. The structure of rules can also be used, for example for discarding rules that are generalizations of others. Discarding rules that are potentially irrelevant or harmful for prediction is called *pruning* [12,13].

Strategy for prediction: Most of the previous work on using association rules for classification has been done on this topic. The simplest approach is to go for the rule with the highest quality, typically measured as confidence, sometimes combined with support [13]. Other approaches combine the rules by some kind of *committee method*, such as voting [10], or weighted voting [12].

Rule selection, or pruning, can be done right after rule generation. However, most of the rule selection techniques can be used before, when the rules are being generated. Pruning techniques rely on the elimination of rules that do not improve more general versions. For example, rule $\{a, b, c\} \to g$, may be pruned away if rule $\{a, c\} \to g$ has similar or better predictive accuracy. CBA [13] uses pessimistic error pruning. Another possibility is to simply use some measure of *improvement* [3] on a chosen rule quality measure. At modeling time we can still reduce the set of rules by choosing only the N-best ones overall, or the N-best ones for each class [10], where N is a user provided parameter. This technique may reduce the number of rules in the model dramatically, but the choice of the best value for N is not clear.

2.2   Combining the Decisions of Rules

In this section we describe the two simplest strategies for using association rule sets as classification models. In the discussion we assume we have a static set R of classification association rules, and a predefined set of classes G and that we want to classify cases with description x, where the description of a case is a set of statements involving independent attributes. The set of rules that apply to the case, or that fire upon the case with description x will be $F(x)$ defined as:

$$\{(x' \rightarrow class = g) \in R \mid x' \subseteq x, g \in G\} \tag{1}$$

Best rule. This strategy classifies using one single rule $bestrule_x$:

$$bestrule_x = arg \max_{r \in F(x)} meas(r) \tag{2}$$

The *meas* used is a function that assigns to each rule a value of its predictive power. *Confidence* is the natural choice when it comes to prediction. It estimates the posterior probability of C given A, and is defined as $confidence(A \rightarrow C) = sup(A \cup C)/sup(A)$.

Conviction is another interest measure [5] somewhat inspired in the logical definition of implication and attempts to measure the degree of implication of a rule. Conviction is infinite for logical implications (confidence 1), and is 1 if A and C are independent, and it sometimes outperforms confidence in terms of prediction [9]. It is defined as $conviction(A \rightarrow C) = (1 - sup(C))/(1 - confidence(A \rightarrow C))$.

The prediction given by the best rule is the best guess we can have with one single rule. When the best rule is not unique we can break ties maximizing support [13]. A kind of best rule strategy, combined with a coverage rule generation method, provided encouraging empirical results when compared with state of the art classifiers on some datasets from UCI [16].

Our implementation of Best Rule prediction follows closely the rules ordering described in CMAR [12]. Thus, R_1 is earlier than R_2 is defined as:

$$R_1 \prec R_2 \quad if \quad meas(R_1) > meas(R_2) \quad or \quad meas(R_1)==meas(R_2) \wedge sup(R1)>sup(R2)$$
$$or \quad meas(R_1)==meas(R_2) \wedge sup(R2)==sup(R2) \wedge ant(R1)<ant(R2).$$

where *meas* is the used interest measure and *ant* is the length of the antecedent.

Weighted voting. This strategy combines the rules $F(x)$ that fire upon a case x. The answer of each rule is a *vote*, and the final decision is obtained by assigning a specific weight to each vote, according to its perceived quality. In the case of association rules, this can be done using one of the above defined measures.

$$prediction_{wv} = arg \max_{g \in G} \sum_{x' \in antecedents(F(x))} vote(x', g). \max meas(x' \rightarrow g) \tag{3}$$

3   Iterative Reordering

In this paper, we propose an approach to increase the accuracy of a CAR set by re-evaluating the interest and support of each rule according to its performance on a specific dataset. This new evaluation works by running the rules on the training set. Then, rule's interest is redefined according to its accuracy on this set. Rule's support is also redefined but as a measure of rule's usage

Algorithm 1. Iterative reordering trial generation

Input: training_set=D, max iterations = $MaxI$
1 Generate rule set R from D;
2 **Trial generation** $Trial_0 = R$;
3 **foreach** i *in 1 to* $MaxI$ **do**
4   **foreach** x *in* D **do**
5   using bestrule_measure approach see which rule r in $Trial_{i-1}$ fires;
6   recomputes interest and support measures of rule in R based on usage and accuracy;
7         $Trial_i$ = rules used from $Trial_{i-1}$, with new interest and support + rules from $Trial_{i-2}$ not used in $i-1$ + rules from $Trial_0$ not used in either $i-1$ or $i-2$;
8   (interest and support measures of rules from trials $i-2$, $i-1$ and 0 are the ones calculated there);
9         $accuracy_i$ = accuracy of $Trial_i$ on D;
10   If $accuracy_i < 0.5$ or $accuracy_i > 0.99$ break for;
11   **end**
12 **end**
Output: $Trials$

in classification. The redefinition yields a new ordering on the original set of rules. This process is applied iteratively, yielding a set of rule models that can be aggregated.

3.1   Ensemble Generation

BestRule prediction is applied to the training dataset using the original CARules. From this application, rule's measures (support and interest) are updated according to usage and accuracy.

For instance, if confidence is used in BestRule prediction in $Trial_{i-1}$ then in $Trial_i$, the confidence of rule $A \rightarrow C$ is:

$$conf(A \rightarrow C, Trial_i, D) = \frac{hits(A \rightarrow C, Trial_{i-1}, D)}{usage(A \rightarrow C, Trial_{i-1}, D)}$$

where

$$hits(A{\rightarrow}C, Trial, D) = \#\{x{\in}D|\ (A{\rightarrow}C) == BestRule(Trial, x) : x \sqsupseteq A \wedge x \sqsupseteq C\}$$

and

$$usage(A{\rightarrow}C, Trial, D) = \#\{x{\in}D|\ (A{\rightarrow}C) == BestRule(Trial, x) : x \sqsupseteq A\}$$

and $BestRule(Trial, x)$ represents the best rule in $Trial$ that applies to x.

Other interest measures can be defined referring to $conf$ and support. For instance, conviction is defined as:

$$conv(A \rightarrow C, Trial_i, D) = \frac{1 - sup(C, D)}{1 - conf(A \rightarrow C, Trial_{i-1}, D)}$$

Notice that the minimal required information to represent trials (rule models) is the *usage* and *hits* associated with each rule. A matrix with $2 \times n$ (for n trials) is enough to represent the ensemble.

Prediction on test cases using the ensemble is obtained using BestRule for each case on each trial. The overal prediction is obtained by weighted trial voting. Different ensemble predictions can be obtained using different weighing strategies. For instance, the weight can be the global accuracy of the trial on the training set. Alternatively, the vote can be the interest measure of the best rule within the trial. In the sequel, the latter will be referred as *IRE.BR.int* and the former as *IRE.V.Acc.int*, where *int* is either confidence or conviction.

On each trial, only rules with *usage* > 0 are considered. During ensemble formation it may be the case that there is no rule with positive usage in a given trial that covers a specific case. In that situation the BestRule prediction consults previous trials looking for rules that may cover the case. Line 8 includes this contingency mechanism.

Similarly to AdaBoost [7] trial construction, in line 11 the algorithm stops earlier either if a very good or very bad trial accuracy is achieved. Algoritm 1 summarizes the IRE ensemble construction process. Given n (number of trials) and a training set, the algorithm derives $Trial_0$ as the set of CAR rules through an association rule engine. Then, it iteratively derives $Trial_i$ applying BestRule to the training set using the rules from $Trial_{i-1}$. After n iterations, the ensemble construction is complete. A test set is evaluated by weight voting using best rule prediction on each trial.

4   Experimental Validation

We have conducted experiments comparing the predictive performance of the *ensemble* approach with *bestrule with AR*, using different prediction measures (for assessing the net effect of this kind of ensembling) and state-of-the-art algorithms (for controlling the results). We have used 17 UCI datasets [16]. The datasets are described in Table 1. As a reference algorithm, we used the decision tree inducer *c4.5* [17]. Due to its availability and ease of use we have also compared the results with *rpart* from the statistical package R [18]. *Rpart* is a CART-like decision tree inducer [4].

For the single model association rule classifiers, we used four *CAREN* variants, by combining two strategies: "Best rule" and "Weighted Voting" with two measures (confidence and conviction). Minimal support was set to 0.01 or 10 training cases. The only exception was the *sat* dataset, where we used 0.02 for computational reasons. Minimal improvement was 0.01 and minimal confidence 0.5. We have also used the χ^2 filter to eliminate potentially trivial rules. For each combination we ran *CAREN* with and without IRE-ensembles. Numerical attributes have been previously discretized using *CAREN*'s implementation of Fayyad and Irani's supervised discretization method [6]. However, both *c4.5* and *rpart* runs used the original raw datasets.

Table 1. Datasets used for the empirical evaluation

Dataset	#examples	#classes	#attr	#numerics
australian	690	2	14	6
breast	699	2	9	8
pima	768	2	8	8
yeast	1484	10	8	8
flare	1066	2	10	0
cleveland	303	5	13	5
heart	270	2	13	13
hepatitis	155	2	19	4
german	1000	2	20	7
house-votes	435	2	16	0
segment	2310	7	19	19
vehicle	846	4	18	18
adult	32561	2	14	6
lymphography	148	4	18	0
sat	6435	6	36	36
shuttle	58000	7	9	9
waveform	5000	3	21	21

An estimation of the error of each algorithm (and *CAREN* variant) was obtained on each dataset with stratified 10-fold cross-validation (Table 2). From the estimated errors we ranked the algorithms separately for each dataset, and used mean ranks as an indication of global rank (Table 3). Besides that, we have tested the statistical significance of the results obtained.

4.1   Analysis of Results

The first strong observation is that the iterative reordering (IRE) approaches rank high, when compared to the other approaches. Of the 10 algorithms tested, the first three employ IRE. Separate experiments, not shown here, indicate that the IRE gains advantage through the ensemble strategy, rather than the filtering of rules. Of the two possibilities for combining the rules in the ensemble, the bestmodel approach works well with confidence but poorly with conviction. Of the two predictive measures used, confidence seems to be preferable for the top strategies, but not in general.

In terms of statistical significance, the IRE ensemble approaches are clearly better than the single model AR classifiers. In a t-test with a significance of 0.01,

Table 2. Average error rates obtained with the algorithms on the datasets (min. sup.=(0.01 or 10 cases, except sat with 0.02), min. conf.=0.5, imp.=0.01). Key: BR=best rule, V=Voting, IRE=Iterative Reordering, cf=confidence, cv=conviction, Acc=trial accuracy voting.

	rpart	c4.5	BR.cf	BR.cv	V.cf	V.cv	IRE.BR.cf[1]	IRE.BR.cv	IRE.V.Acc.cf[2]	IRE.V.Acc.cv
aus	0.1623	0.1392	0.1378	0.1378	0.1871	0.1552	0.1318	0.1392	0.1333	0.1348
bre	0.0615	0.0500	0.0457	0.0428	0.0386	0.0386	0.0386	0.0500	0.0386	0.0357
pim	0.2472	0.2436	0.2278	0.2212	0.2277	0.2264	0.2329	0.2355	0.2316	0.2316
yea	0.4327	0.4427	0.4214	0.4194	0.4301	0.4240	0.4294	0.4435	0.4327	0.4395
fla	0.1773	0.1744	0.1914	0.2025	0.1810	0.1894	0.1932	0.1950	0.1932	0.2026
cle	0.4616	0.5004	0.4570	0.4570	0.4570	0.4570	0.4570	0.4587	0.4570	0.5021
hea	0.2000	0.2109	0.1778	0.1815	0.1741	0.1815	0.1778	0.1963	0.1741	0.1778
hep	0.2600	0.2132	0.1999	0.1870	0.1420	0.1878	0.1682	0.1823	0.1682	0.1761
ger	0.2520	0.3020	0.2820	0.2620	0.2570	0.2630	0.2710	0.2650	0.2730	0.2540
hou	0.0487	0.0325	0.0786	0.0786	0.1266	0.1334	0.0646	0.0668	0.0622	0.0622
seg	0.0831	0.0321	0.1190	0.1190	0.2030	0.1242	0.0779	0.0978	0.0801	0.0801
veh	0.3176	0.2596	0.3673	0.3662	0.3331	0.3342	0.3176	0.3272	0.3222	0.3234
adu	0.1555	0.1361	0.1873	0.1549	0.1735	0.1617	0.1599	0.2113	0.1592	0.1605
lym	0.2527	0.2307	0.1729	0.1800	0.2666	0.1989	0.1595	0.1729	0.1667	0.1667
sat	0.1904	0.1397	0.1975	0.1939	0.3514	0.2309	0.1523	0.1848	0.1510	0.1563
shu	0.0053	0.0005	0.0251	0.0075	0.0569	0.0083	0.0304	0.0168	0.0310	0.0048
wav	0.2664	0.2273	0.1774	0.1770	0.1990	0.1822	0.1654	0.2336	0.1650	0.1654

Table 3. Ranks for each algorithm on each dataset (1 is best, x.5 is a draw). Table lines are sorted by the mean rank, which can be found in the first column.

	mean	aus	bre	pim	yea	fla	cle	hea	hep	ger	hou	seg	veh	adu	lym	sat	shu	wav
IRE.Br.conf	4	1	3.5	7	4	6.5	3.5	4	2.5	7	5	2	2.5	5	1	3	8	2.5
IRE.Vote.Acc.conf	4.06	2	3.5	5.5	6.5	6.5	3.5	1.5	2.5	8	3.5	3.5	4	4	2.5	2	9	1
IRE.Vote.Acc.conv	4.5	3	1	5.5	8	10	10	4	4	2	3.5	3.5	5	6	2.5	4	2	2.5
Best.rule.conv	5.21	4.5	6	1	1	9	3.5	6.5	6	4	7.5	7.5	9	2	6	7	4	4
c4.5	5.53	6.5	8.5	9	9	1	9	10	9	10	1	1	1	1	8	1	1	8
Voting.conv	6.09	8	3.5	2	3	4	3.5	6.5	7	5	10	9	8	7	7	9	5	6
Voting.conf	6.15	10	3.5	3	5	3	3.5	1.5	1	3	9	10	7	8	10	10	10	7
Best.rule.conf	6.21	4.5	7	4	2	5	3.5	4	8	9	7.5	7.5	10	9	4.5	8	7	5
rpart	6.24	9	10	10	6.5	2	8	9	10	1	2	5	2.5	3	9	6	3	10
IRE.Br.conv	7.03	6.5	8.5	8	10	8	7	8	5	6	6	6	6	10	4.5	5	6	9

IRE.Br.conf has 4 significant wins against 0 of Best.rule.conf. When compared to Best.rule.conv, the advantage is of 3 significant wins. Using the same statistical test, we see that the single model AR classifiers tend to be worse than *rpart* (Best.rule.conv looses 3/1), but the IRE ensemble best strategies beat *rpart* (2/0). c4.5 beats the best IRE strategy, in terms of significant wins, by 3/2.

By using Friedman's test on all the data on Table 2, we may reject the hypothesis that all the approaches have equal performance with a confidence of 0.05 (p-value is 0.033).

4.2   Method Behavior

To understand why IR-ensembling improves the results of a bestrule classifier we have performed a bias-variance analysis as described in [11]. For each dataset we proceed as follows. We divide the examples in two sets D and E. This last set is used for evaluation and is a stratified sample, without replacement, with half the size of the original dataset. From the set D we generate 50 simple random samples, without replacement. Each one of these samples is used as training, and the results of the obtained models on E are used to estimate the contribution of the bias and of the variance to the global error. For each dataset we decompose the error into bias and variance for both strategies: bestrule and IR-ensembling. The parameters used were the same as in the experiments reported above, except for the minimum support. In this case, since the training sets were smaller (25% of the original set), we have lifted the admissible support of the rules to values that guarantee that at least 5 cases are covered (instead of 10, as we used above). In any case, support never goes below 0.01.

Figure 1 shows the results of the bias-variance analysis for 12 datasets. Each dataset has two bars, the left for best rule and the right one for IRE. The grey part of each bar corresponds to the bias component and the white part to the variance. In terms of the bias-variance decomposition, we can see that for 10 of the 12 datasets the bias component of the error visibly decreases. For the other

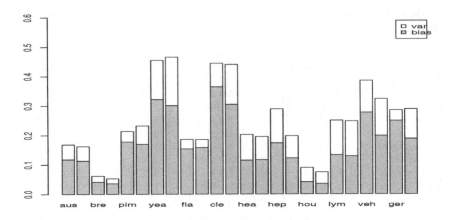

Fig. 1. The decomposition of bias and variance for 12 of the datasets. For each dataset, the bar on the left corresponds to the best rule approach and the bar on the right to the IRE

2 cases (flare and heart) there is at most a small increase in bias. The variance component tends to increase although not in the same proportion as bias.

We can thus hypothesize that the error reduction caused by IRE is mostly due to the reduction of the bias component. Since the variance component will converge to zero with the size of the datasets, IRE seems advantageous for large datasets. We should note that dealing with large datasets is not particularly complicated since the generation of association rules grows linearly with the number of examples [1]. The process of rule reweighting also grows linearly with the number of examples.

In another set of experiments, we have observed how the answers of the models in the ensemble compare with the answers given by the single model. In Figure 2 we can see the result for the *yeast* dataset. The xx axis represents the test examples and the yy axis the percentage of correct answers given by the two strategies for each case. In the case of the best rule, this percentage is either 0 (failure) or 1 (succes). In the case of the ensemble approach we have the percentage of models in the ensemble that gave the correct answer. The examples in the xx axis are sorted by the success of the best rule and than by the percentage of successes of the ensemble.

With this analysis we can see that there is a good number of "easy cases" and of "hard cases". These are the ones at the right and left end of the plot, respectively. The cases in the middle are in a grey area. These are the ones that can be more easily recovered by IRE. To be successful, the IRE approach must recover more examples (improve the answer of the best rule) than the ones it loses (degrades the answer of the best rule). In the case of *yeast*, we can see that many case are recovered (crosses above the 0.5 horizontal line, and to the left of the vertical solid line), although some others are lost (below the horizontal line and to the right of the vertical one). In the case of the *heart* dataset (figure 2), similar observations can be made. Notice the small number of test examples that perform worst in the ensemble method than in the best rule prediction.

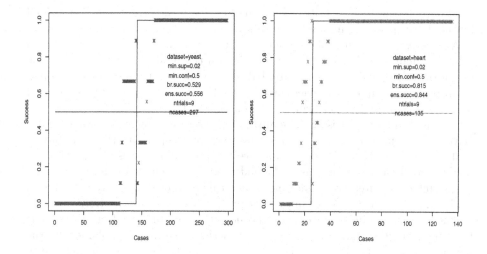

Fig. 2. Left: percentage of correct answers per test case for the *yeast* dataset (crosses) shown against the correct answers given by the best rule approach (solid crisp line). Right: Similar analysis for the *heart* dataset.

5 Discussion

Iterative Reordering Ensembling is a technique that produces replications of an original model without relearning. Each replication is a variant not very different from the original one. The obtained ensemble is thus more than homogeneous. Its elements are rather a result of jittering the original model in the version space. Still, the combined effect of these very similar models reduces the error consistently, when compared to using the single model, and in many datasets the reduction is significant.

The study of the bias variance decomposition indicates that IRE tends to reduce the bias component of the error, more than the variance. This is similar to what happens in boosting and in contrast to the case of bagging, where the reduction of the error is mainly due to the reduction in variance [2].

The intuitive explanation for the reduction of the bias component is that the single model best rule approach is tied to a particular rule ordering, and it is hard to find an ordering that maximizes the number of examples correctly classified. This constraint seems to be softened by combining similar versions of the rule set with different orderings.

6 Related Work

Ensemble learning has concentrated a large number of proposals in the literature. In [2] a study on the performance of several voting methods (including Bagging and Boosting) was presented. A careful analysis of the bias/variance error decomposition is described as means to explain the error reduction yielded by the different voting methods variants.

A novel version of model aggregation obtained from bagging is described in [14]. The main idea is to derive an ordering on the models and to consider the top of the order. This is obtained by halting the bagging process earlier. Only a small part of the models is selected (15% to 30% of the total). This fraction of models are expected to perform best when aggregated.

A similar idea to ours is [8]. The author proposes an iterative version of Naive Bayes. The aim is to boost accuracy by iteratively updating the distribution tables yield from Naive Bayes to improve the probability class distribution associated with each training case. The end product is a single model rather then an ensemble. Our iterative updating of each rule predictive measure within each trial can be seen as a form of improving probability class distribution.

The work in [20] investigates the hypothesis that combining effective ensemble learning strategies leads to the reduction of the test error is explained by the increase of diversity. These authors argue that by trading a small increase in individual test error, a reduction in overall ensemble test error is obtained.

The Post-bagging ensemble method proposed in [9] employs a similar general strategy. Like IRE, Post-bagging derives replications that jitter around the original model. However, the combined effect of the similar models minimizes the test error but mostly due to a reduction on the variance component.

7   Conclusions

Classificaton using association rules can be improved through ensembling. We have proposed Iterative Reordering Ensembling (IRE), which is a procedure that generates multiple models with one single learning step. First, a rule set is obtained from the data. Then, replications of this initial set are obtained by iteratively recalculating the predictive measures of the rules in the set.

Experimental results with 17 datasets suggest that this ensembling technique improves best rule prediction and is competitive when compared to *rpart* and *c*4.5. The bias-variance decomposition indicates that most of the improvement is explained by a reduction of the bias component. This is possibly explained by the ability of the ensembling technique avoiding being tied to one particular ordering of the rules.

This kind of ensemble approach obtains multiple models by perturbing an original one. The resulting models are computationally unexpensive and atend to be similar to each other. Despite that low variety, their combination results in an effective improvement with respect to the single model.

References

1. Agrawal, R., Srikant, R.: Fast algorithms for mining association rules in large databases. In: VLDB '94: Proceedings of the 20th International Conference on Very Large Data Bases, pp. 487–499. Morgan Kaufmann Publishers Inc, San Francisco (1994)

2. Bauer, E., Kohavi, R.: An empirical comparison of voting classification algorithms: Bagging, boosting, and variants. Machine Learning 36(1-2), 105–139 (1999)
3. Bayardo, R.J., Agrawal, R., Gunopulos, D.: Constraint-based rule mining in large, dense databases. In: ICDE, pp. 188–197. IEEE Computer Society, Los Alamitos (1999)
4. Breiman, L., Friedman, J.H., Olshen, R.A., Stone, C.J.: Classification and Regression Trees. Wadsworth (1984)
5. Brin, S., Motwani, R., Ullman, J.D., Tsur, S.: Dynamic itemset counting and implication rules for market basket data. In: Peckham, J. (ed.) SIGMOD Conference, pp. 255–264. ACM Press, New York (1997)
6. Fayyad, U.M., Irani, K.B.: Multi-interval discretization of continuous-valued attributes for classification learning. In: IJCAI, pp. 1022–1029 (1993)
7. Freund, Y., Schapire, R.E.: A decision-theoretic generalization of on-line learning and an application to boosting. In: Vitányi, P.M.B. (ed.) EuroCOLT 1995. LNCS, vol. 904, pp. 23–37. Springer, Heidelberg (1995)
8. Gama, J.: Iterative bayes. Theor. Comput. Sci. 292(2), 417–430 (2003)
9. Jorge, A., Azevedo, P.J.: An experiment with association rules and classification: Post-bagging and conviction. In: Hoffmann, A., Motoda, H., Scheffer, T. (eds.) DS 2005. LNCS (LNAI), vol. 3735, pp. 137–149. Springer, Heidelberg (2005)
10. Jovanoski, V., Lavrac, N.: Classification rule learning with apriori-c. In: Brazdil, P.B., Jorge, A.M. (eds.) EPIA 2001. LNCS (LNAI), vol. 2258, pp. 44–51. Springer, Heidelberg (2001)
11. Kohavi, R., Wolpert, D.: Bias plus variance decomposition for zero-one loss functions. In: ICML, pp. 275–283 (1996)
12. Li, W., Han, J., Pei, J.: Cmar: Accurate and efficient classification based on multiple class-association rules. In: Cercone, N., Lin, T.Y., Wu, X. (eds.) ICDM, pp. 369–376. IEEE Computer Society, Los Alamitos (2001)
13. Liu, B., Hsu, W., Ma, Y.: Integrating classification and association rule mining. In: KDD '98: Proceedings of the fourth ACM SIGKDD International Conference on Knowledge Discovery and Data Mining, pp. 80–86. ACM Press, New York (1998)
14. Martínez-Muñoz, G., Suárez, A.: Pruning in ordered bagging ensembles. In: Cohen, W.W., Moore, A. (eds.) ICML, pp. 609–616. ACM, New York (2006)
15. Meretakis, D., Wüthrich, B.: Extending naïve bayes classifiers using long itemsets. In: KDD, pp. 165–174 (1999)
16. Merz, C.J., Murphy, P.: Uci repository of machine learning database (1996), http://www.cs.uci.edu/~mlearn
17. Quinlan, J.R.: C4.5: Programs for Machine Learning. Morgan Kaufmann, San Francisco (1993)
18. R Development Core Team. R: A language and environment for statistical computing. R Foundation for Statistical Computing, Vienna, Austria (2004) ISBN 3-900051-00-3
19. Schapire, R.E.: The strength of weak learnability. Machine Learning 5, 197–227 (1990)
20. Webb, G.I., Zheng, Z.: Multistrategy ensemble learning: Reducing error by combining ensemble learning techniques. IEEE Trans. Knowl. Data Eng. 16(8), 980–991 (2004)

On Approximating Minimum Infrequent and Maximum Frequent Sets

Mario Boley

Fraunhofer IAIS, Schloss Birlinghoven, Sankt Augustin, Germany
mario.boley@iais.fraunhofer.de

Abstract. The maximum cardinality of a frequent set as well as the minimum cardinality of an infrequent set are important characteristic numbers in frequent (item) set mining. Gunopulos et al. [10] have shown that finding a maximum frequent set is **NP**-hard. In this paper I show that the minimization problem is also **NP**-hard. As a next step I investigate whether these problems can be approximated. While a simple greedy algorithm turns out to approximate a minimum infrequent set within a logarithmic factor one can show that there is no such algorithm for the maximization problem.

1 Introduction

Finding sets of items that appear concurrently in at least a specified number of records in a given database is an important task in data mining. This so-called frequency criterion for sets is used as an additional condition for different interestingness predicates. Examples are association rules [2], correlations [5], or emerging patterns [7].

Algorithms usually perform an exhaustive enumeration of the family of frequent sets or of a reduced family like closed frequent sets or maximal frequent sets. Such an exhaustive enumeration tends to be very time-consuming because both, the search space and the output size, can be exponential in the size of the input database. The running time as well as the semantic significance of the produced output depend on the user-specified frequency parameter. Thus it is of great value to know as much as possible about the results of an exponential time pattern mining algorithm prior to its application. This knowledge can be used to readjust the frequency parameter and thus improve performance and semantic value of the mining algorithm.

For that purpose frequent sets of maximum cardinality resp. infrequent sets of minimum cardinality can be used. Many mining algorithms tend to run exponentially long in the cardinality of a longest pattern, i.e. the size of a maximum frequent set and for level-wise algorithms the size of a minimum infrequent set determines the level where pruning starts. So knowing either of the two would allow to upper bound the running time resp. skip initial search levels. In terms of result quality both indicate whether the chosen frequency threshold provides a significant gain of information for the resulting patterns. If for instance the

V. Corruble, M. Takeda, and E. Suzuki (Eds.): DS 2007, LNAI 4755, pp. 68–77, 2007.
© Springer-Verlag Berlin Heidelberg 2007

minimum cardinality of an infrequent set is 18 in a database containing 20 items this is an indication for a weak parameter choice.

On the one hand, both optimization problems are **NP**-hard. For the maximization problem this was shown by Gunopulos et al. in [10]. For the minimization problem this is shown in Section 3 of this paper. On the other hand, computing approximate solutions would suffice for the described motivations. In this paper I show that not even a reasonable approximation algorithm for a maximum frequent set is likely to exist based on recent results from computational complexity [12], while for a minimum infrequent set a simple greedy algorithm reaches a logarithmic approximation factor. By another recent complexity result [8] this factor cannot be improved substantially. Note that in contrast to approaches that aim at approximating the set of *all* frequent sets (like in [1]) we consider different problems each aiming to compute only *one* set. To the best of my knowledge this is the first investigation on the approximability of these problems.

The rest of the paper is organized as follows: Section 2 introduces basic definitions and notations. In Section 3, the two optimization problems are defined formally and their **NP**-hardness is discussed. Section 4 points out the hardness of approximating the maximization problem, while Section 5 proves the logarithmic performance of the greedy algorithm for the minimization problem. Finally, Section 6 concludes with a summary and ideas for possible future work.

2   Preliminaries

A **hypergraph** is a triple $(V, \mathcal{H}, \mu)$ with V a finite set called **ground set**, $\mathcal{H} \subseteq 2^V$ a family whose elements are called **hyperedges**, and $\mu : \mathcal{H} \to \mathbb{N}$ a mapping representing the multiplicity of each hyperedge. So $\mathcal{H}$ can be seen as a multiset, and thus we mean by its cardinality $|\mathcal{H}|$ the sum $\sum_{H \in \mathcal{H}} \mu(H)$. For the purpose of computational problems we assume a hypergraph to be given as incidence matrix, and thus define $\text{size}((V, \mathcal{H}, \mu)) = |V||\mathcal{H}|$ as the input size. If $\mu(H) = 1$ for all $H \in \mathcal{H}$ we omit μ and $(V, \mathcal{H})$ is called **proper**.

A **graph** is a hypergraph $G = (V, E)$ with $|e| = 2$ for all $e \in E$. The elements of V are called **vertices**, and the elements of E are called **edges**[1]. G is called **bipartite** if V can be partitioned into V_1, V_2 such that all edges are of the form $\{v, w\}$ with $v \in V_1$ and $w \in V_2$. A graph of this form is denoted by (V_1, V_2, E). A set of vertices $X = X_1 \cup X_2$ with $X_1 \subseteq V_1, X_2 \subseteq V_2$ is denoted by (X_1, X_2) and is called a **bipartite clique** if for all $x_1 \in X_1$ and all $x_2 \in X_2$ there is an edge $\{x_1, x_2\} \in E$. It is called **balanced** if $|X_1| = |X_2|$. The size of a balanced bipartite clique (X_1, X_2) is $|X_1| = |X_2|$.

An **optimization problem** is a computational problem formally given by a 4-tuple $P = (X, (S_x)_{x \in X}, c, \text{goal})$ with a set of instances X, a set of feasible solutions S_x for all instances, a target function $c : \bigcup_{x \in X} S_x \to \mathbb{N}$, and goal $\in \{\min, \max\}$. The task is then, given an instance $x \in X$, compute a feasible

---

[1] In this paper we do not consider graphs with parallel edges or loops, i.e. edges with only one element.

solution $y \in S_x$ with $c(y) = \text{goal}\{c(y') : y' \in S_x\}$. If goal = min, P is called a **minimization problem**. If goal = max, P is called a **maximization problem**.

As examples consider the following two well known **NP**-hard optimization problems (see [9]):

MAX BALANCED CLIQUE is the following maximization problem: *Given* a bipartite graph G, *compute* a balanced bipartite clique in G of maximum cardinality. Here the instances are bipartite graphs, the feasible solutions for a graph G are balanced bipartite cliques in G, and the target function maps a balanced bipartite clique (X, Y) to its size $|X|$.

MIN SET COVER is the following minimization problem: *Given* a hypergraph $(V, \mathcal{H})$ with $\bigcup \mathcal{H} = V$, *compute* a family $\mathcal{H}' \subseteq \mathcal{H}$ of minimum cardinality covering V, i.e., $\bigcup \mathcal{H}' = V$.

Let $P = (X, (S_x)_{x \in X}, c, \text{goal})$ be an optimization problem. A deterministic algorithm $\mathcal{A}$ for P can be thought of as a mapping from the instances X to the set of all possible outputs $\bigcup_{x \in X} S_x$. Then $\mathcal{A}$ is called an α-**approximation algorithm** for P with $\alpha : X \rightarrow \mathbb{R}_{\geq 1}$ if for all $x \in X$ with $\text{goal}\{c(y) : y \in S_x\} = \text{OPT}$ it holds that $\mathcal{A}(x) \in S_x$, i.e., the algorithm produces only feasible solutions, $\mathcal{A}$ runs in polynomial time, and

$$\frac{1}{\alpha(x)} \text{OPT} \leq c(\mathcal{A}(x)) \leq \alpha(x)\text{OPT} .$$

For such an algorithm we say that $\mathcal{A}$ approximates P within a factor of α. If $\alpha(x) \equiv 1$, $\mathcal{A}$ solves the problem exactly. Note that the first inequality applies only to maximization problems, while the second applies only to minimization problems. Since we require $\mathcal{A}$ to produce always feasible solutions, it holds that $\mathcal{A}(x) \leq \text{OPT}$ in case goal=max and $\text{OPT} \leq \mathcal{A}(x)$ in case goal=min.

In *frequent set mining* (or frequent itemset mining) [2] the input is a hypergraph $D = (I, \mathcal{T}, \mu)$ called dataset and a positive integer $t \in \{1, \ldots, |\mathcal{T}|\}$ called **frequency threshold**. Sometimes the elements of I are called items and the elements of $\mathcal{T}$ are called transactions. For $X \subseteq I$ the **support set** of X is defined as

$$\mathcal{T}[X] = \{T \in \mathcal{T} : X \subseteq T\} .$$

X is called t-**frequent** in D if $|\mathcal{T}[X]| \geq t$.

3   Problems and Hardness of Exact Solutions

We are now ready to give a formal definition of the problems of interest: *Given* a hypergraph $(I, \mathcal{T}, \mu)$ and a frequency threshold $t \in \{1, \ldots, |\mathcal{T}|\}$ we define

MAX FREQUENT SET as the maximization problem to *compute* a t-frequent set $X \subseteq I$ of maximum cardinality and

MIN INFREQUENT SET as the minimization problem to *compute* a set $X \subseteq I$ of minimum cardinality that is not t-frequent.

Remark 1. In Section 1 we only discussed the use of the maximum resp. the minimum *cardinality* of a frequent resp. infrequent set. Here we require the *construction* of an actual set in each problem. However, these two tasks are polynomially equivalent. In particular a maximum frequent set can be constructed by iteratively trying to remove an element and then checking whether the maximum cardinality has changed.

Next we recall the construction used in [10] to prove hardness of MAX FREQUENT SET. In Section 4 we will reuse this construction, which is a transformation from the **NP**-hard MAX BALANCED CLIQUE problem to MAX FREQUENT SET that uses a canonical correspondence between hypergraphs and bipartite graphs:

For a given bipartite graph $G = (V, U, E)$ construct a hypergraph $D = (V, \mathcal{T}, \mu)$ with

$$\mathcal{T} = \{\Gamma(u) : u \in U\}$$
$$\mu \colon \mathcal{T} \mapsto |\{u \in U \colon \Gamma(u) = \mathcal{T}\}|$$

where $\Gamma(u)$ denotes the set of all neighbors of u, i.e., $\Gamma(u) = \{v \in V \colon \{v, u\} \in E\}$. Note that size$(D) \leq$ size(G). Furthermore, the maximum cardinality of a balanced bipartite clique in G is the maximum t such that there is a t-frequent set X in D with $|X| \geq t$, which can easily be computed from D with an algorithm solving MAX FREQUENT SET. This implies:

Theorem 1 (Gunopulos et al. [10]). MAX FREQUENT SET *is* **NP**-*hard.*

To analyze MIN INFREQUENT SET we define the following generalized version of MIN SET COVER:

MIN GENERAL SET COVER is the following minimization problem: *Given a hypergraph $(V, \mathcal{H})$ and a positive integer $p \in \{0, \ldots, |V| - 1\}$, compute a minimum family of hyperedges $\mathcal{H}'$ covering at least $|V| - p$ elements of V, i.e., $|V \setminus \bigcup \mathcal{H}'| \leq p$.*

MIN GENERAL SET COVER contains the **NP**-hard problem MIN SET COVER as a special case $(p = 0)$, and thus it is itself **NP**-hard. Moreover, we have the following equivalence:

Theorem 2. MIN INFREQUENT SET *is polynomially equivalent to* MIN GENERAL SET COVER.

Proof. Construct a polynomial transformation f from MIN GENERAL SET COVER to MIN INFREQUENT SET by transposing the given incidence matrix and changing 0-entries to 1-entries and vice versa. The frequency parameter t is set to $p + 1$. Note that because of the parameter ranges of t and p this mapping is bijective. For an instance $((V, \mathcal{H}), p)$ this results in:

$$f \colon ((V, \mathcal{H}), p) \mapsto ((\mathcal{H}, \overline{V}, \mu), p + 1)$$
$$\overline{V} = \{\mathcal{H} \setminus \mathcal{H}[\{v\}] \colon v \in V\}$$
$$\mu \colon \mathcal{H}' \mapsto |\{v \in V \colon \mathcal{H}' = \mathcal{H} \setminus \mathcal{H}[\{v\}]\}| \ .$$

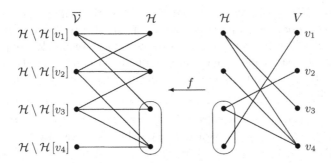

Fig. 1. Construction used to proof Theorem 2. Here the hypergraphs are drawn as bipartite graphs (see proof of Theorem 1) with hyperedges on the left side and ground set on the right side. The marked set is 2-infrequent in the left hypergraph and covers all but 1 element in the right.

So the original hyperedges act as items and every element v of the original ground set becomes a hyperedge, that contains all the sets, which v is not an element of (in the original MIN GENERAL SET COVER instance). Now we claim that an $(p+1)$-infrequent set in $(\mathcal{H}, \overline{V})$ corresponds to a set of hyperedges covering at least all but p elements of $(V, \mathcal{H})$ and vice versa (see Fig. 1). For a subset $\mathcal{H}' \subseteq \mathcal{H}$ it holds that

$$\mathcal{H}' \ (p+1)\text{-infrequent in } (\mathcal{H}, \overline{V}) \Leftrightarrow |\overline{V}[\mathcal{H}']| < p+1$$
$$\Leftrightarrow |\{v \in V : \mathcal{H} \setminus \mathcal{H}[\{v\}] \supseteq \mathcal{H}'\}| < p+1$$
$$\Leftrightarrow |\{v \in V : \forall H \in \mathcal{H}' \ v \notin H\}| < p+1$$
$$\Leftrightarrow |V \setminus \bigcup \mathcal{H}'| < p+1$$
$$\Leftrightarrow \mathcal{H}' \text{ covers at least all but } p \text{ elements .}$$

So an infrequent set of size k corresponds to a subfamily of size k covering sufficient many elements and vice versa. Furthermore, f is a bijection implying polynomial equivalence. □

This implies the main result of this section completing our problem introduction:

Corollary 3. MIN INFREQUENT SET *is* **NP**-*hard.*

4   Hardness of Approximating a Maximum Frequent Set

Since MAX FREQUENT SET is **NP**-hard, the next step is to ask for an approximation algorithm. Proving negative results for the approximation of hard problems has been very successful in recent years. New results have in common that they use so called 'probabilistically checkable proofs' [4] as a characterization of **NP**. As indicated by the proof of Theorem 1 the following result proved by Khot [12] for MAX BALANCED CLIQUE is of particular importance for our purpose:

Unless there are probabilistic algorithms with an arbitrary small exponential time complexity for all problems in **NP** there is no polynomial approximation scheme for MAX BALANCED CLIQUE, i.e., the infimum of all constants k such that there is a k-approximation algorithm for MAX BALANCED CLIQUE is bounded away from 1. It was known before that such a result, once achieved, can be boosted via derandomized graph products (introduced in [3]). So that the result of Khot implies in fact:

Theorem 4 (Khot [12]). *Unless for all $\epsilon > 0$ and all decision problems in* **NP** *there is a probabilistic algorithm $\mathcal{A}$ accepting a YES-instance resp. rejecting a NO-instance of size n with probability at least 2/3 in time 2^{n^ϵ} the following holds: There is a constant $\delta_{BC} > 0$ such that there is no algorithm approximating* MAX BALANCED CLIQUE *within a factor of $size(x)^{\delta_{BC}}$ for instances x.*

Now suppose there is an algorithm $\mathcal{A}$ approximating MAX FREQUENT SET within a factor of $\alpha(size(x))$ for instances x. Then one can construct a hypergraph D from a given bipartite graph G as for Theorem 1 and find t_{APX} the maximum $t \in \{1, \ldots, |T|\}$ for which $|\mathcal{A}(D, t)| \geq t$ by running $\mathcal{A}$ at most $|T|$ times. Let (X, Y) be a maximum balanced bipartite clique in $G = (V, U, E)$ with size t_{OPT}. Any set of transactions corresponding to a subset $Y' \subseteq Y$ contains the t_{OPT} items corresponding to X—in particular those with $|Y'| = t_{\mathrm{OPT}}/\alpha(size(D)) = t^*$. This implies for the maximum cardinality of a t^*-frequent set in D, denoted as $\mathrm{mfs}(D, t^*)$,

$$\mathrm{mfs}(D, t^*) \geq t_{\mathrm{OPT}} \Rightarrow |\mathcal{A}(D, t^*)| \geq t_{\mathrm{OPT}}/\alpha(size(D)) = t^* \ .$$

But then $t_{\mathrm{APX}} \geq t^* = t_{\mathrm{OPT}}/\alpha(size(D)) \geq t_{\mathrm{OPT}}/\alpha(size(G))$, because the transformed instance is of equal or smaller size. Since all necessary computations can be performed in polynomial time, we have a polynomial algorithm approximating MAX BALANCED CLIQUE within a factor of $\alpha(size(x))$ for instances x and hence

Corollary 5. *Under the same assumptions as in Theorem 4 with the same constant $\delta_{BC} > 0$ there is no algorithm approximating* MAX FREQUENT SET *within a factor of $size(x)^{\delta_{BC}}$ for instances x.*

Although stronger than **P** $\neq$ **NP** the stated complexity assumption is still widely believed and thus we have a strong indication that there is no algorithm for MAX FREQUENT SET with a reasonable approximation factor.

5   Greedy Approximation of a Minimum Infrequent Set

The transformation in Theorem 2 maps instances of MIN GENERAL SET COVER to instances of MIN INFREQUENT SET with the same optimum value and vice versa and there is also a bijection between feasible solutions. So an approximation algorithm for either one of the two problems will grant the same approximation factor for the other. To analyze the approximability of the two problems we will use another related coverage problem:

MAX COVERAGE is the following maximization problem: *Given* a hypergraph $(V, \mathcal{H})$ and a positive integer k, *compute* a family $\mathcal{H}' \subseteq \mathcal{H}$ of k hyperedges covering a maximum number of elements.

Using the known fact that the approximation ratio of the greedy algorithm for this problem is $(1 - e^{-1})$ (see for instance [6]), one can analyze the approximation performance of the greedy approach for MIN GENERAL SET COVER.

Theorem 6. MIN GENERAL SET COVER *can be approximated in polynomial time within a factor of* $\lceil \ln(|V| - p) \rceil + 1$ *for instances* $((V, \mathcal{H}), p)$.

Proof. The following algorithm uses the greedy algorithm $\mathcal{G}$ for MAX COVERAGE, to achieve the desired approximation rate for MIN GENERAL SET COVER. Denote with n the number of elements $|V|$ and with $\mathrm{gsc}(V, \mathcal{H}, p)$ the minimum cardinality of a hyperedge set covering at least $n - p$ elements.

1. $i \leftarrow 1$, $S \leftarrow \emptyset$, $V_1 \leftarrow V$, $\mathcal{H}_1 \leftarrow \mathcal{H}$
2. **while** $|V_i| > p$ **do**
3. $k_i \leftarrow \min\{j : |\bigcup \mathcal{G}(V_i, \mathcal{H}_i, j)| \geq e^{1-i}(1 - \frac{1}{e})(|V| - p)\}$
4. $\mathcal{H}_\Delta \leftarrow \mathcal{G}(V_i, \mathcal{H}_i, k_i)$
       $S \leftarrow S \cup \mathcal{H}_\Delta$, $\mathcal{H}_{i+1} \leftarrow \mathcal{H}_i \setminus \mathcal{H}_\Delta$, $V_{i+1} \leftarrow V_i \setminus \bigcup \mathcal{H}_\Delta$
5. $i \leftarrow i + 1$
6. **return** S

Obviously S covers at least $n - p$ elements after termination. We claim that also $|S| \leq (\lceil \ln(n - p) \rceil + 1)\,\mathrm{gsc}(V, \mathcal{H}, p)$. To see this, we first analyze the number of iterations and then the number of hyperedges added to the S in every iteration.

(i) The algorithm terminates after at most $\lceil \ln(n - p) \rceil + 1$ iterations.
Proof of (i):
First show $|V_i| \leq p + e^{1-i}(n - p)$ by induction on i. For $i = 1$ this is true, because $|V| = |V_1| = n$. Now assume that $|V_i| \leq p + e^{1-i}(n - p)$ for a given i. In line 3 k_i is chosen such that at least $e^{1-i}(1 - e^{-1})(n - p)$ elements will be covered. So

$$|V_{i+1}| \leq p + \frac{1}{e^{i-1}}(n - p) - \frac{e-1}{e^i}(n - p) = p + \frac{1}{e^i}(n - p) \ .$$

Since the algorithm terminates when $|V_i| \leq p$ (and $|V_i|$ cannot be fractional), it is for the number of iterations t:

$$t \leq \min\{i \in \mathbb{N} : e^i > n - p\} = \lceil \ln(n - p) \rceil + 1$$

$$(i)\square$$

(ii) For all iterations i it is $k_i \leq \mathrm{gsc}(V, \mathcal{H}, p)$.
Proof of (ii):
By definition there is an optimum cover $\mathcal{O} \subseteq \mathcal{H}$ with

$$\left|\bigcup \mathcal{O}\right| \geq n - p \text{ and } |\mathcal{O}| = \mathrm{gsc}(V, \mathcal{H}, p) \ .$$

So $\mathcal{O}$ covers all but p elements. Let $\mathrm{mc}(V', \mathcal{H}', k)$ denote the maximum number of elements one can cover with k hyperedges in $(V', \mathcal{H}')$. Since in iteration i it is

---

Algorithm 1. $(\lceil \ln(|\mathcal{T}| - t)\rceil + 1)$-approximation for MIN INFREQUENT SET

Require: Dataset $D = (I, \mathcal{T}, \mu)$ and frequency threshold t
Ensure: X infrequent and $|X| \leq (\lceil \ln(|\mathcal{T}| - t)\rceil + 1)$OPT, with OPT the minimum
　　cardinality of a set that is not t-frequent in D

1. $X \leftarrow \emptyset$
2. **while** $|\mathcal{T}| \geq t$ **do**
3. 　　$i \leftarrow i \in I$ with $|\mathcal{T}[\{i\}]| = \min\{|\mathcal{T}[\{i'\}]| : i' \in I\}$
4. 　　$X \leftarrow X \cup \{i\}$
5. 　　$I \leftarrow I \setminus \{i\}$
6. 　　$\mathcal{T} \leftarrow \mathcal{T}[\{i\}]$
7. **return** X

$|V_i| \leq p + e^{1-i}(n - p)$, $|\mathcal{O}|$ elements can still cover at least $e^{1-i}(n - p)$ elements.
It follows

$$\mathrm{mc}(V_i, \mathcal{H}_i, \mathrm{gsc}(V, \mathcal{H}, p)) \geq e^{1-i}(n - p)$$
$$\Rightarrow |\bigcup \mathcal{G}(V_i, \mathcal{H}_i, \mathrm{gsc}(V, \mathcal{H}, p))| \geq (1 - \frac{1}{e})e^{1-i}(n - p)$$

and because k_i is selected in line 3 as the minimum number satisfying this

$$\Rightarrow k_i \leq \mathrm{gsc}(V, \mathcal{H}, p) \ .$$

$$(ii)\square$$

Since k_i sets are added to $\mathcal{S}$ in every iteration i, it follows from (i) and (ii)
that $|\mathcal{S}| \leq (\lceil \ln(n - p)\rceil + 1)\mathrm{gsc}(V, \mathcal{H}, p)$. The polynomial running time is obvious,
because the polynomial time greedy algorithm is called in every iteration at most
$|\mathcal{H}|$ times. 　　　　　　　　　　　　　　　　　　　　　　　　　$\square$

Remark 2. The formulation of the algorithm in the above proof was tailor-made
for the surrounding analysis. In fact it only selects remaining hyperedges covering
a maximum number of remaining elements and thus the simple greedy strategy
stopping, when all but p elements are covered, will select the same hyperedges
or possibly even some less.

Algorithm 1 takes this into account and incorporates the transformation between MIN INFREQUENT SET and MIN GENERAL SET COVER. Note that this
transformation switches the roles of ground set and hyperedges so that the resulting approximation factor does not depend on the number of items but on
the number of transactions. This constitutes the following result:

Corollary 7. MIN INFREQUENT SET *can be approximated within a factor of*
$\lceil \ln(|\mathcal{T}| - t)\rceil + 1$ *for instances* $(I, \mathcal{T}, \mu)$, t.

The approximation ratio achieved above is close to optimal. Otherwise, since
MIN GENERAL SET COVER contains MIN SET COVER as a special case for $p = 0$,
a better ratio would imply the existence of subexponential time algorithms with
extremely small exponents for every problem in **NP** by the following theorem:

Theorem 8 (Feige [8]). *For all $\epsilon > 0$ there is no algorithm approximating* MIN SET COVER *within a factor of* $(1 - \epsilon) \ln |V|$ *for instances* $(V, \mathcal{H})$, *unless for all problems in* **NP** *there is an algorithm running in time* $n^{O(\log \log n)}$ *for instances of size* n.

6   Discussion

In this paper, we have analyzed the algorithmical tasks to approximate a maximum frequent resp. a minimum infrequent set. This investigation is motivated by the need for an efficient parameter evaluation procedure that can be applied before a possibly exponential time pattern mining algorithm. We turned to approximation algorithms because both problems are **NP**-hard. In case of the maximization problem this was well-known. In case of the minimization problem we proved this hardness by showing it to be equivalent to a generalized version of the MIN SET COVER problem.

Using recent results from computational complexity we have argued that a non-trivial approximation algorithm for MAX FREQUENT SET is unlikely to exist. For MIN INFREQUENT SET we gave a polynomial time greedy algorithm, which was proven to compute an infrequent set of cardinality smaller than $\lceil \ln(m - t) \rceil + 1$ times the minimum cardinality of an infrequent set for instances with frequency threshold t and m transactions. Slavík proved in [14] that the approximation ratio of the greedy algorithm for MIN SET COVER can in fact be bounded by $\ln n - \ln \ln n + 0.79$. It is likely that his tight analysis can be transferred to MIN GENERAL SET COVER, which is a task for possible future work. The fact that the approximation factor depends on the number of transactions and not on the number of items indicates that the algorithm is useful for gene expression data [13], which can contain up to 100,000 items but typically only about 1000 transactions. In general, knowing the approximation factor allows valuable conclusions. If the cardinality of the returned set is c this implies that all sets of cardinality smaller than $c/(\lceil \ln(m - t) \rceil + 1)$ are frequent. In turn, this provides a lower bound on the number of frequent sets and for level-wise algorithms determines an earliest level where pruning can occur so that search need not to be started before this level.

Other important characteristics that can be used for parameter evaluation are the number of frequent resp. closed or maximal frequent sets resulting from a given parameter, all of which are hard *counting* problems [15,10]. It is an interesting question whether the positive results from computing the permanent of a 0-1 matrix can be transferred to those problems. For 0-1-PERMANENT the existence of a fully polynomial randomized approximation scheme has been shown [11]. Another question is, how quick parameter evaluation can be done in other domains with similar problems as frequent set mining (exponential output size and even greater search space). Examples for such domains are pattern mining tasks with structured data like sequences or graphs.

Acknowledgments. I would like to thank Prof. Bhaskar DasGupta who pointed me towards the maximum coverage approach.

References

1. Afrati, F., Gionis, A., Mannila, H.: Approximating a collection of frequent sets. In: ACM SIGKDD Int. Conf. on Knowledge Discovery and Data Mining (2004)
2. Agrawal, R., Mannila, H., Srikant, R., Toivonen, H., Verkamo, A.I.: Fast discovery of association rules. In: Advances in Knowledge Discovery and Data Mining, pp. 307–328. AAAI/MIT Press, Cambridge (1996)
3. Alon, N., Feige, U., Wigderson, A., Zuckerman, D.: Derandomized graph products. Computational Complexity 5(1), 60–75 (1995)
4. Arora, S., Safra, S.: Probabilistic checking of proofs: A new characterization of NP. J. ACM 45(1), 70–122 (1998)
5. Brin, S., Motwani, R., Silverstein, C.: Beyond market baskets: Generalizing association rules to correlations. In: SIGMOD Conference, pp. 265–276 (1997)
6. Cornuejols, G., Fisher, M., Nemhauser, G.: Location of bank accounts to optimize float: an analytic study of exact and approximate algorithms. Management Science (23), 789–810 (1977)
7. Dong, G., Li, J.: Efficient mining of emerging patterns: Discovering trends and differences. In: KDD, pp. 43–52 (1999)
8. Feige, U.: A threshold of ln n for approximating set cover. J. ACM 45(4), 634–652 (1998)
9. Garey, M.R., Johnson, D.S.: Computers and Intractability: A Guide to the Theory of NP-Completeness. W. H. Freeman, (1979) ISBN 0-7167-1044-7
10. Gunopulos, D., Khardon, R., Mannila, H., Saluja, S., Toivonen, H., Sharm, R.S.: Discovering all most specific sentences. ACM Trans. Database Syst. 28(2), 140–174 (2003)
11. Jerrum, M., Sinclair, A.: Approximating the permanent. SIAM J. Comput. 18(6), 1149–1178 (1989)
12. Khot, S.: Ruling out ptas for graph min-bisection, densest subgraph and bipartite clique. In: FOCS, pp. 136–145 (2004)
13. Pan, F., Cong, G., Tung, A.K.H., Yang, J., Zaki, M.J.: Carpenter: Finding closed patterns in long biological datasets. In: ACM SIGKDD Int. Conf. on Knowledge Discovery and Data Mining, ACM Press, New York (2003)
14. Slavík, P.: A tight analysis of the greedy algorithm for set cover. Journal of Algorithms 25, 237–254 (1997)
15. Valiant, L.G.: The complexity of computing the permanent. Theor. Comput. Sci. 8, 189–201 (1979)

A Partially Dynamic Clustering Algorithm for Data Insertion and Removal

Haytham Elghazel, Hamamache Kheddouci, Véronique Deslandres,
and Alain Dussauchoy

LIESP Laboratory, Lyon 1 University, 43 Bd du 11 novembre 1918,
69622 Villeurbanne cedex, France
{elghazel,hkheddou,deslandres,dussauchoy}@bat710.univ-lyon1.fr

Abstract. We consider the problem of *dynamic* clustering which has been addressed in many contexts and applications including dynamic information retrieval, Web documents classification, etc. The goal is to efficiently maintain homogenous and well-separated clusters as new data are inserted or existing data are removed. We propose a framework called *dynamic b-coloring clustering* based solely on pairwise dissimilarities among all pairs of data and on cluster dominance. In experiments on benchmark data sets, we show improvements in the performance of clustering solution in terms of quality and computational complexity.

Keywords: Dynamic clustering, graph *b-coloring*, dissimilarity, dominance.

1 Introduction

Cluster analysis is one of the most important aspects in the data mining process for discovering groups and identifying interesting distributions or patterns over the considered data sets [1]. Clustering algorithms are widely used in many areas including information retrieval, image segmentation and so on.

In [2] a new partitioning clustering scheme is introduced. It is based on the *b-coloring of graph* [3]. This technique consists in coloring the vertices of a graph G with the maximum number of colors such that (i) no two adjacent vertices (vertices joined by an weighted edge representing the *dissimilarity* between objects) have the same color (*proper coloring*), and (ii) for each color c, there exist at least one vertex with this color which is adjacent (has a sufficient dissimilarity degree) to all other colors. This vertex is called *dominating vertex*, there can have many within the same class. This specific vertex reflects the properties of the class and also guarantees that the class has a distinct separation from all other classes of the partitioning. The *b-coloring based clustering method* in [2] enables to build a fine partition of the data set (*numeric* or *symbolic*) in clusters when the number of clusters is not specified in advance.

In dynamic information environments, such as the World Wide Web, it is usually desirable to apply adaptive methods for document organization such as clustering. Incremental clustering methods are of great interest in particular when we examine their ability to cope with a high rate of dataset update. In this paper, we consider the problem of *online clustering* in the form of data insertion and removal. The difference

V. Corruble, M. Takeda, and E. Suzuki (Eds.): DS 2007, LNAI 4755, pp. 78–90, 2007.
© Springer-Verlag Berlin Heidelberg 2007

between these *learning approaches* and the traditional ones in particular is the ability to process instances as they are added (new data) or deleted (outmoded or inefficient data) from the data collection, eventually with an updating of existing clusters without having to frequently performing complete re-clustering.

In the dynamic setting, instances arrive or leave one by one, and we need to deal with an arriving or removed data before seeing any future instances. Problems faced by such algorithms include how to find the appropriate cluster to assign for a new object, how to deal with deletion of an existing object, and how to reassign objects to other clusters

Many algorithms are proposed to investigate the dynamic clustering problem. The *Single-Pass clustering* algorithm basically processes instances sequentially, and compares each instance to all existing clusters. If the dissimilarity between the instance and any cluster[1] is above a certain threshold, then the instance is added to the closest cluster; otherwise it forms its own cluster. The *k-Nearest Neighbor clustering* [4] algorithm computes for each new instance its dissimilarity to every other instance, and chooses the top k instances. The new instance is assigned to the most frequent class label among the k nearest training.

In this paper, a dynamic algorithm is proposed for the *b-coloring based clustering* approach presented in [2]. It depends only on pairwise dissimilarities among all pairs of data and on *dominance property* of vertices.

The paper is structured as follows: in Section 2, the *b-coloring* technique is introduced in broad outline. Section 3 is devoted to the dynamic algorithm. Some experiments using relevant benchmarks data set are shown in Section 4. Further works and applications linked with dynamic clustering will be proposed in conclusion.

2   Clustering with Graph b-Coloring

In this section, we briefly introduce the *b-coloring based clustering approach* and we refer the reader to [2] for more details.

When the dissimilarities among all pairs of data to be clustered $\{x_1,...,x_n\}$ are specified, these can be summarized as a weighed dissimilarity matrix D in which each element $D(x_i; x_j)$ stores the corresponding dissimilarity. Based on D, the data can also be conceived as a weighted linkage graph $G = (V, E)$, where $V = \{v_1,v_2,...,v_n\}$ is the vertex set which correspond to the data (vertex v_i for data x_i), and $E = V \times V$ is the edge set which correspond to a pair of vertices $(v_i; v_j)$ weighted by their dissimilarities $D(v_i; v_j)$. It must be noticed that the possibility of a complete graph would not be interested for clustering problem because in such a case, the *b-coloring* algorithm would provide the trivial partition where each cluster is a singleton. Hence, our algorithm starts from a subgraph (non complete graph) from the original graph. The subgraph is a *superior threshold graph* which is commonly used in graph theory. Let $G_{>\theta}=(V,E_{>\theta})$ be the superior threshold graph associated with threshold value θ chosen among the dissimilarity matrix D. In other words, $G_{>\theta}$ is given by $V=\{v_1,...,v_n\}$ as vertex set and $\{(v_i,v_j)|\ D(v_i,v_j)>\theta\}$ as edge set.

The data to be clustered are now depicted by a *non-complete edge-weighted graph* $G_{>\theta}=(V,E_{>\theta})$.   In   order   to   divide   the   vertex   set V into   a   partition

---

[1] The dissimilarity between an instance x and a cluster C is the average of dissimilarities between x and instances of C.

$P=\{C_1,C_2,..,C_k\}$ where for $\forall$ $C_i,C_j \in P$, $C_i \cap C_j=\varnothing$ for $i{\neq}j$ (when the number of clusters k is not pre-defined), our *b-coloring based clustering algorithm* performed on the graph $G_{>\theta}$ consists of two steps: 1) generate an initial coloring of vertices using a maximum number of colors, and 2) removing each color that has no dominating vertices yet using a greedy algorithm. Step 2 is performed until the coloring is stable, i.e. each color of $G_{>\theta}$ has at least one dominating vertex.

Let illustrate the *b-coloring* algorithm on one example. $\{A,B,C,D,E,F,G,H,I\}$ is the data set to analyse for which dissimilarity matrix D is given in table 1. Figure 1 shows the *superior threshold graph* for θ =0.15. Therefore here, the *b-coloring* of $G_{>0.15}$ (cf. Fig.2) gives four classes, namely: $C_1=\{\mathbf{B}\}$, $C_2=\{\mathbf{A,D}\}$, $C_3=\{\mathbf{C,E,G,H,I}\}$ and $C_4=\{\mathbf{F}\}$. Bold characters show dominating vertices.

The clustering algorithm is iterative and performs multiple runs, each of them increasing the value of the dissimilarity threshold θ. Once all threshold values passed, the algorithm provides the optimal partitioning (corresponding to one threshold value θ_o) which maximizes *Dunn's generalized index* ($Dunn_G$) [5]. $Dunn_G$ is designed to offer a compromise between the *intercluster separation* and the *intracluster cohesion*. So, it is the more appropriated to partition data set in *compact* and *well-separated* clusters. As an illustration, successive threshold graphs are constructed for each threshold θ selected from the dissimilarity Table 1, and our approach is used to give the *b-coloring* partition of each graph. The value of the *Dunn's generalized index* is computed for the obtained partitions. We conclude that the partition $\theta=0.15$ has the maximal $Dunn_G$ among other ones with different θ.

Table 1. Dissimilarity matrix

v_i	A	B	C	D	E	F	G	H	I
A	0								
B	0.20	0							
C	0.10	0.30	0						
D	0.10	0.20	0.25	0					
E	0.20	0.20	0.10	0.40	0				
F	0.20	0.20	0.20	0.25	0.65	0			
G	0.15	0.10	0.15	0.10	0.10	0.75	0		
H	0.10	0.20	0.10	0.10	0.05	0.05	0.05	0	
I	0.40	0.075	0.15	0.15	0.15	0.15	0.15	0.15	0

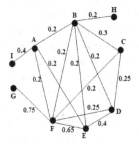 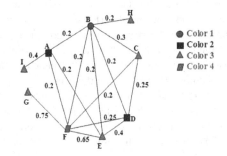

Fig. 1. *Superior threshold graph $G_{>0.15}$ (θ =0.15)*

Fig. 2. *b-coloring* of graph $G_{>0.15}$: four classes are identified

3 Online *b-Coloring* Based Clustering Algorithm

We now present the *online* clustering algorithm based on the above scheme (the *b-coloring* based clustering). The algorithm works incrementally by receiving a new document or removing existing data. The principle is; once the best partition (associated to the optimal threshold θ_o) returned from the *b-coloring based-clustering algorithm*, working to assign new instances to their respective clusters as they arrive or to rearrange the partition when existing instances leave the system. Let suppose the data set $X=\{x_1,...,x_n\}$ depicted by the *optimal threshold graph* $G=(V,E)$ and divided into $P=\{C_1,C_2,..,C_k\}$ The adding of new instance x_{n+1} transforms the vertex set V on $V\cup\{v_{n+1}\}$ and the edge set E on $E\cup\{(v_i,v_{n+1})|\ v_i\in V$ and $D(v_i,v_{n+1})>\theta_o\}$. The deletion of one instance $x_m\in X$ transforms the vertex set V on $V-\{v_m\}$ and the edge set E on $E-\{(v_i,v_m)|\ v_i\in V$ and $D(v_i,v_m)>\theta_o\}$. The main problem is to find the appropriate color to assign for v_{n+1} (*i.e.* in the case of insertion) or to rearrange the coloring of G (*i.e.* in the case of deletion) which is constrained to incrementally maintain the *b-coloring* of G and the clustering performances in terms of quality (*Dunn$_G$ value*) and runtime.

Assuming that the vertices of G are colored, the following notations will be used:

- Δ: the maximum degree of G.
- $c(v_i)$: the color (integer value) of the vertex v_i in G.
- $N(v_i)$: the neighborhood of vertex v_i in G.
- $N_c(v_i)$: the neighborhood colors of vertex v_i.
- $Dom(v_i)$: the dominance of v_i. $Dom(v_i)=1$ if v_i is one dominant vertex of $c(v_i)$ and 0 otherwise.
- k: the current number of colors (clusters) in G.

3.1 Adding a New Instance x_{n+1}

When a new instance x_{n+1} is introduced which corresponds to the vertex and edges adding in G, the following update operations on G are allowed:

- v_{n+1} is assigned to one of the existing k colors of G.
- v_{n+1} forms its own color.
- The insertion of v_{n+1} in G launches the merge of some colors in G.

As mentioned above, our *dynamic algorithm* relies only on the knowledge of the dissimilarity matrix and the dominating vertices of each color. Under this hypothesis, the following scenarios are to be considered:

3.1.1 Scenario 1: v_{n+1} Is Adjacent to at Least One Dominating Vertex of Each Color

When the neighborhood of v_{n+1} contains at least one dominating vertex from each k colors, v_{n+1} forms its own color $(k+1)^{th}$. Otherwise, the next *Scenario 2* is performed.

Proposition 1. After the creation of the new $(k+1)^{th}$ color, the coloring of G is a *b-coloring*.

Proof. $\forall C_h\in P=\{C_1,C_2,..,C_k\}\ \exists v\in(C_h\cap N(v_{n+1}))$ such that $Dom(v)=1$. Thus, $Dom(v_{n+1})=1$ and the vertex v remains dominating of its color $c(v)$ (*i.e.* $Dom(v)=1$). Consequently, $\forall\ C_h\in P=\{C_1,C_2,..,C_k,C_{k+1}\}\ \exists\ v\in C_h$ such that $Dom(v)=1$: the coloring of G using $k+1$ colors is a *b-coloring*.

In order to improve the quality of the new partition $P=\{C_1,C_2,..,C_k,C_{k+1}\}$ in terms of $Dunn_G$ value, the color of some vertices can be changed providing that the coloring of G remains a *b-coloring*. For that, the following definitions are introduced:

Definition 1. A vertex v_s is called "*supporting vertex*" if v_s is the only vertex colored with $c(v_s)$ in the neighborhood $(N(v_d))$ of one dominating vertex v_d. Thus, v_s cannot be re-colored.

Definition 2. A vertex v_c is called "*critical vertex*" if v_c is a *dominating* or a *supporting* vertex. Thus, v_c cannot be re-colored.

Definition 3. A vertex v is called "*free vertex regarding a color C*" if v is a non *critical vertex* and C is not in the neighborhood colors of v (*i.e.* $C \notin N_c(v)$). Thus, the color C can be assigned to v.

In order to evaluate the efficiency in the color change for *one free vertex v regarding one color C*, we compute the dissimilarity degree from the vertex v to the color C which is defined as the average dissimilarity from v to all vertices colored with C (eq.(1)). If this latter is lower to the dissimilarity degree from v to its current color, the color C is assigned to v.

$$d(v,C) = \frac{1}{|C|}\sum_{y \in C} D(v,y) \tag{1}$$

Due to this re-coloring, the *intraclass dissimilarity* can decrease which can maximally increase $Dunn_G$ by decreasing its numerator.

Suppose that the vertex v was initially assigned to $c(v)$ and re-colored with C. Since, the re-coloring of v causes to change the dissimilarity values $d(v_i, c(v))$ and $d(v_i,C)$ for each vertex v_i of G. Furthermore, although naive calculation of $d(v_i, c(v))$ and $d(v_i,C)$ takes $O(n^2)$, it can also be reduced to $O(n)$ using their old values as defined in the following equations:

$$d^{new}(v_i,C) = \frac{|C|d^{old}(v_i,C) + D(v_i,v)}{|C|+1} \tag{2}$$

$$d^{new}(v_i,c(v)) = \frac{|c(v)|d^{old}(v_i,c(v)) - D(v_i,v)}{|c(v)|-1} \tag{3}$$

Procedure Scenario 1()

```
BEGIN
c(vₙ₊₁):=k+1 ;
For each free vertex vᵢ regarding the color k+1 do
  If (d(vᵢ,k+1)<d(vᵢ,c(vᵢ)) then
    for each vertex vⱼ from G do
      Update(d(vⱼ,k+1);// using eq.(2)
      Update (d(vⱼ,c(vᵢ));// using eq.(3)
    Enddo
    c(vᵢ):=k+1;
  EndIf
Enddo
find_dominating();
END.
```

Therefore, after the re-coloring of every *free vertex* v_i *regarding the color k*, the method *find_dominating()* of order $O(n)$ tries to identify the new dominating vertices in *G*.

Proposition 2. The procedure *Scenario 1()* runs in $O(n^2)$.
Proof. After the coloring of v_{n+1} using the $(k+1)^{th}$ color, the algorithm for *Scenario 1* verifies if the color of each free vertex v_i regarding the color $k+1$ (at most n) can be changed by $(k+1)$. In this case, for each vertex v_j from *G* we update the dissimilarities $d(v_j,k+1)$ and $d(v_j,c(v_i))$ using the formulas eq.(2,3) in $O(n)$. Therefore, *Scenario 1* uses at most $(n*n)$ instructions, and the complexity is $O(n^2)$.

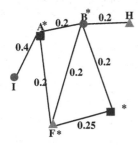

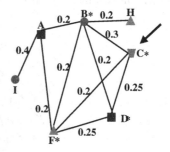

Fig. 3. Optimal Partition of $\{A,B,D,F,H,I\}$ on 3 clusters for $\theta=0.15$. "*" is used to denote the dominating vertices.

Fig. 4. Insertion of vertex *C* using *Scenario 1*: the neighborhood of *C* contains at least one dominating vertex of each color

3.1.2 Scenario 2: Neighborhood of v_{n+1} Has No Dominating Vertex of *m* Colors
The neighborhood of v_{n+1} does not contain any dominating vertex from *m* colors among the *k* current colors. These colors are called "*available to receive v_{n+1}*". Two cases are then considered:

❖ *Scenario 2.1 : m_1 colors ($m_1 \leq m$) are not present in v_{n+1} neighborhood colors*

The neighborhood colors of v_{n+1} does not contain m_1 among the *m* current colors (*cf.* Fig.5). This means that there is no significant dissimilarity between vertex v_{n+1} and these m_1 colors. Among m_1 colors, the one having the smaller dissimilarity with v_{n+1} will color it. Otherwise, the *Scenario 2.2* is performed.

Procedure Scenario 2.1()

```
BEGIN
H := {h | h∉N_c(v_{n+1})};
c(v_{n+1}):= {C| d(v_{n+1},C)=min_{h∈H}(d(v_{n+1},h))};
For each vertex v_i from G do
    Update(d(v_i, c(v_{n+1}))  ;// using eq.(2)
Enddo
For each vertex v_i ∈ N(v_{n+1}) do
    test_dominance(v_i) ;
Enddo
END.
```

After the insertion of v_{n+1} in the selected color (among m_1), some vertices from the neighborhood of v_{n+1} became dominating vertices. These vertices needed only one neighbor within the selected color to become dominating. In order to verify this situation, we need to recall the method *test_dominance(vertex)* which is in order $O(1)$.

Proposition 3. The procedure *Scenario 2.1()* performs in $O(n)$.

Proof. After the coloring of v_{n+1} using the selected color, the procedure *Scenario 2.1()* tries to update the dissimilarity $d(v_i,c(v_{n+1}))$ for each vertex v_i from G using the formula eq.(2) ($O(n)$). Afterward, it verifies the dominance property of the neighbors of v_{n+1} (at most Δ) using *test_dominance method* ($O(1)$). Therefore, the procedure *Scenario 2.1()* uses at most ($\Delta+n*1$) instructions, and the complexity is $O(n)$.

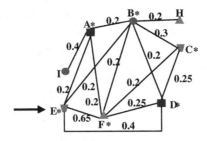

Fig. 5. Insertion of vertex E using *Scenario 2.1*: the color of C does not belong to the neighborhood colors of E. Consequently, the color of C is assigned to E.

❖ **Scenario 2.2: v_{n+1} is neighbor to at least one vertex in each m colors.**

Contrary to the previous scenario, v_{n+1} has at least one *non dominating vertex* per color in its neighborhood. We distinguish here the two following complementary sub-cases:

    o   *Scenario 2.2.1 : number of colors m=1*

If $m=1$ that is only one color C *available to receive* v_{n+1}, we assign this color C to v_{n+1}. Since this assignment generates a *non proper coloring* of G due to the presence of some neighbors of v_{n+1} in C, the colors of these vertices must be changed. For each vertex v_i among the latter the transformation is feasible because it is *non dominating*. As our objective is to find a partition such that the sum of vertex dissimilarities within each class is minimized, the color whose dissimilarity with v_i is minimal (eq.(1)) will be selected if there is a choice between many colors for v_i.

Procedure Scenario 2.2.1()

```
BEGIN
c(v_{n+1}):= C;// C the color available to receive v_{n+1}
For each vertex v_i from G do
Update(d(v_i, c(v_{n+1})) ;// using eq.(2)
Enddo
For each vertex v_i ∈ N(v_{n+1}) such that c(v_i)=C fo
    H := {h | h∉N_c(v_i)};
    k := {color| d(v_i,color)=min_{h∈H}(d(v_i,h))};
```

```
For each vertex vⱼ from G do
    Update (d(vⱼ,k);// using eq.(2)
    Update (d(vⱼ,c(vᵢ));// using eq.(3)
Enddo
c(vᵢ):=k ;
Enddo
For each vertex vᵢ ∈ N(vₙ₊₁) do
    test_dominance(vᵢ) ;
Enddo
END.
```

Proposition 4. The new coloring given from *Scenario 2.2.1* is a *b-coloring*.

Proof. $\forall v_i$ one vertex from G such that $c(v_i)=C$ and $v_i \in N(v_{n+1})$ we have $Dom(v_i)=0$. By the *dominance property*, $\exists h \in \{1,2,..,k\}$ such that $C_h \neq C$ and $C_h \notin N_c(v_i)$. Therefore, the color C_h will be assigned to v_i which guarantees *proper coloring*. In addition, $\forall h \in \{1,2,..,k\}$ such that $C_h \neq C$, $\exists v \in (C_h \cap N(v_{n+1}))$ having $Dom(v)=1$. Thus, v remains a dominating vertex of its color (*i.e.* $Dom(v)=1$) and likewise for v_{n+1} (*i.e.* $Dom(v_{n+1})=1$ in its color C). Consequently, there is at least one dominating vertex for each color ($\forall C_h \in P=\{C_1,C_2,..,C_k\}$ $\exists$ v such that $c(v)=C_h$ and $Dom(v)=1$): the *dominance property* is satisfied in P. The coloring of G is a *b-coloring*.

Proposition 5. The procedure *Scenario 2.2.1()* performs in $O(n\Delta)$.

Proof. When the color C is assigned to v_{n+1}, the neighbor vertices of v_{n+1} colored with C (at most Δ) change their colors which require the updates of the dissimilarities values in $O(n)$. Afterward, the dominance property of the neighbors of v_{n+1} (at most Δ) is verified using the *test_dominance method* ($O(1)$). Therefore, the procedure *Scenario 2.2.1()* uses at most ($\Delta*n+\Delta*1$) instructions, and the complexity is $O(n\Delta)$.

      o   *Scenario 2.2.2 : number of colors m>1*

In this case, several colors are *available to receive* v_{n+1} (*m>1*). The following definition of *color transformation* is required:

Definition 4. A color C among the m candidate colors to receive v_{n+1} is called "*transformation subject*" if its transformation does not violates the *b-coloring* constraints for the (*m-1*) remaining colors. In other words, the color C is a *non transformation subject* if it exists at least one color C' (among m) such that all the neighbors in C for the dominating vertices of C' are in the neighborhood of v_{n+1}.

Example: As an illustration, the figure 6 shows two colors C_1 and C_2 *available to receive* the vertex **F** (*m=2*). The unique neighbor in C_1 to the dominating vertex of C_2 (the vertex **B**) is the vertex **A** (called a *supporting vertex*) which belong to the neighborhood of **F**. Thus, the color C_1 is a *non transformation subject*. In fact, if the color C_1 is affected to the vertex **F**, the vertex **A** (dissimilar to **F**) must be re-colored. Due to this transformation, the color C_2 is removed from the neighborhood colors of **B** which becomes a non dominating vertex and the color C_2 without dominating vertices. Consequently, the transformation of C_1 is forbidden. Contrary to C_1, C_2 is a *transformation subject* and it is hence available to receive **F**.

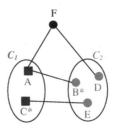

Fig. 6. A *transformation subject* colors identification

This shows that a color can undergo some transformations when a new vertex is presented (exclusion of vertices, change of dominating). Only the colors maintaining the *b-coloring* constraints are transformable. A relevant stage in the incremental approach will consist to identifying the number (m_2 among m) of *transformation subject* colors. The following sub-cases are then considered:

- *Scenario 2.2.2.1 : one color as a transformation subject*

In this case, only one color ($m_2=1$) is identified as a *transformation subject*. Therefore, the vertex v_{n+1} is assigned to this color and its transformation is allowed alike the previous *Scenario 2.2.1*.

- *Scenario 2.2.2.2 : m2>1 colors as a transformation subject*

The actual scenario considers the presence of a number m_2 ($1<m_2\leq m$) *transformation subject* colors. The color whose dissimilarity with v_{n+1} is minimal (eq.(1)) will be selected to receive it. Since the neighbor vertices of v_{n+1} in these m_2 colors must change their colors behind the inclusion of v_{n+1}, these vertices do not contributes to compute the dissimilarity values. Once the color available to receive v_{n+1} being selected, we transform it alike the previous *Scenario 2.2.1*.

- *Scenario 2.2.2.3 : no color as a transformation subject*

If any color is selected as *transformation subject* among the m colors, v_{n+1} forms its own color $(k+1)^{th}$ it becomes its dominating vertex (*i.e. $Dom(v_{n+1})=1$*). Due to this transformation, the m colors becomes without dominating vertices. Regarding this problem, we define a procedure which tries to find a *b-coloring* of G where all colors are *dominating*. The idea is the following: each *non dominating* color C among the m *no subject transformation* colors can be changed. In fact, after removing C from the graph G, for each vertex v_i colored with C (i.e. $c(v_i)=C$), a new color is assigned to v_i which is different from those of its neighborhood. As our objective is to find a partition such that the sum of vertex dissimilarities within each class is minimized, the color whose distance with v_i is minimal will be selected if there is a choice between many colors for v_i. Before starting again with another *non dominating* color C' the procedure verifies if the remaining colors have now a dominating vertex (in such a case, these colors are identified as a dominating color).

Discussion

In order to process new data instances as they are arrived, the learning algorithm has two steps: initialization and cluster update. It initially adopts the *b-coloring partition* associated to the optimal dissimilarity threshold θ_o and works to update it. In the initialization step it is better if we have a sample of the data set that is significant overall the feature space as that we can get a significant clustering, but we can work as well with a normal data set. If the data set used for initialization step does not reflect the true clusters structure, the online approach allows an eventual updating of existing clusters by re-coloring certain instances. Due to this re-assignment strategy, the *intraclass dissimilarity*, an increasing monotonous function of threshold θ, can decrease by improving the partition quality and monotonically decreasing the optimal dissimilarity threshold θ_o during the incremental process.

For more improving the partition quality, we propose an additional operation to optimize the groups of existing clusters called *color merging*. Typically, two colors are merged when the dissimilarity between them is below the optimal dissimilarity threshold θ_o. Consequently, the optimal threshold θ_o can increase although the *b-coloring constraints* are violated. To solve this problem, the procedure used in Scenario 2.2.2.3 to find a *b-coloring* of G is applied for every color without dominating vertices.

3.2 Removal of an Existing Instance v_m

When an instance x_m is introduced which corresponds to the vertex and edges deleting in G, we must rearrange the coloring of G in order to maintain the *b-coloring* properties and a high quality clustering. Likewise to the previous scenarios, our idea is based only on the knowledge of the dissimilarity matrix and the dominating vertices of each color. Under this hypothesis, the following scenarios are to be considered:

3.2.1 Scenario 3: v_m Is the Sole Dominating Vertex of Its Color $c(v_i)$

In this case, v_m is the only one dominating vertex of its color. Therefore, by removing v_m, the color $c(v_m)$ becomes without dominating vertices and the coloring of G is not a *b-coloring*. Consequently, the colors of the remaining vertices of $c(v_m)$ must be changed. For each vertex v_i among the latter, the transformation is feasible because it is *non dominating*. As our objective is to find a partition such that the sum of vertex dissimilarities within each class is minimized, the color whose dissimilarity with v_i is minimal (eq.(1)) will be selected if there is a choice between many colors for v_i. In the opposite case of this scenario, the following *Scenario 4* is performed.

3.2.2 Scenario 4: v_m Is a Supporting Vertex of All Dominating Vertices from at Least One Color

In this situation, v_m is the sole vertex colored with $c(v_m)$ in the neighborhood of all dominating vertices from at least one color C. As a result, the deletion of v_m, pushes these vertices to become *non dominating* and C without dominating vertices. To solve this problem, the procedure used in Scenario 2.2.2.3 to find a *b-coloring* of G is applied for every color C without dominating vertices.

If v_m does not verify any of the previous *Scenarios* 3 and 4, the dynamic algorithm process the deletion of v_m without any rearrangement and the new coloring of G is a *b-coloring*.

4 Experimental Results

Experiments have been made using three relevant benchmark data sets chosen from UCI database [6]. The first data set (*Zoo data*) is a collection of 100 animals with 17 features (1 quantitative, 1 nominal and 15 boolean). The second data set (*Auto import data*) consists of 193 instances of cars with 24 features (14 quantitative and 10 nominal) and the third data set (*Tic-tac-toe data*) contains 958 instances, each described by 9 categorical attributes.

In order to examine the effectiveness of the *online b-coloring* algorithm, the experimental methodology in conducted as follows: for each data set, the *b-coloring partition* id firstly generated upon a data sample (which contains 50 instances for Zoo, 100 for Auto import and 700 for tic-tac-toe) from the original data sets; then the partition is updated by adding sequentially the remaining points and the value of $Dunn_g$ index is computed as more instances are included.

For an interesting assess of the results gained on these data set, our algorithm was compared against *original b-coloring*, *Single-Pass*[2] and *k-NN* (*k=5*). The *original b-coloring* consists in performing complete re-clustering using *b-coloring clustering algorithm* [2].

We note that the Euclidian distance is applied to define the dissimilarity level between two instances characterized with m features a_f ($f \in \{1...m\}$) as given by the following formula:

$$d_{i,j} = D(x_i, x_j) = \left(\sum_{f=1}^{m} \left(\frac{g_f(a_{i,f}, a_{j,f})}{m_f} \right)^2 \right)^{1/2} \tag{4}$$

where m_f is the *normalized coefficient* for the attribute a_f and g_f is the comparative dissimilarity function between the two attribute values $a_{i,f}$ and $a_{j,f}$ corresponding respectively to the instances x_i and x_j.

For *numeric* attributes, g_f is: For *categorical* attributes, g_f is:

$$g_f(a_{i,f}, a_{j,f}) = |a_{i,f} - a_{j,f}| \qquad g_f(a_{i,f}, a_{j,f}) = \begin{cases} 1 & iff \ a_{i,f} \neq a_{j,f} \\ 0 & iff \ a_{i,f} = a_{j,f} \end{cases} \tag{5}$$

The figures 7, 8 and 9 show the evolution of $Dunn_g$ values according to the number of instances. The curves comparisons of the different clustering algorithms show the performance of the *online b-coloring algorithm*. It appears clearly that the online algorithm achieves better results than *k-NN* (*k=5*) and more significant results than *Single-Pass*. It appears that the incremental algorithm slightly improves the performance of the *original b-coloring algorithm* (except the runtime profit) especially due to the efficiency of the re-assignment strategy (re-coloring) which improves the partition quality in terms of $Dunn_g$ value. Finally, one can see that similar experiments may be done for removal instances.

---

[2] The dissimilarity threshold used in the *Single-Pass* algorithm is the optimal threshold θ_o.

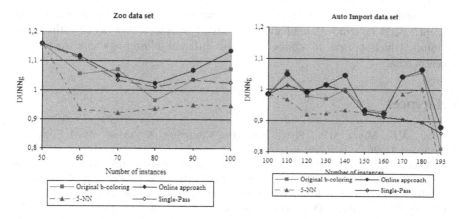

Fig. 7. Performances on *Zoo*   **Fig. 8.** Performances on *Auto Import*

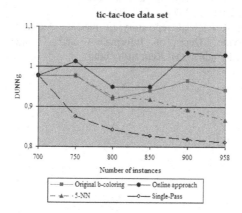

Fig. 9. Performances on *tic-tac-toe*

5 Conclusion

We proposed a dynamic *version* for the *b-coloring based clustering approach* which relies only on *dissimilarity matrix* and *cluster dominating vertices* in order to cluster new data as they are added to the data collection or to rearrange a partition when an existing data is removed. A real advantage of this method is that it performs a dynamic classification that correctly satisfies the *b-coloring* properties and the clustering performances in terms of quality ($Dunn_G$ *value*) and runtime, when the number of clusters is not pre-defined and without any exception on the type of data. The results obtained over three *UCI data sets* have illustrated the efficiency of our algorithm to generate good results than *Single-Pass* and *k-NN* algorithms.

There are many interesting issues to pursue: (1) leading additional experiments on a larger medical data set where a patient stay typology is required and an inlet patient stay is regular and has to be incorporate to the typology, (2) extending the incremental concept to add or remove simultaneously sets of instances, and (3) to define some operators which permit to combine easily different clusterings constructing on different data.

References

[1] Jain, A.K., Murty, M.N., Flynn, P.J.: Data Clustering: A Review. ACM Computing Surveys 31, 264–323 (1999)
[2] Elghazel, H., et al.: A new clustering approach for symbolic data and its validation: Application to the healthcare data. In: Esposito, F., Raś, Z.W., Malerba, D., Semeraro, G. (eds.) ISMIS 2006. LNCS (LNAI), vol. 4203, pp. 473–482. Springer, Heidelberg (2006)
[3] Irving, W., Manlove, D.F.: The b-chromatic number of a graph. Discrete Applied Mathematics 91, 127–141 (1999)
[4] Cover, T., Hart, P.: Nearest neighbor pattern classification. IEEE Transactions on Information Theory 13(1), 21–27 (1967)
[5] Kalyani, M., Sushmita, M.: Clustering and its validation in a symbolic framework. Pattern Recognition Letters 24(14), 2367–2376 (2003)
[6] Blake, C.L., Merz, C.J.: UCI repository of machine learning databases. University of California, Irvine, Dept. of Information and Computer Sciences (1998), http://www.ics.uci.edu/ mlearn/MLRepository.html

Positivism Against Constructivism: A Network Game to Learn Epistemology

Hélène Hagège[1,*], Christopher Dartnell[2], and Jean Sallantin[2]

[1] Université Montpellier II, Laboratoire Interdisciplinaire de Recherche en Didactique
Education et Formation, CC 077, place Eugène Bataillon, 34 095 Montpellier cedex 5, France
hhagege@univ-montp2.fr
[2] Laboratoire d'Informatique de Robotique et de Microélectronique de Montpellier, UMR 5506,
161 rue Ada, 34392 Montpellier Cedex 5, France
dartnell, sallantin@lirmm.fr

Abstract. As mentioned in French secondary school official texts, teaching science implies teaching scientific process. This poses the problem of how to teach epistemology, as traditional science teaching is mostly dogmatic and based on contents. Previous studies show that pupils, science students and teachers mostly own positivist and realist spontaneous conceptions of science and scientific discovery. Here, we present the evaluation of the didactic impact of a network game, Eleusis+Nobel, on third year biology students who aim at becoming teachers. This cards game, based on a Popperian epistemology, has been designed to reproduce the scientific discovery process in a community. In the limits of our study, results obtained with classical social psychology tools indicate that students who played this game specifically assimilated the subjective dimension of knowledge and the role of the community in their conception of science, on the contrary to negative control students, who did not play.

Keywords: epistemology, positivism, constructivism, science education.

1 Introduction

Scientific discovery is a complex process including psychological, social and historical dimensions. As far as the cognitive psychological dimension is concerned, research made an advance since both science products (concept or knowledge) and process (experimental design and evidence evaluation skills) have been integrated in the descriptive framework of Scientific Discovery as Dual Search [13]. Simulated science discovery tasks have then been focalised on domain specific discoveries integrating science process consideration (for a review see [21]). However, few simulations take into account the social dimension of scientific discovery, which is considered as central by epistemologists (*e.g.* [14]). Here, we are interested in one of those: Eleusis+Nobel network game (E+N; [5,6]).

---

* Corresponding author.

V. Corruble, M. Takeda, and E. Suzuki (Eds.): DS 2007, LNAI 4755, pp. 91–103, 2007.
© Springer-Verlag Berlin Heidelberg 2007

1.1 Problematics

All these interesting developments raise the question of their social utility: what do people who take part in such simulations really learn? Are they able to transfer what they learn to their conception of science? We try here to provide elements to answer these questions, which have not yet occurred, as far as we know.

In other words, the matter is to know whether such simulations can be used as tools to teach epistemology. In France, all secondary science teachers enter a classroom very soon, without having entered a research laboratory. That is to say, they do not know what science looks like. New French official texts contain the explicit obligation for them to teach science process [3], although traditional education always put the emphasis on contents. Given the little practical place devoted to epistemology during the formation of teachers, and the contradiction of teaching epistemology in a dogmatic way, we were willing to evaluate alternative ways of teaching. We propose here an evaluation of the impact of E+N game on third year general biology students, who aim at becoming secondary biology teachers.

1.2 E+N Game

E+N is a card game inspired from Abott's game [9], and was designed in collaboration with cognitive scientists [5] to simulate scientific discovery and to study the players' strategies during a collective process of proof and refutation.

Players have to discover a set of hidden rules, each determining the valid card sequences that can be formed during the game. A hidden rule is a set of clauses as "A red card can be followed by a black card, and a black card by a red card" using the colour and the form of a card (red, heart …) and/or its rank (ace, figure, pair…).

Each player has access to private experimentation spaces corresponding to each hidden rule, in which he/she can play cards and form sequences which are classified as true or false by the hidden rule. Players can publish their own theories explaining hidden rules, read the ones submitted by other players and possibly refute them when they find a sequence which is irrelevant with what was published. This game is based on a Popperian conception of science, where validation goes through conjectures and refutations. Publications and refutations are sanctioned by the following score system. A player scores n points when publishing a theory, and n' points by refuting an existing one, in which case the publisher of this theory looses n' points. The game ends after a fixed duration (two or three hours), and the player with the highest score wins the Nobel Prize. The ratio n/n' can be changed from one game to another to study the variations in player's strategies. We refer to previous paper [6] for a more detailed description of the game's rules and interface.

1.3 Theoretical Frame of the Study

In science education and epistemology, a constructivist vision of building knowledge has been developed (*e.g.* [8, 14, 19]), to which a majority of research workers in these domains seem to adhere [15]. According to constructivism, all knowledge is linked to a subject who knows [8]. So, its profound nature is subjective. Thus conviction, point of views and beliefs are part of science and learning [1, 14]. On the other hand, all knowledge is issued from a construction process. This process consists in qualitative reorganisation of initial knowledge structure [17], and can be assimilated to change of conceptions [19]. Conceptions play an organisational role in thinking and learning [19], but affects and values also do [10].

Here, we refer to *personal epistemology* as a system of interacting attitudes related to knowledge construction objects (such as *error, science...*). Attitudes are composed of a cognitive and an affective component (*i.e.* conception of an object, and affective relation to this object; [11]). They interact together and norms and associated values emerge from this epistemic attitudes system [10]. Norms are rules telling how the subject should behave in a particular situation and values consist in general principles which justify the corresponding norms.

Most studies on epistemology learning and teaching concern conceptions, *i.e.* what we call the cognitive component of attitudes. Science teachers and students do not own constructivist spontaneous science conceptions (*e.g.* [2, 16, 20]). For instance, to future biology teachers, knowledge is an "external truth that can be discovered through observation, discussion, sense-making" and also a collection of additive facts [16]. In that sense, experiment can constitute a supreme arbitrary to verify theories. This naïve, *positivist* labelled epistemology also contains a realist view, given which the world is intimately knowledgeable (in opposition to an idealist conception), so that scientific knowledge tells us about a truth: the world as it is. This positivist and realist vision is coherent with *naïve* [18] and *traditionalist* [4] epistemologies evaluated by other authors, in the sense that knowledge would be composed of information units which are progressively added, thus allowing knowledge progress. In fact, secondary teachers define teaching as a "maximum information transfer" and learning as an "every information absorption" [2, 20].

In the following, we evaluate E+N playing impact on science conceptions, values, and to a less extent, affects. We used the standard pre-test/post-test procedure. The test was mostly composed of a Likert-type scale and of Osgood's semantic differentiators (OSD). Values are considered to be implicit in all adjectives, but some of those explicitly refer to values, such as *good* and *beautiful*. Affects correspond to pleasure and pain domain. Conceptions are here considered as moving from a positivist and realist extremity to an idealist and constructivist one. One has to notice that we refer to philosophical corresponding notions, to be able to characterise students' undifferentiated epistemology. These students initially had no deep thought about scientific process. E+N implements the Popperian *intersubjective construction of objectivity* concept, which is a central point of what became constructivism. That is why we expected E+N game to favour constructivist epistemology development.

2   Methodology

2.1   Procedure and Subjects

The study has been realized in South France, in the University Montpellier II. In January 2007, 43 third year general biology students filled up the initial test (= *initial experiment*). All these students aimed at becoming life and earth science secondary school teachers and were registered to follow the same science education and epistemology courses. One and a half month later, 14 of them (=Pl for *Players*) played E+N then filled up the final test (6 days later), whereas 14 others (=NC for *Negative Controls*) filled up the final test without having played. The final test corresponds to the initial test plus some additive questions. For both Pl and NC groups, the initial experiment is called the *pre-test* and the final one the *post-test*. Players have been told that this game mimics scientific discovery as it occurs – in community. During the game, Pl was mixed together with 24 other students and the whole sample was split into 16 teams of 2 or 3 players. All 16 computers were in the same room. The game lasted 2 hours and the winner team won a 1kg candy box (the Nobel Price). There was a non desired function in the program: players could refute themselves and win points whereas they should normally have lost the points gained during publication. Particularly, two teams concentrated on this strategy and this provoked a revolt atmosphere at the end of the session.

2.2 Measuring Tools

Classical tools of socio-psychology have been used for this study. A Likert-type scale is a group of propositions which measure the same psychometric variable. Subjects have to indicate their degree of agreement for each proposition (see Appendix 1). For OSD relative to a term, subjects have to choose a position between two opposite adjectives, depending on the one that best describes the term from their point of view. For each scale and each individual, we calculated a score, which corresponds to the average answer to the scale's items.

The pre-test is composed of a questionnaire and an OSD series. The questionnaire aims at assessing positivist and realist conceptions in opposition with constructivist and idealist ones. It is composed of two subscales: "Realism and truth status" (RTS) and "Research worker's status" (RWS) subscales (Appendix 1). OSD were designed to evaluate conceptions (C1 to C3 scores), values (V1 to V4 scores) and affects (A score) related to five terms, considered as epistemic objects (Appendix 2).

The post-test contains additional OSD, relative to conceptions of *proof* and *refutation* (Appendix 3) and two open questions: "1) Give 3 terms you associate to the communication of results in a scientific community" and "2) Give 3 terms you associate to scientific discovery". We respectively expected the occurrence of *publication* and *refutation* terms specifically for the Pl group. Since these parts did not appear in the pre-test, we are not able to observe any change in conceptions. Consequently, the results are only indicative.

2.3 Results Analysis

Data were collected, reported in Excel^R and analyzed with SPSS^R 9.0 software. Non-parametric tests were used to compare item *per* item (Wilcoxon signed ranks test on paired samples[*] and Mann-Whitney test on independent samples). Independent or paired samples[*] T-tests allowed scores comparison. ([*]for pre-test/post-test comparison in a given subpopulation – Pl or NC)

3 Results

3.1 Homogeneity of Pl and NC Subpopulations at the Pre-test

To verify that Pl and NC subpopulations were comparable, we first looked at social variables (Table 1A). Both subpopulations were significantly the same average age and were composed of the same number of males and females. Concerning parents' socio-professional category, we cannot know much since the majority of subjects answered *other*, although our sampling do not seem to be biased relatively to professions linked to scientific research or scientific education.

Table 1. Comparison of Pl and NC Subpopulations Pre-test Variable Means and Average Variable Means for Pooled Pl and NC.

sd : standard deviation
for T-test, df = 26
t is obtained after an independent samples T-test

A) Social Variables

	age	sexe	spc of parents
t (Pl *vs* NC)	-1.381	0.000[a]	-0.801[a]
Mean (Pl+NC); (sd)	20.86 (1.11)	1.29 (0.46)	5.11 (1.64)

spc : socio-professional category
[a] concerning these ordinal variables, Wilcoxon test also leads to the conclusion of population homogeneity
sex : 1 female, 2 male
spc of parents : 1 scientific education, 2 scientific research , 3 agriculture, 4 industry, 5 health and 6 other

B) Pre-test Scores

Score	RTS	RWS	V1	V2	V3	V4	A	C1	C2	C3
t (Pl *vs* NC)	-0.245	0.217	2.664*	0.137	0.113	1.168	0.437	-0.303	-0.625	0.077
Mean (Pl+NC)	0.18	-0.20	1.23	1.00	1.33	1.28	-1.09	0.31	-0.08	0.23
(sd)	(0.60)	(0.77)	(0.65)	(0.54)	(0.55)	(0.62)	(0.64)	(0.62)	(0.57)	(0.73)

statistical significance: *$p<0.05$

Secondly, we compared epistemology scores between each subpopulation through a T-test (Table 1B). We can notice that with the exception of the esthetical value V1 score, all scores can be assumed as similar. Some means have an absolute value superior to 1, whereas other means are closer to 0. The former, clear-cut epistemology scores, concern the positive values associated to scientific knowledge, science, error, teaching and knowledge (V2, V3 and V4) and the negative affects associated to error (A). Relatively to the latter, which does not reflect a shared tendency between individuals, population is more heterogeneous. Positive RTS, C1 and C3 scores means correspond to a dominant positivist and realist epistemology, whereas negative C2 and RWS scores means indicate a constructivist tendency.

V1 score is significantly higher in Pl subpopulation (see Table 2). However, we can notice that all Pl and NC subjects have a null or a positive V1 score (not shown), which suggests that if the quantity of this value is not comparable, the quality is the same: it is positive. Item *per* item analysis through Mann-Whitney test indicates that only 4 items among 106 initial items were statistically different between Pl and NC subpopulations (not shown). Two of those items enter V1 score, one has been excluded from the analysis and the last one is part of V4 score.

We conclude that for all considered scores but V1, NC subpopulation constitutes a satisfying negative control for Pl subpopulation.

3.2 Pl Subpopulation Specific Scores Changes of Answers in the Post-test

Table 2 shows that only two scores means (RWS and C3) significantly changed in Pl's post-test. For this subpopulation, RWS scores mean is more negative in the post-test than in the pre-test, whereas C3 scores mean becomes negative in the post-test. Among nine RWS subscale items, six concern the role of a research worker's subjectivity in science (Appendix 1). Moreover, all C3 semantic differentiators focus on subjectivity (or creativity and imagination) relatively to different epistemology objects. So it seems that a major change in players' conception deals with the central role of *subjectivity* – of subjects – in building knowledge.

Table 2. Evaluation of E+N Specific Effect on Pl and NC Subpopulations Scores

sd : standard deviation
t is obtained following a paired-samples T-test comparing pre-test and post-test scores means
for T-test, df = 13
statistical significance: *$p < 0.05$, **$p < 0.01$

Score (sd)	NC			Pl		
	Mean at the pre-test	Mean at the post-test	t	Mean at the pre-test	Mean at the post-test	t
RTS	0.21 (0.68)	0.04 (0.65)	1.201	0.15 (0.55)	-0.16 (0.52)	1.967
RWS	-0.23 (0.78)	-0.56 (0.48)	1.469	-0.17 (0.79)	-0.72 (0.54)	3.016**
V1	0.93 (0.66)	0.81 (0.69)	0.563	1.52 (0.52)	1.31 (0.59)	1.188
V2	0.99 (0.54)	0.83 (0.38)	1.230	1.01 (0.56)	0.96 (0.53)	0.328
V3	1.32 (0.60)	1.25 (0.54)	0.479	1.35 (0.51)	1.29 (0.40)	0.359
V4	1.14 (0.73)	1.10 (0.64)	0.268	1.41 (0.47)	1.59 (0.45)	-1.075
A	-1.14 (0.53)	-0.79 (0.64)	-1.859	-1.04 (0.75)	-0.86 (0.77)	-1.439
C1	0.34 (0.76)	0.50 (0.50)	-0.962	0.29 (0.47)	0.09 (0.61)	1.129
C2	-0.01 (0.68)	-0.11 (0.56)	0.490	-0.15 (0.46)	-0.25 (0.33)	0.766
C3	0.22 (0.67)	0.00 (0.49)	1.223	0.25 (0.80)	-0.32 (0.56)	2.543*

Item *per* item analysis revealed only few differences between Pl post-test and pre-test answers (Table 3). We can notice that among seven significantly changing items, four deal with subjectivity (Q2, Q4, D1, D3), and always in the sense of enhancing subjectivity integration in their conceptions. The fact that Q2 and Q4 are part of RTS score reinforces the previous result obtained with RWS score (Table 2). An interesting result is obtained with Q1 item; it seems that the game has convinced a third of players (not shown) that an isolated research worker cannot do science.

Table 3. Evaluation of E+N Specific Effect on Pl and NC Subpopulations Item Answers

	Z^a	NC Mean at the pre-test	Mean at the post-test	Z^b	Pl Mean at the pre-test	Mean at the post-test	Z^b
Q1	-0.026	0.71 (1.49)	0.29 (1.59)	-1.540	0.69 (1.70)	-0.57 (1.40)	-2.401*
Q2	-0.951	0.31 (1.38)	0.07 (1.27)	-0.666	0.79 (1.58)	-0.38 (1.04)	-2.476*
Q3	-0.171	-1.46 (0.88)	-1.21 (0.70)	-1.000	-1.57 (0.64)	-0.86 (1.17)	-2.309*
Q4	-1.278	-1.08 (1.32)	-1.14 (1.03)	-0.520	-0.29 (1.64)	-1.36 (0.84)	-2.324*
D1	-0.025	0.36 (1.08)	0.14 (0.53)	-0.918	0.38 (1.04)	-0.57 (1.09)	-2.220*
D2	-1.524	0.36 (0.84)	0.07 (0.73)	-1.265	-0.29 (1.20)	0.64 (1.01)	-2.804**
D3	-1.135	-0.21 (1.12)	0.14 (1.03)	-0.905	0.29 (0.99)	-0.50 (1.02)	-1.995*

[a] Mann-Whitney test variable is issued from comparison of Pl and NC answers at the pre-test.
[b] Z is issued from Wilcoxon signed ranks test on paired samples comparing pre-test and post-test items answers means. All items of initial experiment that present a significant difference between Pl and NC subpopulations at the post-test are presented here.
Q1: "An isolated research worker can do science." Po. **Q2:** "Scientific theories are inventions." Co (RTS). **Q3:** "There is always more than one way to interpret an experiment result." Co (RTS). **Q4:** "Researchers do not use their beliefs to do science" Po. **D1:** "scientific knowledge": subjective/objective (C3). **D2:** error; awful-beautiful. **D3:** learning; subjective/objective.
Po indicates that a total agreement is counted as +2 and *Co* that the answer is reversed (total agreement as -2). When the item enters a scale, it is mentioned (Q1, Q4, D2 and D3 have not been retained in the scales presented in this paper).

Another promising result concerns D2; to players, error has significantly become more beautiful. This is the only result of our study concerning the change of a value after playing E+N. Finally, an unexpected result is found in Q3 answers change.

3.3 Putative Pl Subpopulation Specific Changes of Conceptions

Answers to additional open questions (Table 4) indicate that our expectations concerning the occurrence of the term *publication* – which corresponds to E+N nomenclature – in subpopulation Pl have not been satisfied : not only did Pl subjects mention *article* instead of *publication*, but they also did it nearly as much as NC subjects. Also, *refutation* is not mentioned. The only two clear-cut answers specific to subpopulation Pl, which were not predicted, are *discussion* and *subjective*. As these questions were not in the pre-test, we cannot be sure that this specificity appeared through the game. However, this result contributes to reinforce previous ones concerning subjectivity and the role of community in science.

Specific OSD relative to *proof* and *refutation* in the post-test (Table 5) indicate that *proof* is significantly more relative, temporary, statistic and collective for Pl than NC subjects. Again for Pl subjects, both *proof* and *refutation* are more collective, experimental and complex. It is tempting to think that this corresponds to a game effect. Item *per* item analysis (not shown) reveals that changes concern *complexity* for both proof and refutation and on the experimental property of refutation.

Table 4. Number of Five Selected Terms' Occurrence and Number of Subjects Concerned by these Occurrences in Answers to both Additional Open Questions in the Post-test

	NC		Pl	
	N (occurrence)	N (subjects)	N (occurrence)	N (subjects)
article	4	3	5	5
eurêka[b]	1	1	1	1
experiment[a,b]	3	3	3	3
discussion	1	1	6	5
subjective[a]	0	0	4	4

[a] or related term : *experimentation, subjectivity, ...*
[b] both terms are chosen as negative controls and were not particularly expected.
Discussion is the term with the highest overall occurrence. Other terms which are not indicated here are very disparate and seem to come under heterogeneous categories.

Table 5. Comparison of Pl and NC Additional OSD Post-test Scores Means

	RePr	Re	Pr
NC score mean (sd)	-0.21 (0.30)	-0.11 (0.49)	0.11 (0.57)
Pl score mean (sd)	-0.91 (0.55)	-0.07 (0.87)	-0.33 (0.48)
t (NC *vs* Pl)	-4.201***	0.134	-2.124*

sd : standard deviation
t is issued from independent samples T-test
statistical significance : *** $p < 0.0005$, * $p < 0.05$

4   Discussion

4.1   Population Initial Epistemology

We proposed a pre-test and a post-test to students who played E+N for two hours and we compared changes in answers with the ones of negative controls (non-players). The test evaluates conceptions, values and affects concerning scientific process.

Before the game, initial Pl's and NC's epistemology where similar, except from esthetical values, which were higher for Pl. This heterogeneity effect underlies a limit of our study: the smallness of our samples. Future experiment will be done with greater samples. Otherwise, positive values were expected from students who aim to become science teachers. The negative affective dimension of attitude towards error had already been characterized [7] and is explained, together with general conception tendencies elsewhere [10]. Slightly negative scores means (RWS and C2) – indicating constructivist conceptions – are interpreted as concerning on-going science: these students know that error takes part of science and that scientists can have "wrong" interpretations or theories. But they think that once the error is detected, knowledge which is kept is true. This last point would explain slightly positive scores means (RTS, C1 and C3) and correspond to a realist and positivist point of view.

4.2   Conception Change Through E+N Playing

We tried to evaluate several aspects linked to constructivism. Among these, the aspect which is recurrently and significantly changed – specifically to Pl – concerns the role of subjectivity in scientific process. These results are reinforced by those obtained with additional specific post-test questions. Additionally, Q1 item and answers that indicate putative conception changes focus on the role of community in scientific process. Thus, to us, the game allowed Pl to become aware of these central aspects of constructivism, so that they specifically assimilated them in the cognitive components of their epistemic attitudes. The only one result which was not predicted is the change of Q3 answer; Pl are in fact less likely to believe that several interpretations are possible in front of a given result. Maybe they assimilated *possible*, in the sense of what a research worker can propose, with *right*, in the sense of what is acceptable given a theory. This could be due to the strict formalism of the game, in which theories are predetermined and perfectly knowledgeable.

Because of the difficulty to find volunteers, we organised this experiment with our students, who were supposed to follow epistemology courses. This could explain why NC's scores also change between the pre-test and the post-test. However, statistics give us a clear limit and the significance levels that we use are absolutely standard. So no statistically significant score change has been observed in NC subpopulation.

4.3 Suitability of the Game for Epistemology Teaching

In the game, hidden rules represent what would be in reality "facts resistance to experimentation". Thus, the conventional law constructed by players' community do not necessarily correspond to the hidden rule. In that way, the game partly modelises construction of knowledge by a research worker community. Although we did not wanted that auto-refutation could allow point winnings, we noticed that this could modelise an existing scientific strategy. It is possible that this parameter greatly influenced Pl in their consideration of science as relying on subjectivity; the one who wins can do it through cheating! As all observed answers changes do not focus on themes that are explicitly dealt with in the game, but just practiced, we infer that this constructivist conception has been subconsciously assimilated, in the Piagetian sense. We cannot exclude that this effect occurred synergistically with traditional epistemology courses. Even so, observed changes are very encouraging, because they would have been caused by only two hours of playing.

An important factor for such a teaching tool is users' pleasure. Open questions in the post-test treated of the matter of feelings during playing (not shown). We noticed that answers extremely differed: either players liked it much, or they got "very frustrated because of cheats". This highlights what we also observed during the game: they really got involved into it. Previous experiments with 13 or 20-year-old pupils lead to the same conclusion. When time was out, a majority was disappointed and wanted to continue (that rarely happens with a traditional course!).

Altogether, it indicates that E+N game can constitute a very interesting complementary tool to teach epistemology. In this report, we did not address the evaluation of what ability players learn through the game. It would be interesting to evaluate students' skills to apply the refutation principle, to manipulate hypothesis and to propose experiences in front of a problem. We shall go deeper into this question, which will be dealt with in future investigations.

References

1. Bachelard, G.: Epistémologie (1971). Presses Universitaires de France (2001)
2. Boulton-Lewis, G.M., Smith, D.J.H., McCrindle, A.R., Burnett, P.C., Campbell, K.J.: Secondary Teachers' Conceptions of Teaching and Learning. Learn. Instr. 11, 35–51 (2001)
3. Bulletin Officiel: Les compétences professionnelles des maîtres. MENS0603181A (2007)
4. Chan, K.-W., Elliott, R.G.: Relational Analysis of Personal Epistemology and Conceptions About Teaching and Learning. Teach. Teach. Educ. 20, 817–831 (2004)
5. Chavalarias, D.: La thèse de Popper est-elle réfutable? Mémoire de DEA CREA-CNRS/Ecole Polytechnique (1997)
6. Dartnell, C., Sallantin, J.: Assisting Scientific Discovery with an Adaptive Problem Solver. Discov. Sci. (2005)
7. Favre, D.: Conception de l'erreur et rupture épistémologique. Rev. Fr. Pédagog 111, 85–94 (1995)
8. Fourez, G., Englebert-Lecomte, V., Mathy, P.: Nos savoirs sur nos savoirs, DeBoeck Université (1997)
9. Gardner, M.: Mathematical Games. Sci. Am. (1959)
10. Hagége, H.: Jugement de valeurs, affects et conceptions sur l'élaboration du savoir scientifique: à la recherche d'obstacles à l'enseignement des questions vives. In: Giordan, A., Martinand, J.-L. (eds.) XXVIII^{émes} journées internationales sur la communication, l'éducation et la culture scientifiques, techniques et industrielles (under press, 2007)
11. Hagège, H., Reynaud, C., Caussidier, C., Favre, D.: New Conceptualisation of Environmental Attitudes: Cut, Relatedness and Fusion Towards the Nonhuman Environment - Preliminary Measure. Environ. Behav. (submitted)
12. Howard, B.C., McGee, S., Schwartz, N., Purcell, S.: The Experience of Constructivism: Transforming Teacher Epistemology. J. Res. Comput. Educ. 32, 455–465 (2000)
13. Klahr, D., Dunbar, K.: Dual Search Space During Scientific Reasoning. Cogn. Sci. 12, 1–48 (1988)
14. Kuhn, T.: La structure des révolutions scientifiques. Champs Flammarion (1962)
15. Lederman, N.G., Abd-El-Khalick, F., Bell, R.L., Schwartz, R.S.: Views of Nature of Science Questionnaire: Toward Valid and Meaningful Assessment of Learners' Conceptions of Nature of Science. J. Res. Sci. Teach. 39, 497–521 (2002)
16. Lemberger, J., Hewson, P.W., Park, H.-J.: Relationships between Prospective Secondary Teachers' Classroom Practice and Their Conceptions of Biology and of Teaching Science. Sci. Educ. 83, 347–371 (1999)
17. Lonka, K., Joram, E., Brysin, M.: Conceptions of Learning and Knowledge: Does Training Make a Difference? Contemp. Educ. Psychol. 21, 240–260 (1996)
18. Schommer, M.: Synthesizing Epistemological Belief of Research: Tentative Understandings and Provocative Confusions. Educ. Psychol. Rev. 6, 293–319 (1994)
19. Strike, K.A., Posner, G.J.: A Revisionist Theory of Conceptual Change. In: Duschl, R.A., Hamilton, R.J. (eds.) Philosophy of Science, Cognitive Psychology, and Educational Theory and Practice, pp. 147–176 (1992)
20. Waeytens, K., Lens, W., Vandenberghe, R.: Learning to Learn: Teachers Conceptions of Their Supporting Role. Learn. Instr. 12, 305–322 (2002)
21. Zimmerman, C.: The Development of Scientific Reasoning Skills. Dev. Rev. 20, 99–149 (2000)

Appendix

For OSD and questionnaires, we proposed, for each item, five intermediate possible choices. As our test is prospective, and given the small size of our samples, we calculated scores. Principal Components Analysis (PCA) was made on initial experiment results to check and if necessary uncover items that seemed to measure the same dimension. Based on these results, we grouped correlated items into scales and checked again the internal consistency of these scales by calculating the Cronbach's α.

A.1 Composition of Questionnaire Subscales Used in Pre-test and Post-test

The questionnaire used for initial experiment (N=43) questionnaire was composed of 39 items. Based on these results, we chose 19 items which constitute a robust scale (Chronbach's α = 0.823). PCA allowed to distinguish two subscales. We named the subscales according to repartition specificity of observed items, although each subscale also contains items assessing comparable themes.

Research Worker Status (RWS) Subscale	
Po	Objectivity is intrinsic to scientific activity.
Co	Subjectivity is intrinsic to scientific activity.
Po	Scientific progress consists in a gradual accumulation of knowledge.
Po	Every scientific observation is neutral.
Po	Every scientific observation is objective.
Co	Every scientific theory is likely to be questioned in the future.
Po	Research workers do not use their beliefs to do science.
Po	If an experimental result is not compatible with a scientific theory, then this theory will necessarily be questioned.
Co	Even advice from experts should often be questioned.[a]
Realism and Truth Status (RTS) Subscale	
Po	Science produces knowledge which progressively accumulates.
Co	Scientific theories are inventions.
Co	The notion of atom is an invention.[b]
Po	The notion of atom is a discovery.[b]
Po	The result of an experimentation imposes a conclusion.
Co	There are always several possible interpretations for an experimental result.
Po	There is some scientific knowledge considered as acquired and which will never be questioned.
Po	We can say about a part of scientific knowledge that it is true.
Po	Before, there were theories which were false, but now we tend more and more towards truth.
Co	Sometimes I don't believe the facts in textbooks written by authorities.[a]

[a] propositions which belong to the same epistemological belief scale published elsewhere [4]
[b] propositions inspired from an open questionnaire published elsewhere [15]

Po indicates a positivism (or realism) measuring item and *Co* a constructivist (or idealist) one. Answers are counted as follow:

Po item :	Agree	2	1	0	-1	-2	Disagree
Co item :	Agree	-2	-1	0	1	2	Disagree

For each subscale and each student, we calculate a score between -2 (constructivist/idealist extremity) and +2 (positivist/realist, *i.e. naïve* extremity), which is their average answers to corresponding subscale items. Chronbach's α of RWS and RTS subscales are 0.750 and 0.738, respectively.

A.2 Composition and Internal Consistency of Osgood's Semantic Differentiators (OSD) Subscales of Pre-test and Post-test

We classified antagonistic adjectives into three registers. We refer to *explicit register* of values, conceptions and affects because we made the *a priori* hypothesis that this adjectives mostly appeal to the corresponding dimension. However, no term has a pure connotation.

A) Composition of V1, V2, V3, V4, A, C1, C2 and C3 subscales

Items Explicit register of values		scientific knowledge[a]	science	error	teaching	knowledge[b]
negative pole (-2)	positive pole (+2)					
awful	beautiful	V1	V1		V1	V4
false	true		V2			
bad	good	V2	V2	V3	V3	V4
negative	positive			V3		V3
useless	useful			V2		V3
not interesting	interesting			V3		V4
Explicit register of affects						
painful	pleasant			A		V4
scaring	tempting			A		V4
Explicit register of conceptions						
non dogmatic pole (-2)	dogmatic pole (+2)					
approximate	exact	C1	C1			
imprecise	precise	C1	C1			
contextual	universal	C2	C2			
relative	absolute	C2		C2		C2
temporary	definitive	C2				
subjective	objective	C3	C2	C3	C3	C3
stemming from imagination	stemming from reason	C3				
created	given	C3				

[a] "*savoir scientifique*" in French. [b] "*connaissance*" in French

On an initial amount of 42 differentiators, comprising 3 types of explicit registers, we kept this 37 differentiators. From initial experiment, they were shown by PCA to be organized into two values groups and two conceptions groups, except for error affects which where apart. Then we defined, through two other PCA (one for values and one for conceptions), subgroups of differentiators for each category. We can notice that explicit registers of knowledge affects work as knowledge specific values. Apart from these last differentiators and for those we removed, our *a priori* explicit registers were consistent.

B) Internal Consistency (Chronbach's α)

subscale	V1	V2	V3	V4	A	C1	C2	C3
α	0.7494	0.7126	0.6853	0.7239	0.5837	0.7034	0.7137	0.6844

A.3 Composition and Internal Consistency of Post-test Specific OSD Subscales

A) Composition

explicit register of conceptions		proof	Refutation
non dogmatic pole (-2)	dogmatic pole (+2)		
relative	absolute	Pr	
temporary	definitive	Pr	Re
statistic	logic	Pr	Re
collective	individual	Pr	RePr
experimental	theoretical	RePr	RePr
complex	simple	RePr	RePr

B) Internal Consistency

subscale	Pr	Re	RePr
Chronbach's α	0.4179	0.6047	0.4267

Learning Locally Weighted C4.4 for Class Probability Estimation

Liangxiao Jiang[1], Harry Zhang[2], Dianhong Wang[1], and Zhihua Cai[1]

[1] Faculty of Computer Science, China University of Geosciences
Wuhan, Hubei, P.R. China, 430074
ljiang@cug.edu.cn
[2] Faculty of Computer Science, University of New Brunswick
P.O. Box 4400, Fredericton, NB, Canada E3B 5A3
hzhang@unb.ca

Abstract. In many real-world data mining applications, accurate class probability estimations are often required to make optimal decisions. For example, in direct marketing, we often need to deploy different promotion strategies to customers with different likelihood (probability) of buying some products. When our learning task is to build a model with accurate class probability estimations, C4.4 is the most popular one for achieving this task because of its efficiency and effect. In this paper, we present a locally weighted version of C4.4 to scale up its class probability estimation performance by combining locally weighted learning with C4.4. We call our improved algorithm locally weighted C4.4, simply LWC4.4. We experimentally tested LWC4.4 using the whole 36 UCI data sets selected by Weka, and compared it to other related algorithms: C4.4, NB, KNN, NBTree, and LWNB. The experimental results show that LWC4.4 significantly outperforms all the other algorithms in term of conditional log likelihood, simply CLL. Thus, our work provides an effective algorithm to produce accurate class probability estimation.

Keywords: class probability estimation, C4.4, locally weighted C4.4, locally weighted learning, conditional log likelihood.

1 Introduction

Classification has been extensively studied and various learning algorithms have be developed, such as decision tree, Bayesian network, and k-nearest-neighbor. The predictive performance of a classifier is usually measured by its classification accuracy on the testing instances. In fact, most classifiers, including decision tree, Bayesian network, and k-nearest-neighbor, can also produce probability estimations or "confidence" of the class prediction. Unfortunately, however, this information is completely ignored in classification. This is often taken for granted since the true probability is unknown for the test instances anyway.

In many real-world data mining applications, however, classifiers' classification accuracy are not enough, because they can't express the information how "far-off" (be it 0.45 or 0.01?) is the prediction of each instance from its target.

V. Corruble, M. Takeda, and E. Suzuki (Eds.): DS 2007, LNAI 4755, pp. 104–115, 2007.
© Springer-Verlag Berlin Heidelberg 2007

For example, in direct marketing, we often need to deploy different promotion strategies to customers with different likelihood (probability) of buying some products. To accomplish these tasks, we need more than a mere classification of buyers and non-buyers, namely an accurate class probability estimation of customers in terms of their likelihood of buying. It is obvious that an accurate probability estimation of class membership is much more desirable than just an accurate classification in these cases.

This fact raises the question of wether is there another better criterion than the classification accuracy to evaluate classifiers that also produce class probability estimation, if we are aiming at an accurate class probability estimation from a classifier? Recent research show that the answer is the conditional log likelihood, simply CLL [1,2]. Now, given a built classifier G and a set of test instances $D = \{e_1, e_2, \ldots, e_i, \ldots, e_N\}$, where $e_i = < a_{i1}, a_{i2}, \ldots, a_{in} >$, N is the number of test instances, n is the number of attributes, and c_i is the true (ideal) class label of the test instance e_i. Then, the conditional log likelihood $CLL(G|D)$ of the built classifier G on the set of test instances D is:

$$CLL(G|D) = \sum_{i=1}^{N} log P_G(c_i|a_{i1}, a_{i2} \ldots, a_{in}) \tag{1}$$

In this paper, we firstly conduct an extensive experiment to compare some state-of-the-art algorithms such as C4.4 [3], NB (naive Bayes) and KNN (k-nearest-neighbor) in terms of class probability estimation (measured by CLL). The experimental results show that C4.4 performs significantly better than NB and KNN. This results indicate that C4.4 is an attractive model for class probability estimation. Motivated by the success of locally weighted linear regression and locally weighted naive Bayes [4], we present a locally weighted version of C4.4 to scale up its class probability estimation performance by combining locally weighted learning (LWNB) [5] with C4.4. We call our improved algorithm locally weighted C4.4, simply LWC4.4.

The rest of the paper is organized as follows. In Section 2, we summarize some related algorithms can be used to class probability estimation. In Section 3, we present our improved algorithm called Locally Weighted C4.4. In Section 4, we describe the experimental setup and results in detail. In Section 5, we make a conclusion and outline our main directions for future research.

2   Related Work

Just as discussed in section 1, many classification models such as decision tree, Bayesian network, and k-nearest-neighbor can also be used for class probability estimation. Now, we simply look back them in this section.

2.1   Decision Tree

Decision tree is one of the most widely used classification models. It classifies an instance by sorting it down the tree from the root node to one leaf node, which provides the classification of this instance. Each node in the tree specifies a test of one attribute of the instance, and each branch descending from that node corresponds to one of the possible values for this attribute. An instance is classified by starting at the root node of the tree, testing the attribute specified by this node, then moving down the tree branch corresponding to the value of the attribute in the given instance. This process is then repeated for the subtree rooted at the new node. After the tree is built, a manipulation called tree pruning is performed to scale up the classification accuracy of the learned tree.

Unfortunately, traditional decision tree algorithms, such as C4.5 [6], have been observed to produce poor class probability estimation [7]. Aiming at this fact, Provost and Domingos [3] presented an improved algorithm simply called C4.4 to improve C4.5's performance on class probability estimation. In C4.4, two techniques are used to improve C4.5's class probability estimation:

1. Smooth class probability estimation by Laplace estimation: Assume that there are p instances of the class at a leaf, N total instances, and C total classes. The frequency-based estimation calculates the estimated probability of class membership as $\frac{p}{N}$. The Laplace estimation calculates the estimated probability of class membership as $\frac{p+1}{N+C}$.
2. Turn off pruning: Provost and Domingos show that pruning a large tree damages the probability estimation. Thus, a simple strategy to improve the probability estimation is to build a large tree without tree pruning.

2.2 Bayesian Network

A Bayesian network consists of a structural model and a set of conditional probabilities. The structural model is a directed acyclic graph in which nodes represent attributes and arcs represent attribute dependencies. Attribute dependencies are quantified by conditional probabilities for each node given its parents. Bayesian networks are often used for classification problems, in which a learner attempts to construct a classifier from a given set of training instances with class labels. Assume that $A_1, A_2, \cdots, A_n$ are n attributes (corresponding to attribute nodes in a Bayesian network). A test instance e is represented by a vector $(a_1, a_2, , \cdots, a_n)$, where a_i is the value of A_i. Let C represent the class variable (corresponding to the class node in a Bayesian network). We use c to represent the value that C takes and $c(e)$ to denote the class of e. The Bayesian network classifier represented by a Bayesian network is defined in Equation 2.

$$c(e) = \arg\max_{c \in C} P(c)P(a_1, a_2, \cdots, a_n|c). \tag{2}$$

Assume all attributes are independent given the class. Then, the resulting classifier is called naive Bayes, simply NB:

$$c(e) = \arg\max_{c \in C} P(c) \prod_{i=1}^{n} P(a_i|c). \tag{3}$$

In NB, each attribute node has the class node as its parent, but does not have any parent from attribute nodes. Although naive Bayes is easy to construct, the attribute conditional independence assumption made by the naive approach harms the classification accuracy of naive Bayes when it is violated. In order to relax this assumption effectively, an appropriate language and efficient machinery to represent and manipulate independence assertions are needed [8]. Both are provided by Bayesian networks [9]. Unfortunately, however, it has been proved that learning an optimal Bayesian network is NP-hard [10]. In order to avoid the intractable complexity for learning Bayesian networks, learning improved naive Bayes has attracted much attention from researchers. For example, Kohavi [11] presented an algorithm called naive Bayes tree, simply NBTree. It uses decision trees to scale up the classification accuracy of naive Bayes. Learning an NBTree is similar to C4.5 [6] except for its score function of evaluating split attributes. After a tree is grown, a naive Bayes is constructed for each leaf using the data associated with that leaf. NBTree classifies a test instance by sorting it to a leaf and applying the naive Bayes in that leaf to assign a class label to it.

2.3   K-Nearest-Neighbor

KNN (k-nearest-neighbor) has been widely used in classification problems. KNN is based on a distance function that measures the difference or similarity between two instances. The standard Euclidean distance $d(x, y)$ between two instance x and y is often used as the distance function, defined as follows.

$$d(x, y) = \sqrt{\sum_{i=1}^{n}(a_i(x) - a_i(y))^2} \qquad (4)$$

Given a test instance x, KNN assigns the most common class of x's k nearest neighbors to x, as shown in Equation 5. KNN is a typical example of lazy learning, which just stores training data at training time and delays its learning until classification time.

$$c(x) = \arg\max_{c \in C} \sum_{i=1}^{k} \delta(c, c(y_i)) \qquad (5)$$

where y_1, y_2, $\cdots$, y_k are the k nearest neighbors of x, k is the number of the neighbors, and $\delta(c, c(y_i)) = 1$ if $c = c(y_i)$ and $\delta(c, c(y_i)) = 0$ otherwise.

KNN uses a simple voting to produce the class probability estimation. That is say that the class labels of instances in the neighborhood are treated equally. So, an obvious improved method is to weight the vote of k nearest neighbors differently according to their distance to the test instance x. The resulting classifier is called k-nearest-neighbor with distance weighted defined as follows.

$$c(x) = \arg\max_{c \in C} \sum_{i=1}^{k} \frac{\delta(c, c(y_i))}{d(y_i, x)^2} \qquad (6)$$

Another most efficient approach is deploying a local probability-based classification model within the neighborhood of the test instance consisting of the k nearest neighbors. Talking of the local probability-based classification models, naive Bayes is absolutely necessary. The idea of combining KNN with naive Bayes is quite straightforward. Whenever a test instance is classified, a local naive Bayes is trained using the k nearest neighbors of the test instance, with which the test instance is classified. Locally weighted naive Bayes [4], simply LWNB, is a state-of-the-art example, which implements the locally weighed manipulation using locally weighted learning [5]. In LWNB, k nearest neighbors of a test instance are firstly found and each of them is weighted in terms of its distance to the test instance. Then a local naive Bayes is built from the locally weighted training instances.

3   Locally Weighted C4.4

Thinking of C4.5's bad performance of class probability estimation, Provost and Domingos [3] presented an improved algorithm simply called C4.4 to improve its class probability estimation performance. In C4.4, two techniques called Laplace correction and turning of tree pruning are used.

Locally weighted learning [5] is meta method, which has been successfully used to improve some efficient and effective algorithms. For example, locally weighted linear regression, which is a locally weighted version of linear regression. It uses a local linear regression to fit to a subset of the training instances that is in the neighborhood of the test instance. The training instances in this neighborhood are weighted according to its distance from the test instance, with less weight being assigned to instances that are further from the test instance. A regression prediction is then obtained from linear regression taking the attribute values of the test instance as input. Similar to locally weighted linear regression, Li [12] uses locally weighted learning to improve SMOreg (a support vector machine algorithm for Regression) for Regression.

For another example, Frank et al. [4] presented a hybrid algorithm called locally weighted naive Bayes, simply LWNB, by combining locally weighted learning with naive Bayes. When call upon to classify a test instance, LWNB firstly finds the k nearest neighbors of this test instance. Then, LWNB assigns different weights to different instances in the neighborhood according to its distance from the test instance. At last, a local naive Bayes is built on these locally weighted training instances, with which this test instance is classified.

For solving the regression problems, the linear regression algorithm is the most popular one. Its locally weighted version demonstrates great improvement. In the same way, naive Bayes performs well in classification [8], its improved algorithm called locally weighted naive Bayes significantly outperforms it in terms of classification accuracy. Thus, we can draw a conclusion: a remarkable character in applying locally weighted learning is that local models all need to be efficient and effective. Fortunately, C4.4 exactly meets this character. These facts raise the question of whether such a locally weighted learning can be used to improve the class probability estimation performance of C4.4.

Responding to this question, we present a locally weighted version of C4.4 by combining locally weighted learning with C4.4. We call our improved algorithm locally weighted C4.4 , simply LWC4.4. LWC4.4 use C4.4 in exactly the same way as naive Bayes is used in locally weighted naive Bayes: a local C4.4 is built on the subset of the training instances that is in the neighborhood of the test instance whose probability of class membership is to be estimated. The training instances in this neighborhood are weighted according to the inverse of its distance from the test instance, with less weight being assigned to instances that are further from the test instance. A class probability estimation is then obtained from C4.4 taking the attribute values of the test instance as input.

The subset of the training instances used to training each locally weighted C4.4 are determined by a k-nearest-neighbor algorithm. A user-specified parameter k controls how many instances are used. So, like locally weighted linear regression and locally weighted naive Bayes, our locally weighted C4.4 also is a k-related algorithm. Fortunately, we get almost same experimental results with LWNB: LWC4.4 is not particularly sensitive to the choice of value of k as long as it is not too small. This makes it a very attractive alternative to the k-related algorithms, which requires fine-tuning of k to achieve good results.

Although our experimental results show that LWC4.4 significantly outperforms the original C4.4 measured by CLL. Our improvements turn an eager learning algorithm into a lazy learning algorithm. Like all the other lazy learning algorithms, LWC4.4 simply stores training instances and defers the effort involved in learning until prediction time. When called upon to predict a test instance, LWC4.4 constructs an C4.4 using a weighted set of training instances in the neighborhood of the test instance. In a word, an obvious disadvantage with LWC4.4 is that it has relatively higher time complexity. So, enhancing LWC4.4's efficiency is one main direction for our future research.

4   Experimental Methodology and Results

We ran our experiments on 36 UCI data sets [13] selected by Weka [14], which represent a wide range of domains and data characteristics. In our experiments, we adopted the following four preprocessing steps.

1. Replacing missing attribute values: We don't handle missing attribute values. Thus, we used the unsupervised filter named *ReplaceMissingValues* in Weka to replace all missing attribute values in each data set.
2. Discretizing numeric attribute values: We don't handle numeric attribute values. Thus, we used the unsupervised filter named *Discretize* in Weka to discretize all numeric attribute values in each data set.
3. Removing useless attributes: Apparently, if the number of values of an attribute is almost equal to the number of instances in a data set, it is a useless attribute. Thus, we used the unsupervised filter named *Remove* in Weka to remove this type of attributes. In these 36 data sets, there are only three such attributes: the attribute "Hospital Number" in the data set "colic.ORIG", the attribute "instance name" in the data set "splice" and the attribute "animal" in the data set "zoo".

4. Sampling large data sets: For saving the time of running experiments (three lazy algorithms are used in our experiments), we used the unsupervised filter named *Resample* with the size of 20% in Weka to randomly sample each large data set having more than 5000 instances. In these 36 data sets, there are only three such data sets: "letter", "mushroom", and "waveform-5000".

We conducted extensive experiments to compare LWC4.4 on CLL with other related algorithms: C4.4, NB, KNN, NBTree, and LWNB. In our experiments, the CLL score of each classifier is computed using Equation 1. We use the implementation of C4.4 (J48 with Laplace smoothing but without tree pruning), NB (NaiveBayes), KNN (IBk without distance weighting), NBTree, and LWNB (LWL with NaiveBayes as the basic classifier) in Weka system. We use the LWL with C4.4 as the basic classifier for the implement of LWC4.4. Besides, we set

Table 1. The detailed experimental results on CLL and standard deviation. C4.4: C4.5 with laplace correction and without tree pruning; NB: naive Bayes; KNN: K-Nearest-Neighbor; NBTree: naive Bayes tree; LWNB: locally weighted naive Bayes; LWC4.4: locally weighted C4.4.

Dataset	C4.4	NB	KNN	NBTree	LWNB	LWC4.4
anneal	-7.72±2.17	-14.08±8.72	-8.25±3.33	-16.61±18.73	-11.95±8.13	-6.5±1.63
anneal.ORIG	-22.25±4.16	-23.85±7.26	-28.34±4.13	-29.84±12.46	-22.18±6.96	-20.25±4.3
audiology	-14.62±3.68	-65.71±21.12	-31.82±7.66	-87.29±33.27	-59.9±21.59	-14.47±3.47
autos	-12.83±2.52	-44.52±20.45	-18.89±6.38	-31.58±12.78	-39.65±19.65	-11.5±2.82
balance-scale	-52.4±3.99	-31.77±1.44	-66.96±2.28	-31.77±1.44	-31.63±1.4	-49.79±5.54
breast-cancer	-18.47±3.22	-18.12±6.01	-18.05±4.99	-18.89±4.99	-18.01±6.03	-19.49±3.63
breast-w	-11.23±4.59	-18.2±16.18	-9.18±5.19	-15.41±12.81	-18.21±16.23	-10.89±4.67
colic	-16.84±4.29	-30.29±9.6	-18.49±4.71	-35.19±15.06	-29.61±9.35	-17.65±5.2
colic.ORIG	-18.07±3.48	-20.41±5.55	-25.63±5.35	-33.22±11.57	-19.96±5.35	-18.73±3.11
credit-a	-27.72±3.16	-28.52±7.77	-30.82±7.76	-32.74±12.46	-28.44±7.83	-29.26±3.67
credit-g	-61.8±6.98	-52.42±7.29	-61.16±8.28	-58.95±17.18	-52.26±7.35	-63.4±6.9
diabetes	-43.91±4.94	-40.86±8.11	-44.54±6.41	-40.86±8.11	-40.78±8.08	-44.29±4.98
glass	-20.31±2.12	-24.16±4.21	-23.46±4.83	-33.31±10.14	-23.65±4.17	-20.21±2.03
heart-c	-15.7±4.69	-13.66±5.08	-14.52±5.9	-14.73±4.39	-13.64±5.14	-16.38±4.95
heart-h	-14.75±4.4	-13.69±5.2	-14.1±5.56	-14.87±5.61	-13.69±5.24	-15.14±5.2
heart-statlog	-13.95±3.45	-12.17±4.52	-12.04±4.46	-15.6±5.99	-12.21±4.53	-13.87±3.87
hepatitis	-5.7±2.13	-8.57±4.11	-7.43±4	-7.38±4.33	-8.55±4.08	-5.94±2.28
hypothyroid	-90.86±7.72	-97.44±19.4	-133.81±29.3	-97.81±19.8	-96.61±20.2	-91.6±9.1
ionosphere	-11.04±2.34	-35.01±13.73	-13.53±6.16	-24.27±12.88	-34.92±13.58	-10.65±2.51
iris	-3.67±1.33	-2.56±2.77	-3.1±2.63	-2.76±2.97	-2.53±2.87	-3.53±1.41
kr-vs-kp	-8.61±3.69	-93.21±8.36	-58.71±7.16	-34.67±19.94	-85.73±7.67	-7.45±4.11
labor	-2.47±1.45	-0.96±1.11	-1.67±0.99	-1.63±2.95	-1.01±1.2	-2.47±1.45
letter	-320.96±8.1	-564.72±52.8	-429.4±42.2	-618.49±64.8	-505.48±52.3	-294.91±6.7
lymph	-7.57±3.03	-6.43±3.16	-7.05±3.21	-9.67±7.7	-6.3±3.21	-7.12±2.67
mushroom	-2.53±0.87	-34.7±16.35	-0.55±0.78	-2.61±5.85	-20.64±11.49	-2.13±0.9
primary-tumor	-51.58±2.82	-65.27±10.04	-94.05±11.89	-73.04±15.37	-65.98±10.48	-50.58±2.97
segment	-49.66±6.05	-124.26±38.23	-56.84±7.02	-115.81±62.16	-109.49±35.34	-41.97±6.01
sick	-20.57±3.31	-45.74±11.62	-26.2±3.84	-41.58±13.58	-42.21±11.12	-20.49±4.61
sonar	-11.98±2.28	-20.87±12.2	-9±1.94	-34.68±23.7	-20.5±11.91	-12.21±2.1
soybean	-17.84±2.47	-26.41±9.7	-15.51±4.38	-30.65±15.5	-23.74±8.92	-16.61±2.45
splice	-66.6±8.6	-46.67±8.63	-178.56±18.2	-46.67±8.63	-45.69±8.82	-66.09±7.68
vehicle	-53.61±2.85	-169.76±27.29	-60.78±9.34	-131.69±26.73	-160.81±24.86	-52.37±3.81
vote	-7.31±4.78	-27.08±12.99	-10.03±4.48	-5.43±5.23	-24.33±11.9	-7.22±5.14
vowel	-70.91±4.74	-87.41±8.91	-61.38±5.2	-42.52±11.4	-66.63±6.95	-64.9±4.94
waveform-5000	-67.31±6.21	-74.37±17.55	-69.42±8.55	-104.36±47.13	-69.95±16.39	-66.48±6.32
zoo	-2.96±1.56	-1.21±1.12	-1.61±1.06	-0.96±0.89	-0.99±1.04	-2.67±1.56
Mean	**-34.62±3.84**	-55.14±11.62	-46.25±7.21	-53.82±15.52	-50.77±11.15	**-33.31±4.02**

Table 2. The compared results of two-tailed t-test on CLL with a 95% confidence level. An entry w/t/l in the table means that the algorithm at the corresponding row wins in w data sets, ties in t data sets, and loses in l data sets, compared to the algorithm at the corresponding column.

	C4.4	NB	KNN	NBTree	LWNB
NB	4/19/13				
KNN	2/23/11	13/17/6			
NBTree	4/21/11	4/29/3	6/21/9		
LWNB	5/19/12	19/17/0	5/19/12	5/27/4	
LWC4.4	9/26/1	13/18/5	10/25/1	11/21/4	12/19/5

the weighting kernel function to the inverse of their distance in LWNB[1] and LWC4.4, the number of neighbors to 5 in KNN, 50 in LWNB and LWC4.4. The CLL of each classifier on each data set was obtained via 10-fold cross validation. Run with the various algorithms were carried out on the same training sets and evaluated on the same test sets. In particular, the cross-validation folds are the same for all the experiments on each data set. Finally, we compare each pair of algorithms via two-tailed t-test with a 95% confidence level. According to the statistical theory, we speak of two results for a data set as being "significantly different" only if the probability of significant difference is at least 95%.

Table 1 shows the CLL and standard deviation of each algorithm on the test sets of each data set, the average CLL and standard deviation are summarized at the bottom of the table. Table 2 shows the results of two-tailed t-test with a 95% confidence level between each pair of algorithms in terms of CLL. each entry w/t/l in Table 2 means that the algorithm at the corresponding row wins in w data sets, ties in t data sets, and loses in l data sets, compared to the algorithm at the corresponding column.

The detailed results displayed in Table 1 and Table 2 show that our improved algorithm LWC4.4 significantly outperforms all the other algorithms used to compare measured by CLL. Now, let's summarize the highlights as follows:

1. C4.4 significantly outperforms NB. In the 36 data sets we test, C4.4 wins in 13 data sets, only loses in 4 data sets. C4.4's average CLL is -34.62, much higher than that of NB (-55.14). This fact proves that C4.4 is an attractive alternative for class probability estimation.
2. C4.4 significantly outperforms KNN. In the 36 data sets we test, C4.4 wins in 11 data sets, only loses in 2 data sets. C4.4's average CLL is -34.62, much higher than that of KNN (-46.25). This fact also proves that C4.4 is an attractive alternative for class probability estimation.
3. LWC4.4 significantly outperforms C4.4. In the 36 data sets we test, LWC4.4 wins in 9 data sets, surprisingly loses in 1 data sets. LWNB's average CLL is -33.31, much higher than that of C4.4 (-34.62). This fact proves that locally weighted learning is an effective method for scaling up the class probability estimation performance of C4.4.
4. LWC4.4 significantly outperforms other two algorithms: NBTree (11 wins and 4 losses) and LWNB (12 wins and 5 losses). This fact is another

---

[1] It is a little different from LWNB published in UAI 2003.

Table 3. The detailed experimental results on AUC and standard deviation.
C4.4: C4.5 with laplace correction and without tree pruning; NB: naive Bayes; KNN: K-Nearest-Neighbor; NBTree: naive Bayes tree; LWNB: locally weighted naive Bayes; LWC4.4: locally weighted C4.4.

Dataset	C4.4	NB	KNN	NBTree	LWNB	LWC4.4
anneal	93.78±2.9	95.9±1.3	93.66±5.92	96.45±0.28	96.1±1.2	96.1±1.31
anneal.ORIG	92.69±3.15	94.49±3.67	93.95±1.44	94.71±3.74	94.63±3.61	94.27±2.1
audiology	70.58±0.63	70.96±0.73	70.59±0.66	71.14±0.71	71.03±0.73	70.88±0.59
autos	90.73±4.52	89.18±4.93	89.29±3.84	93.93±2.68	90.77±5.1	94.1±3.3
balance-scale	63.06±6.18	84.46±4.1	65.84±2.89	84.46±4.1	84.01±4.4	62.24±5.37
breast-cancer	59.3±12.03	69.71±15.21	62.14±13.51	68.95±11.27	69.37±14.79	57.86±12.17
breast-w	97.85±1.86	99.19±0.87	98.71±1.38	99.21±0.73	99.21±0.86	98.29±1.6
colic	85.02±7.03	83.71±5.5	85.3±5.09	85.92±6.3	83.98±5.41	83.21±9.56
colic.ORIG	80.56±8.94	80.67±6.98	71.35±7.56	80.06±8.69	81.45±6.19	79.66±6.75
credit-a	89.42±3.1	92.09±3.43	91±3.14	91.48±3.52	92.22±3.41	88.24±2.89
credit-g	69.62±5	79.27±4.74	74.36±5	77.75±5.97	79.5±4.65	69.06±4.71
diabetes	75.5±5.76	82.31±5.17	77.57±3.98	82.31±5.17	82.44±5.19	75.46±5.87
glass	82.36±4.38	80.5±6.65	83.36±5.86	82.53±8.46	82.23±6.2	85.1±4.48
heart-c	83.1±1.19	84.1±0.54	83.85±0.84	83.96±0.51	84.1±0.56	83±1.24
heart-h	83.04±0.85	83.8±0.7	83.47±0.99	83.78±0.62	83.8±0.71	83.15±0.95
heart-statlog	81.36±9.15	91.3±4.19	89.79±4.36	89.66±3.42	91.06±4.24	82.76±9.13
hepatitis	82.03±14.04	88.99±8.99	83.14±12.51	88.03±8.29	88.99±8.99	81.7±12.83
hypothyroid	81.58±8.8	87.37±8.52	83.12±11.13	87.01±9.1	87.52±8.61	81.85±9.9
ionosphere	93.1±3.76	93.61±3.36	93.85±3.99	96.84±2.16	94.24±3.14	93.06±4.42
iris	97.33±2.63	98.58±2.67	97.75±3.22	98.08±2.67	98.58±2.67	99.25±1.39
kr-vs-kp	99.95±0.06	95.17±1.29	99.33±0.36	99.17±0.68	96.18±1.08	99.96±0.07
labor	74.17±31.04	98.33±5.27	92.5±7.03	100±0	98.33±5.27	88.33±31.48
letter	88.83±1.12	95.51±0.78	96.38±0.58	96.38±0.76	96.35±0.69	90.83±0.89
lymph	87.26±3.75	89.69±1.49	88.41±3.09	89.08±2.08	89.77±1.34	88.63±3.05
mushroom	99.98±0.02	99.59±0.18	99.97±0.02	99.97±0.1	99.86±0.09	100±0
primary-tumor	75.48±2.33	78.85±1.35	77.1±2.08	78.26±1.75	79.08±1.45	76.62±2.3
segment	98.85±0.32	98.51±0.46	99.01±0.16	99.09±0.43	98.73±0.39	99.36±0.2
sick	99.07±0.35	95.91±2.35	98.55±0.54	94.46±2.65	96.46±2.07	99.11±0.5
sonar	77.01±8.59	85.48±10.82	88.32±7.39	79.72±12.51	85.48±10.82	77.64±7.29
soybean	91.43±2.6	99.53±0.6	96.16±1.8	99.33±0.64	99.54±0.61	99.2±0.87
splice	98.14±0.72	99.41±0.22	96.99±0.97	99.41±0.22	99.43±0.22	98.16±0.63
vehicle	86.5±2.28	80.81±3.51	88.48±2.05	85.83±2.9	81.94±3.41	87.24±2.78
vote	96.77±2.96	96.56±2.09	97.39±1.49	98.82±1.61	96.77±1.92	98.21±1.8
vowel	91.28±2.46	95.81±0.84	97.58±0.64	98.66±0.68	97.46±0.55	93.28±2.02
waveform-5000	79.22±3.91	95.26±1.4	85.49±3.18	91.3±4.35	95.83±1.17	80.27±3.06
zoo	88.88±4.5	89.88±4.05	89.7±4.17	89.88±4.05	89.88±4.05	89.88±4.05
Mean	**85.69±4.80**	89.57±3.58	87.87±3.69	89.88±3.44	89.90±3.49	**86.83±4.49**

Table 4. The compared results of two-tailed t-test on AUC with a 95% confidence level. An entry w/t/l in the table means that the algorithm at the corresponding row wins in w data sets, ties in t data sets, and loses in l data sets, compared to the algorithm at the corresponding column.

	C4.4	NB	KNN	NBTree	LWNB
NB	12/20/4				
KNN	8/25/3	6/24/6			
NBTree	13/21/2	6/30/0	11/24/1		
LWNB	14/18/4	13/23/0	8/25/3	3/30/3	
LWC4.4	**8/28/0**	5/19/12	7/23/6	2/24/10	5/19/12

evidence to prove that LWC4.4 is an effective algorithm for addressing the class probability estimation problems.

In our experiments, we also observe the ranking performance of LWC4.4 in term of AUC (the area under the Receiver Operating Characteristics curve)

Table 5. The detailed experimental results on classification accuracy and standard deviation. C4.4: C4.5 with laplace correction and without tree pruning; NB: naive Bayes; KNN: K-Nearest-Neighbor; NBTree: naive Bayes tree; LWNB: locally weighted naive Bayes; LWC4.4: locally weighted C4.4.

Dataset	C4.4	NB	KNN	NBTree	LWNB	LWC4.4
anneal	99±0.98	94.32±2.38	96.88±2.15	98.33±1.6	95.65±2.13	99.11±0.88
anneal.ORIG	91.76±3.07	87.53±4.69	87.31±3.35	90.98±4.46	87.64±3.57	92.31±3.22
audiology	78.3±8	71.23±7.03	60.57±7.87	79.66±6.6	74.74±6.4	78.28±8.51
autos	81.45±7.48	64.83±11.18	66.29±8.28	78.12±7.02	69.17±8.96	82.9±9.22
balance-scale	69.3±4.25	91.36±1.38	83.84±4.71	91.36±1.38	91.36±1.38	72.02±4.69
breast-cancer	68.57±7.49	72.06±7.97	73.78±4.38	74.53±8.37	71.71±8.35	65.05±7.6
breast-w	92.99±3.66	97.28±1.84	94.99±2.81	96.99±1.85	97.28±1.84	92.99±3.72
colic	80.17±5.95	78.81±5.05	80.68±6.65	83.42±4.49	79.62±4.95	79.08±8.36
colic.ORIG	76.08±8.74	75.26±5.26	70.63±5.06	76.07±5.03	75.53±5.04	75.55±6.1
credit-a	83.19±3.5	84.78±4.28	85.07±3.62	85.07±3.81	85.22±4.36	80.72±4.21
credit-g	68.6±4.3	76.3±4.76	71.5±2.42	75.9±4.48	76.2±4.59	67.2±4.16
diabetes	69.54±5.12	75.4±5.85	69.14±1.84	75.4±5.85	75.4±5.38	69.8±3.85
glass	58.83±7.73	60.32±9.69	58.92±7.8	56.99±10.66	60.35±8.98	56.95±8.11
heart-c	74.26±11.46	84.14±4.16	81.41±12.65	82.16±3.66	84.14±4.16	72.59±11.54
heart-h	72.78±11	84.05±6.69	81.36±6.65	82.36±7.71	84.05±6.07	73.11±9.83
heart-statlog	75.93±8.95	83.7±5	80.74±6	82.59±6.06	83.7±5	74.44±7.29
hepatitis	81.25±11.52	83.79±8.79	84.46±6.25	83.79±9.91	83.13±8.22	79.33±11.15
hypothyroid	92.5±0.58	92.79±1.02	93.03±0.89	93.08±1	92.79±0.99	92.07±0.92
ionosphere	84.63±4.45	90.89±3.49	89.44±3.34	91.45±3.3	90.89±3.49	85.2±5.3
iris	92.67±5.84	94.67±8.2	93.33±6.29	94±7.98	95.33±8.34	92.67±5.84
kr-vs-kp	99.41±0.45	87.89±1.81	96.03±1.19	97.09±2.38	88.86±1.35	99.41±0.43
labor	77.67±15.64	93.33±11.65	91.67±11.79	91.67±11.79	93.33±11.65	79.33±15.22
letter	70.3±1.67	66.15±2.15	73.3±2.24	73.9±1.69	69.7±2.34	72.58±2.28
lymph	74.29±12.56	85.67±9.55	82.33±9.81	83.05±8.01	86.33±8.8	75.67±9.55
mushroom	99.75±0.32	93.84±2.02	99.82±0.3	99.88±0.26	95.57±2.16	99.75±0.32
primary-tumor	38.91±4.97	46.89±4.32	41.26±8.05	46.9±6.22	48.37±4.08	38.03±3.83
segment	92.86±1.39	88.92±1.95	90.74±1.61	92.51±1.77	90±2.14	93.51±1.71
sick	97.83±0.61	96.74±0.53	97.51±0.59	97.96±0.73	96.85±0.48	97.69±0.66
sonar	67.69±10.94	77.5±11.99	80.79±10.06	73.62±13.8	77.98±12.03	66.74±9.25
soybean	92.68±1.56	92.08±2.34	90.76±3.76	92.24±2.08	92.96±2.5	92.67±1.72
splice	91.57±1.37	95.36±1	79.81±2.81	95.36±1	95.42±1	90.78±1.5
vehicle	69.03±2.63	61.82±3.54	70.57±3.02	68.1±5	62.41±4.04	69.5±3.71
vote	94.96±3.83	90.14±4.17	94.03±2.69	95.41±4.03	90.6±3.93	94.95±4.41
vowel	75.66±5.18	67.07±4.21	81.31±1.73	88.59±2.74	75.56±5.08	77.47±4.37
waveform-5000	65.8±3.77	79.7±4	70.4±4.09	79.4±3.31	80.7±3.68	65.8±3.08
zoo	92.18±8.94	94.18±6.6	92.09±6.3	95.09±5.18	96.18±6.54	92.18±8.94
Mean	**80.34±5.55**	82.24±5.02	81.55±4.81	84.53±4.87	83.19±4.83	**80.21±5.43**

Table 6. The compared results of two-tailed t-test on classification accuracy with a 95% confidence level. An entry w/t/l in the table means that the algorithm at the corresponding row wins in w data sets, ties in t data sets, and loses in l data sets, compared to the algorithm at the corresponding column.

	C4.4	NB	KNN	NBTree	LWNB
NB	11/15/10				
KNN	4/26/6	8/23/5			
NBTree	10/26/0	10/26/0	9/27/0		
LWNB	11/17/8	7/29/0	7/23/6	0/28/8	
LWC4.4	**2/33/1**	10/16/10	6/24/6	0/24/12	9/16/11

[15,16,17] shown in Table 3 and Table 4. Fortunately, LWC4.4 also significantly outperforms C4.4. In the 36 data sets we test, LWC4.4 wins in 8 data sets, surprisingly loses in 0 data sets, and ties all the other data sets. Besides, an interested observation is that LWC4.4 almost ties C4.4 in term of classification accuracy shown in Table 5 and Table 6. So, we can draw a conclusion that locally weighted learning can be used to improve C4.4 for class probability estimation and ranking but not for classification.

5   Conclusions and Future Work

C4.4 is one of the most popular algorithms for addressing the class probability estimation problems. C4.4 is an improved version of C4.5, in which two techniques respectively called Laplace correction and turning of tree pruning are used. Motivated by the success of using locally weighted learning to improve linear regression for regression and using locally weighted learning to improve naive Bayes for classification, we present to apply locally weighted learning to C4.4 to scale up its class probability estimation performance. We call our improved algorithm locally weighted C4.4, simply LWC4.4. Our experimental results show that LWC4.4 is surprisingly effective in class probability estimation and significantly outperforms all the other algorithms used to compare.

Aiming at accurate classification, Friedman et al. [18] presented another lazy decision tree learning algorithm, simply called LazyDT. LazyDT creates a path in a tree for a test instance instead of a neighborhood. According to the experimental results in [18], LazyDT is certainly effective in classification. However, it is not clear whether LazyDT also is effective in class probability estimation. In our future work, we will compare LWC4.4 with LazyDT.

References

1. Grossman, D., Domingos, P.: Learning Bayesian Network Classifiers by Maximizing Conditional Likelihood. In: Proceedings of the Twenty-First International Conference on Machine Learning, Banff, Canada, pp. 361–368. ACM Press, New York (2004)
2. Guo, Y., Greiner, R.: Discriminative Model Selection for Belief Net Structures. In: Proceedings of the Twentieth National Conference on Artificial Intelligence, pp. 770–776. AAAI Press (2005)
3. Provost, F.J., Domingos, P.: Tree Induction for Probability-Based Ranking. Machine Learning 52(3), 199–215 (2003)
4. Frank, E., Hall, M., Pfahringer, B.: Locally Weighted Naive Bayes. In: Proceedings of the Conference on Uncertainty in Artificial Intelligence, pp. 249–256. Morgan Kaufmann, San Francisco (2003)
5. Atkeson, C.G., Moore, A.W., Schaal, S.: Locally Weighted Learning. Artificial Intelligence Review 11(1-5), 11–73 (1997)
6. Quinlan, J.R.: C4.5: Programs for Machine Learning. Morgan Kaufmann, San Mateo, CA (1993)
7. Provost, F., Fawcett, T., Kohavi, R.: The case against accuracy estimation for comparing induction algorithms. In: Proceedings of the Fifteenth International Conference on Machine Learning, pp. 445–453. Morgan Kaufmann, San Francisco (1998)
8. Friedman, Geiger, Goldszmidt: Bayesian Network Classifiers. Machine Learning 29, 131–163 (1997)

9. Pearl, J.: Probabilistic Reasoning in Intelligent Systems. Morgan Kaufmann, San Francisco, CA (1988)
10. Chickering, D.M.: Learning Bayesian networks is NP-Complete. In: Fisher, D., Lenz, H. (eds.) Learning from Data: Artificial Intelligence and Statistics V, pp. 121–130. Springer, Heidelberg (1996)
11. Kohavi, R.: Scaling Up the Accuracy of Naive-Bayes Classifiers: A Decision-Tree Hybrid. In: Proceedings of the Second International Conference on Knowledge Discovery and Data Mining (KDD-96), pp. 202–207. AAAI Press (1996)
12. Li, C., Jiang, L.: Using Locally Weighted Learning to Improve SMOreg for Regression. In: Yang, Q., Webb, G. (eds.) PRICAI 2006. LNCS (LNAI), vol. 4099, pp. 375–384. Springer, Heidelberg (2006)
13. Merz, C., Murphy, P., Aha, D.: UCI repository of machine learning databases. In Dept of ICS, University of California, Irvine (1997),
    http://www.ics.uci.edu/mlearn/MLRepository.html
14. Witten, I.H., Frank, E.: Data Mining: Practical machine learning tools and techniques, 2nd edn. Morgan Kaufmann, San Francisco (2005),
    http://prdownloads.sourceforge.net/weka/datasets-UCI.jar
15. Bradley, A.P.: The use of the area under the ROC curve in the evaluation of machine learning algorithms. Pattern Recognition 30, 1145–1159 (1997)
16. Hand, D.J., Till, R.J.: A simple generalisation of the area under the ROC curve for multiple class classification problems. Machine Learning 45, 171–186 (2001)
17. Ling, C.X., Huang, J., Zhang, H.: AUC: a statistically consistent and more discriminating measure than accuracy. In: Proceedings of the International Joint Conference on Artificial Intelligence IJCAI03, Morgan Kaufmann, San Francisco (2003)
18. Friedman, J., Kohavi, R., Yun, Y.: Lazy decision trees. In: Proceedings of the Thirteenth National Conference on Artificial Intelligence, pp. 717–724. The AAAI Press, Menlo Park, CA (1996)

User Preference Modeling from Positive Contents
for Personalized Recommendation

Heung-Nam Kim[1], Inay Ha[1], Jin-Guk Jung[1], and Geun-Sik Jo[2]

[1] Intelligent E-Commerce Systems Laboratory,
Department of Computer Science & Information Engineering, Inha University
{nami, inay, gj4024}@eslab.inha.ac.kr
[2] School of Computer Science & Engineering, Inha University,
253 Yonghyun-dong, Incheon, Korea 402-751
gsjo@inha.ac.kr

Abstract. With the spread of the Web, users can obtain a wide variety of information, and also can access novel content in real time. In this environment, finding useful information from a huge amount of available content becomes a time consuming process. In this paper, we focus on user modeling for personalization to recommend content relevant to user interests. Techniques used for association rules in deriving user profiles are exploited for discovering useful and meaningful patterns of users. Each user preference is presented the frequent term patterns, collectively called PTP (Personalized Term Pattern) and the preference terms, called PT (Personalized Term). In addition, a content-based filtering approach is employed to recommend content corresponding with user preferences. In order to evaluate the performance of the proposed method, we compare experimental results with those of a probabilistic learning model and vector space model. The experimental evaluation on *NSF research award* datasets demonstrates that the proposed method brings significant advantages in terms of improving the recommendation quality in comparison with the other methods.

1 Introduction

Thanks to technological developments related to the Internet and the World Wide Web, anyone living in today's information society can access a wealth of content and information on the web. However, in accordance with the massive growth of the Internet, users have to contend with an immense and huge amount of content, and often waste time trying to find content relevant to their interests. In addition, with the advent of blogs and RSS (Really Simple Syndication), a tremendous amount of content is generated overnight. Even if a user subscribes to content of interest, failing to read subscribed content for even a single day makes users feel overwhelmed the following day. Recommender systems have been issued as a solution to the problem of information overload [10]. In addition, user modeling for efficient personalization has become a key technique in recent information filtering systems [7, 9].

In this research, we focus on user modeling for personalization to recommend contents relevant to user interests. We exploit the techniques of data mining in deriving user preferences for discovering useful and meaningful patterns of users, collectively

V. Corruble, M. Takeda, and E. Suzuki (Eds.): DS 2007, LNAI 4755, pp. 116–126, 2007.
© Springer-Verlag Berlin Heidelberg 2007

called PTP (**P**ersonalized **T**erm **P**attern). By capturing users' contents of interest, we mine the frequent term patterns and the preference terms existing in the user's contents of interest. The main objective of this research is to develop an effective method that provides high-quality recommendations of content relevant to user interests. In addition, we employ a content-based filtering approach to recommend content that is similar to personalized term patterns.

The subsequent sections of this paper are organized as follows: The next section contains a brief overview of some related work. In section 3, we describe our approach for modeling user preference and filtering contents. A performance evaluation is presented in section 4. Finally, conclusions are presented and future work is discussed in section 5.

2 Related Work

This section briefly explains previous studies related to user modeling and personalized recommendation. Two approaches for recommender systems have been discussed in the literature, *i.e.*, a content-based filtering approach and a collaborative filtering approach. The traditional task in the collaborative filtering is to predict the utility of a certain item for the target user from the opinions of other similar users, and thereby make appropriate recommendations [10]. Instead of computing the similarities between the users, the content-based filtering systems recommend only the items that are highly relevant to the single user profile by computing the similarities between the items and the user preference [9]. This research focuses only on the content-based filtering for personalized recommendations. Personalized recommender systems based on a single user have been developed learning procedures and need to use training data to identify personal preference from information object and their contents. *Webmate* tracks user interests from his positive information only (i.e., documents that the user is interested in) and exploits the vector space model using TF-IDF method [3]. A classification approach has been explored to recommend articles relevant user profile, such as *NewsDude* and *ELFI* [4, 5]. In *NewsDude*, two types of the user interests are used: short-term interests and long-term interests. To avoid recommendations of very similar documents, short-term profile is used. For the long-term interests of a user, the probabilities of a document are calculated using Naïve Bayes to classify a document as interesting or not interesting. Instead of learning from users' explicit information, *PVA* learns a user profile implicitly without user intervention, such as relevance feedback, and represents it as keyword vector in the form of a hierarchical category structure [8], similar to *Alipes* [6]. In *Newsjunkie*, novelty-analysis algorithm is employed to present novel information for users by identifying novelty of articles in the contexts of articles they have already reviewed [12]. Although these systems have their own method to building a user model, they do not deliberate on concurrence of terms and offer the ability to identify meaningful or useful patterns, which are important features for representing articles or contents [13]. For example, when content contains 'apple Macintosh computer', the semantic of 'apple' are discriminated from those of apple in 'apple pie'. Likewise, mouse in 'optical mouse' implies not an animal but an input device of computers. Therefore, our motivation is to develop a learning algorithm which supports the identification of useful patterns of a user.

3 User Preference Modeling for Content Recommendation

The proposed method is divided into three phases: an observation phase, a user modeling phase, and a content filtering phase. Fig. 1 provides a brief overview of the proposed approach.

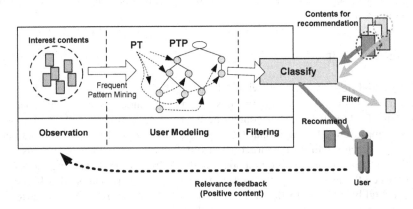

Fig. 1. Overview of the proposed method for personalized content recommendations

3.1 Modeling User Preference

The capability to learn users' preferences is at the heart of a personalized recommender system. Additionally, since every user has different interests, feature selection for representing users' interests should be personalized and be performed individually for each user [9]. In this section, we describe our approach to modeling user preference, which is mined from the user's preferred contents (positive contents).

The first step in user modeling is the extraction of the terms from positive contents that have been preprocessed by: removing stop words and stemming words [15]. After extracting terms, each positive content C_j is represented as a vector of attribute-value pairs as follows:

$$C_j = \{(T_1, w_{1,j}), (T_2, w_{2,j}), ..., (T_m, w_{m,j})\}$$

where T_i is the extracted term in C_j and $w_{i,j}$ is the weight of T_i in C_j, which is computed by static TF-IDF term-weighting scheme [1] and defined as follows:

$$w_{i,j} = \frac{f_{i,j}}{\max_l f_{l,j}} \times \log \frac{n}{n_i}$$

where $f_{i,j}$ is the frequency of occurrence of term T_i in content C_j, n is the total number of contents in the collections, and n_i is the number of contents in which term T_i occurs. The weight indicates the importance of a term in representing the content. All weight values of terms, $w_{i,j}$, in a positive content C_j are sorted in descending order. The first K terms (Top-K terms) are selected as content C_j features and used to mine frequent term patterns that occur at least as frequently as a predetermined minimum support, i.e., $PS > min\_sup$ [16].

Definition 1 (Pattern Support, PS). Let $T = \{T_1, T_2, \ldots, T_m\}$ be a set of terms, I_u be a set of contents of interest of user u where each content C is a set of terms such that $C \subseteq T$. Let pattern P_k be a set of terms. A content C is said to contain pattern P_k if and only if $P_k \subseteq C$. *Pattern support* for pattern P_k, $PS(P_k)$, in I_u is the ratio of contents in I_u that contain pattern P_k.

In this paper, each transaction corresponds to a positive content of a user and items in transaction are terms extracted from the content. For effective mining of the term patterns, we should choose a minimum support threshold. A high *min_sup* discards more patterns, and thus remaining term patterns may not be sufficient to represent user preference. In contrast, a low *min_sup* includes many noise patterns. Therefore, the threshold is chosen heuristically through experiments.

Once the patterns are mined, a model for user u is defined as a tuple $M_u = (PTP_u, PT_u)$ where PTP_u models the interest patterns (Definition 2) and PT_u models the interest terms (Definition 3). And the model is stored in a prefix tree structure to save memory space and explore relationships of terms.

Definition 2 (Personalized Term Patterns, PTP). If the pattern support of pattern P_k, that is composed of at least l different terms ($l \geq 2$), satisfies a pre-specified minimum support threshold (*min_sup*), then pattern P_k is a frequent term pattern. *Personalized term patterns* for user u, PTP_u, is defined as a set of frequent term patterns.

Definition 3 (Personalized Term, PT). *Personalized term* is a term that occurs within *personalized term patterns*. The set of *personalized terms* for user u is denoted as PT_u, $PT_u \subseteq T$. In addition, The vector for PT_u is represented by $\overrightarrow{PT_u} = (\mu_{1,u}, \mu_{2,u}, \ldots, \mu_{t,u})$, where t is the total number of personalized terms and $\mu_{i,u}$ is the mean of term weight for term T_i and is computed as follows:

$$\mu_{i,u} = \frac{1}{|I_u(i)|} \times \sum_{j \in I_u(i)} w_{i,j}$$

where $I_u(i)$ is a set of contents of interest for user u containing term T_i and $w_{i,j}$ is the term weight of term T_i in content C_j.

For example, if five personalized term patterns are found, as shown in Table 1, after mining content of interest for user u, a tree structure of a model for user u is then constructed as follows.

All PT_u are stored in header table and sorted according to descending their frequency. First, create the root of the tree, labeled with "null". For the first term pattern, $\{T_1, T_2, T_3\}$ is insert into the tree as a path from root node where T_2 is linked as child of the root, T_1 is linked to T_2, and T_3 is linked to T_1. And PS and *length* of the pattern $PS(P_1)=0.56$, *length*=3) are then attached to the last node T_3. For the second pattern, since its term pattern, $\{T_1, T_2, T_3, T_4\}$, shares common prefix $\{T_2, T_1, T_3\}$ with the existing path for the first term pattern, a new node T_4 is created and linked as a

Table 1. After mining content of interest of user u, five personalized term patterns are found

Pattern-id	PTP	PS	Length
P_1	$\{T_1, T_2, T_3\}$	0.56	3
P_2	$\{T_1, T_2, T_3, T_4\}$	0.51	4
P_3	$\{T_1, T_2, T_5\}$	0.47	3
P_4	$\{T_4, T_5\}$	0.41	2
P_5	$\{T_2, T_3, T_4\}$	0.32	3

child of node T_3. Thereafter, $PS(P_2)$ and $length(P_2)$ are attached to the last node T_4. (The third, fourth, and fifth patterns are inserted in a manner similar to the first and second patterns. To facilitate tree traversal, header table is built in which each term points to its occurrence in the tree via a *Node-link*. Nodes with the same *term-name* are linked in sequence via such *node-links*. Finally, a model for user *u* is constructed as shown in Fig. 2.

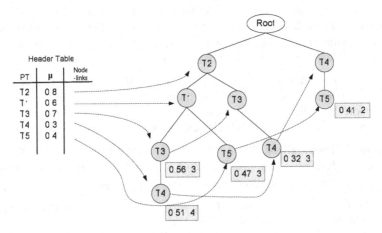

Fig. 2. A tree structure of M_u for personalized term patterns in Table 1

3.2 Personalized Content Filtering

In this paper, we consider two aspects for judging whether content is relevant or irrelevant to the user based on user preference. First, cosine similarity [13, 15], which quantifies the similarity of two vectors according to their angle, is employed to measure the similarity values between new content and **PT** for a user. As noted in Definitions 4, the personalized terms of user *u*, PT_u, are represented as the vector of attribute-value pairs. Further, the term vector for the new content C_n is represented by $\vec{c_n} = (w_{1,n}, w_{2,n}, \ldots, w_{t,n})$, where the weight $w_{i,n}$ is the TF-IDF value of term T_i in content C_n. Therefore, content C_n and **PT** of user *u*, PT_u are represented as t-dimensional vectors, and the cosine similarity for theses two vectors, $\vec{PT_u}$ and $\vec{c_n}$ is measured by equation (1) [15].

$$Sim(u,C_n) = \frac{\vec{PT_u} \cdot \vec{c_n}}{|\vec{PT_u}| \times |\vec{c_n}|} = \frac{\sum_{k=1}^{t} \mu_{k,u} \times w_{k,n}}{\sqrt{\sum_{k=1}^{t} \mu_{k,u}^2} \times \sqrt{\sum_{k=1}^{t} w_{k,n}^2}} \qquad (1)$$

The second approach considers matched patterns between the new contents and **PTP** for a user. Formally, the similarity between content C_n and user u is defined in equation (2).

$$Sim(u,C_n) = \frac{|MP|}{|PTP_u|} \times \sum_{P_k \in MP} length(P_k) \cdot PS(P_k) \qquad (2)$$

where MP is a set of matched patterns between PTP_u and content C_n. $PS(P_k)$ and $length(P_k)$ refer to the pattern support value and the length of matched pattern P_k, respectively. The main concept of the second scheme dictates that patterns with numerous occurrences in user preference present a greater contribution with regard to similarity than patterns with a smaller number of occurrences.

Definition 4 (Matched Pattern). Let $TP_k = \{T_1, T_2, \dots , T_n\}$ be a set of terms contained in pattern P_k such that TP_k is a subset of personalized terms for user u, $TP_k \subseteq PT_u$. If all terms in contained P_k appear content C_n, $TP_k \subseteq C_n$, then pattern P_k is deemed a *matched pattern* between PTP_u and content C_n.

Each similarity value, which is obtained by using the equation (1) and (2), is normalized to [0, 1] and divided by the maximum similarity value, i.e., $sim(u, C_n)/max_l$ $sim(u, C_l)$. Once the similarities between user u and the new contents, which the user u has not yet read, are computed, the contents are sorted in order of descending similarity value. Two strategies can then be used to select the relevant contents to user u. First, if the similarity values are greater than a reasonable threshold value (i.e., $sim(u, C_n)/max_l \; sim(u, C_l) > \theta$), the contents are recommended to user u [3, 5]. Second, a set of N rank contents that have obtained higher similarity values are identified for user u, and then those contents are recommended to user u (Top-N recommendation) [10]. We choose the second approach for filtering the personalized contents.

Definition 5 (Top-N recommendation). Let C be a set of all contents, I_u be a content list that user u has already collected or added to his preference list (positive contents), and NI_u be a content list that user u has not yet read, $NI_u = C - I_u$ and $I_u \cap NI_u = \varnothing$. Given two contents C_i and C_j, $C_i \in NI_u$ and $C_j \in NI_u$, content C_i will be of more interest to user u than content C_j if and only if a similarity value $sim(u, C_i)$ between user u and content C_i is higher than that of content C_j, $sim(u, C_i) > sim(u, C_j)$. Top-N recommendations for user u identifies an ordered set of N contents, $TopN_u$, that will be of interest to user u such that $|TopN_u| \leq N$, $TopN_u \cap I_u = \varnothing$, and $TopN_u \subseteq NI_u$.

4 Experimental Evaluation

In this section, experimental results of the proposed approaches are presented. All experiments were carried out on a Pentium IV 3.0GHz with 2GB RAM, running a

MS-Window 2003 server. In order to mine personalized term patterns, FP-growth software implemented by Frans Coenen[1] was used.

The experimental data is taken from NSF (National Science Foundation) research award abstracts [14]. The original data set contains 129,000 abstracts describing NSF awards for basic research from 1900 to 2003. However, the set is too large to be used for experiments, and thus we selected award abstracts from 2000 to 2003, i.e. the selected data set contained 30,384 abstracts and 3,086,090 terms as obtained from the abstracts (cf. 22,236 distinct terms). 10 users participated in the experiments by scrapping only contents relevant to their interests from the total contents (30,384 contents). Whenever they found the content related to their own preferences, they added that content to their preference list. Each user added at least 700 content items. To evaluate the performance of the proposed approaches, we divided the preference contents of the users into *a test set* with exactly 100 contents per user in the test set and *a training set* with the remaining contents. A model $\mathbf{M}_u$ of each user was then constructed using only the *training set*. We assume that each user does not change his/her interests during the experiments if a user preference is learned (static user profile) [9].

The performance was measured by looking at the number of *hits*, and their *ranking* within the *top*-N contents and the overall contents that were recommended by a particular scheme. We computed three quality measures that are defined as follows.

Hit Rate (HR). In the context of *top*-N recommendations, *hit-rate*, a measure of how often a list of recommendations contains contents that the user is actually interested in, was used for the evaluation metric [6, 10]. The *hit-rate* for user u is defined as:

$$HR(u) = \frac{\left| Test_u \cap TopN_u \right|}{\left| Test_u \right|}$$

where $Test_u$ is the content list of user u in the test data and $TopN_u$ is a *top*-N recommended content list for user u. Finally, the overall *HR* of the *top*-N recommendation for all users is computed by averaging these personal *HR* in test data.

Reciprocal Hit Rank (RHR). One limitation of the *hit-rate* measure is that it treats all hits equally regardless of the ranking of recommended contents. In other words, content that is recommended with a top ranking is treated equally with content that is recommended with an Nth ranking [10]. To address this limitation, therefore, we adopted *the reciprocal hit-rank* metric described in [10]. The *reciprocal hit-rank* for user u is defined as:

$$RHR(u) = \sum_{C_n \in (Test_u \cap TopN_u)} \frac{1}{rank(C_n)}$$

where $rank(C_n)$ refers to a recommended ranking of content C_n within the *hit set* of user u. That is, hit contents that appear earlier in the *top*-N list are given more weight than hit contents that occur later in the list. Finally, the overall *RHR* for all users is computed by averaging the personal *RHR(u)* in test data. The higher the *RHR*, the more accurately the algorithm recommends contents.

---

[1] The software is available at http://www.csc.liv.ac.uk/~frans/KDD/Software/

Reciprocal Total Rank (RTR). This metric is similar to *the reciprocal hit-rank* but instead of only using the ranking of the *hit set* it uses the ranking of all test data for user *u*. We refer to this as *the reciprocal total rank* for user *u* and is defined as follows:

$$RTR(u) = \sum_{C_n \in Test_u} \frac{1}{rank(C_n)}$$

where *rank(C_n)* refers to a recommended ranking of content C_n for user *u* in the test data. Likewise, the overall *RTR* for all users is also computed by averaging the personal *RTR(u)* in test data.

Benchmark Algorithms. In order to compare the performance of the proposed scheme, a probabilistic learning algorithm, which applies a *naïve Bayesian classifier* (denoted as *NB*) [4, 5], and a TF-IDF vector-based algorithm, which is employed in the *Webmate* system (denoted as *Webmate*) [3], were implemented. To make the comparison fair, both of the algorithms were designed to learn users' preferences from positive examples only. For the content filtering process, in the case of *NB*, contents are ranked using the calculated probability value whereas they are ranked using the calculated cosine similarity for *Webmate*. The *top*-N recommendation of our strategy was then evaluated in comparison with the benchmark algorithms.

4.1 Experimental Results

In this section, we present the experimental results of the proposed algorithms. In our algorithms, *SimPT* denotes when equation (1) is used for the similarity method, whereas *SimPTP* denotes the case of equation (2). The performance evaluation is divided into two dimensions. The sensitivity of the two parameters *minimum support* and *Top-K terms* were first determined, and then the quality of the *top*-N recommendations is evaluated.

4.1.1 Experiments with Minimum Support

As noted previously, minimum support controls the size of $\mathbf{M_u}$. In general, if the size of $\mathbf{M_u}$ is too small, some information may be lost. On the other hand, if it is too large, some noise patterns may be included. Therefore, different *min_sup* values were used for mining personalized term patterns: 5%, 8%, 10%, and 20%. In addition, we selected all terms as the content feature during the mining process (*K*=all). Examining the average number of patterns in the users' $\mathbf{M_u}$, in the case of *min_sup*=5%, we found that 2667 patterns had been mined, whereas the average number was 1049, 490, and 58 in the case of *min_sup*=8%, *min_sup*=10%, and *min_sup*=20%, respectively. The recommendation performance obtained by changing *min_sup* in terms of RTR is shown in Fig. 3 (a). The results demonstrate that, at all *min_sup* levels, *SimPTP* provides more accurate recommendations than *SimPT*. For example, when *min_sup* is set to 10%, *SimPTP* yields a RTR of 1.75, which is the best value, whereas *SimPT* gives a RTR of 1.05. It is observed from the graph that the performance of *SimPTP* is slightly affected by *min_sup* relative to that of *SimPT*. These results indicate that even for a small size of $\mathbf{M_u}$, *SimPTP* provides reasonably accurate recommendations. Note that a suitable size should be selected for vector-based similarity approaches such as *SimPT*.

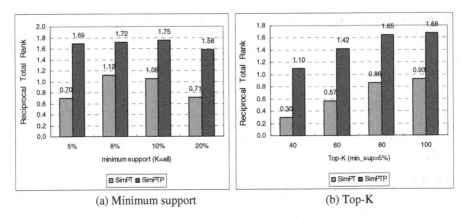

(a) Minimum support   (b) Top-K

Fig. 3. Reciprocal total rank (RTR) according to variation of *min_sup* (a) and *K* (b)

4.1.2 Experiments with Top-K Terms

Theoretically, all terms extracted from the contents can be applied immediately to mine the personalized term patterns. However, the complexity of the learning process is increased by the content feature size. In order to reduce the feature size and refine noise terms, the *K* highest weight terms are selected instead of selecting all terms. In these experiments, *min_sup*=5% was chosen because the sufficient patterns for representing user preference were not discovered at high thresholds of *min_sup* (i.e., *min_sup*=10%, 20%). To evaluate the sensitivity to the value of *K* we performed an experiment with *K* values of 40, 60, 80, and 100. As in the previous experiments, we analyze the average number of patterns mined for users. As a result, patterns of 21, 84, 344, and 1064 were discovered on average when *K* was set to 40, 60, 80, and 100, respectively. That is, the mined patterns were clearly reduced as compared with the number of patterns discovered in the previous experiment (*min_sup*=5%). Fig. 3 (b) depicts the variation of RTR according to the value of *K*. It can be observed from the graph that *SimPTP* yields better RTR than *SimPT*. When we compare the results of RTR achieved by *SimPTP* using *K*=all and *K*=100, *SimPTP* in the case of *K*=100 (RTR of 1.68) offers reasonable performance comparable to that of *K*=all (RTR of 1.69). On the contrary, RTR of the *SimPT* using *K*=100 (RTR of 0.93) is superior to that of *SimPT* using *K*=all (RTR of 0.70). This is particularly important since a small amount of content features leads to low computational requirements.

4.1.3 Comparisons of Performance

For evaluating the *top*-N recommendation, the number of recommended contents (the value of *N*) was increased, and we calculated the hit rate (HR) and the reciprocal hit rank (RHR) achieved by *SimPT*, *SimPTP*, *Webmate*, and *NB*. Table 2 summarizes the results of RHR while Table 3 summarizes the HR of the algorithms as the value of *N* increased from 100 to 500. In general, with the growth of recommended items *N*, HR, and RHR tend increase. Although HR for all algorithms is unsatisfactorily low at a small number of *N*, *SimPTP* provides considerably improved HR on all occasions compared to the benchmark algorithms. Similar conclusions can be made by looking

Table 2. Comparison of the reciprocal hit rank (RHR) as the value of N increases

Algorithms	100	200	300	400	500	Average
SimPT	0.8550	0.9300	0.9813	1.0088	1.0238	0.9598
SimPTP	1.5088	1.6525	1.7250	1.7325	1.7338	1.6705
Webmate	1.3031	1.3942	1.4213	1.4406	1.4512	1.4003
NB	1.1575	1.2413	1.2775	1.2863	1.2925	1.2510

Table 3. Hit rate (HR) as the value of N (number of recommended contents) increases

Algorithms	100	200	300	400	500	Average
SimPT	0.17	0.28	0.41	0.50	0.57	0.38
SimPTP	0.28	0.38	0.57	0.68	0.69	0.52
Webmate	0.14	0.27	0.35	0.42	0.47	0.33
NB	0.16	0.27	0.36	0.39	0.43	0.32

(K=all and *min_sup*=10% for *SimPT* and *SimPTP*)

at the RHR results as well. In addition, comparing the results achieved by *SimPT* and the benchmark algorithms, HR of the former found to be superior to that of the benchmark algorithms. However, with respect to RHR, *SimPT* is worse than that of the benchmark algorithms. Overall, *SimPTP* achieves 19% and 36% improvement in terms of RHR on average, compared to *Webmate* and *NB*, respectively, whereas *SimPT* brings 33% and 23% degradation of RHR, respectively. We conclude from this experiment that the proposed strategy for top-N recommendation is effective in terms of improving the performance, although RHR is diminished in the case of *SimPT*.

5 Conclusions

The capability to model users' preferences is at the heart of a personalized recommender system that discriminates interesting information from uninteresting data. In this paper, a new and effective method for learning and modeling user preferences and for filtering contents relevant user interests is proposed. The major advantage of the proposed learning method is that it supports the identification of useful patterns of each user. In order to evaluate the effectiveness of the approach, we compare our experimental results with those of probabilistic learning model and vector space model. The experimental results demonstrate that the proposed method offers significant advantages in terms of improving recommendation quality as compared to the traditional learning algorithms. A research area that is attractive attention at present is collaborative modeling of user preferences among users with similar interest. In addition, we are currently extending our algorithm to allow for changing user interests. Therefore, we plan to further study the techniques of adaptive and incremental learning [6, 13].

References

1. Salton, G., Buckley, C.: Term Weighting Approaches in Automatic Text Retrieval. Information Processing and Management, Vol.24 (1988) 513-523
2. Agrawal, R., Srikan, R.: Fast Algorithms for Mining Association Rules. In Proc. of the 20th Int. Conf. on Very Large Data Bases (1994)
3. Chen, L., Sycara, K.: WebMate: Personal Agent for Browsing and Searching. In Proc. of the 2nd Int. Conf. on Autonomous Agents and Multi Agent Systems (1998) 132-139
4. Billsus, D., Pazzani, M.J.: A hybrid user model for News story classification. In Proc. of the 7th Int. Conf. on User Modeling (1999) 99–108
5. Schwab, I. Pohl, W., Koychev, I.: Learning to Recommend from Positive Evidence. In Proc. of Int. Conf. on Intelligent User Interfaces (2000)
6. Widyantoro, D.H., Ioerger, T., Yen, J.: Learning User Interest Dynamics with a Three-Descriptor Representation. Journal of the American Society for Information Science and Technology, Vol. 52 (2001) 212-225
7. Aggarwal, C.C., Philip, S.Y.:An Automated System for Web Portal Personalization. In Proc. of the 28th VLDB Conference (2002) 1031-1040
8. Chen, C.C., Chen, M.C., Sun, Y.: PVA: A Self-Adaptive Personal View Agent. Journal of Intelligent Information Systems, Vol.18 (2002) 173-194
9. Eirinaki, M., Vazirgiannis, M.: Web Mining for Web Personalization, ACM Transactions on Internet Technology, Vol. 3 (2003) 1-27
10. Deshpande, M., Karypis, G.: Item-based Top-N Recommendation Algorithms. ACM Transac-tions on Information Systems, Vol. 22 (2004) 143–177
11. Han, J., Pei, J., Yin, Y.: Mining Frequent Patterns without Candidate Generation: A Frequent-Pattern Tree Approach. Data Mining and Knowledge Discovery, Vol. 8 (2004) 53-87
12. Gabrilovich, E., Dumais, S., Horvitz, E.: Newsjunkie: Providing Personalized News-feeds via Analysis of Information Novelty. In proc. of the 13th Int. Conf. on World Wide Web (2004) 482-490
13. Chung, S., McLeod, D.: Dynamic Pattern Mining: An Incremental Data Clustering Approach. Lecture Notes in Computer Science, Vol. 3360 (2004) 85-112
14. Pazzani, M.J., Meyers, A.: NSF Research Awards Abstracts 1990-2003. http://kdd.ics.uci.edu/databases/nsfabs/nsfawards.html
15. Baeza-Yates, R., Ribeiro-Neto, B.: Modern Information Retrieval. Addison -Wesley (1999)
16. Han, J., Kamber, M.: Data Mining: Concepts and Techniques, 2nd ed. Morgan Kaufmann Publishers (2006)

Reducing Trials by Thinning-Out in Skill Discovery

Hayato Kobayashi[1], Kohei Hatano[2], Akira Ishino[1], and Ayumi Shinohara[1]

[1] Graduate School of Information Science, Tohoku University, Japan
[2] Department of Informatics, Kyushu University, Japan
kobayashi@shino.ecei.tohoku.ac.jp, hatano@i.kyushu-u.ac.jp,
ishino@ecei.tohoku.ac.jp, ayumi@ecei.tohoku.ac.jp

Abstract. In this paper, we propose a new concept, thinning-out, for reducing the number of trials in skill discovery. Thinning-out means to skip over such trials that are unlikely to improve discovering results, in the same way as "pruning" in a search tree. We show that our thinning-out technique significantly reduces the number of trials. In addition, we apply thinning-out to the discovery of good physical motions by legged robots in a simulation environment. By using thinning-out, our virtual robots can discover sophisticated motions that is much different from the initial motion in a reasonable amount of trials.

1   Introduction

Skill discovery is a task to obtain an ability to perform well in a target domain, through *trial and error*. It can be formulated as an optimization problem, in which the goal is to find a solution x in a vast search space X that maximizes (or minimizes) a score function $f : X \rightarrow \mathbf{R}$. When the shape of f is not known and there is no satisfactory problem-specific algorithm or heuristic, meta-heuristic guides to pick up the next candidate, based on the previous candidates and their evaluated scores. Various meta-heuristics are proposed and experimented, such as Genetic Algorithms (GA) and Simulated Annealing (SA). These meta-heuristics in general contribute to reduce the number of trials, in order to find a good solution x in X. Nevertheless, for some problem domains, the number of trials is still too large, especially when each trial consumes a considerable amount of time and costs. Skill discovery in robot movements, which we treat as a target application in this paper, is an instance of them.

When GA or SA picks up the next candidate to try, it is often duplicated: it could be already experimented in the past trials. Ratle [1] proposed an efficient method to avoid such duplications using function approximation. He showed that the method can reduce the number of actual function calls by creating an approximate model of the score function using kriging interpolation and using the model instead of the original score function for evaluating some of the next generations. When the score function can not be regarded as a deterministic function because of the noisy environment, duplicated trials are meaningful to increase the certainty of the evaluations. Sano et al. [2] proposed Memory-based

V. Corruble, M. Takeda, and E. Suzuki (Eds.): DS 2007, LNAI 4755, pp. 127–138, 2007.
© Springer-Verlag Berlin Heidelberg 2007

Fitness Evaluation GA for noisy environment. They estimated more proper fitness values (or scores) by weighted average of neighboring scores, so as to reduce the number of trials more than multiple sampling methods (i.e., to evaluate fitness values several times in each trial.)

In this paper, we take another approach to reduce the trials, based on the idea that we can theoretically determine whether or not the selected candidates are worth evaluating, if the gradient of the score function is given. If the candidate x is unlikely to improve the results obtained so far, we do not perform the trial and just skip it. We call this method *thinning-out*, which contrasts to *pruning* in a search tree. One advantage of our method is that we can naturally combine the thinning-out with any search methods including GA and SA as well as random search. In preliminary experiments, we combined it with a simple random search method, and observed that the resulting number of trials is usually reduced to logarithmic with respect to the number of candidates. In this paper, we show that our thinning-out method significantly reduces the number of trials with a small failure rate with a combination with GA.

We address skill discovery by legged robots in a simulation environment as its application. For legged robots to function in the real world, they must need the ability to acquire such basic skills as walking, running, pushing, kicking, and so on. The ability for robots to learn some skills is known as *skill learning*, and is regarded as important. For several years, there have been many studies conducted on skill learning by legged robots. Kim and Uther [3] studied the learning of fast quadruped locomotion skills by modeling the locus of their gait as a quadrangle. Kohl and Stone [4] also studied the learning of stable quadruped locomotion by modeling the locus of their gait as a semi-ellipsoid. Fidelman and Stone [5] proposed a learning method for acquiring the ball-grasping skill. The learning task consists of two layers, the first for walking and the second for pinching the ball by its chin. Kobayashi et al. [6] studied the reinforcement learning to trap a moving ball. The goal of the learning was to acquire a good timing to initiate the catch motion, depending on the distance and the speed of the ball, whose movement was restricted to one dimension.

In this paper, we make robots to discover good shot motions, as Zagal and Solar [7] also addressed in their work. Compared with other tasks mentioned above, discovery of good shot motions could be more challenging, because it is difficult to construct a good model for shot motions. Although our work is similar to Zagal and Solar's, our parameterization is more flexible. On the other hand, flexibility of parameters implies that it takes many trials in the discovery process. Lee et al. [8] successfully realized flexible movements of legged robots to climb over a variety of obstacles by reinforcement learning with supervised information. Since we can not prepare supervised information for good motions other than an initial motion, we must find good motions in large search spaces. Our thinning-out method can reduce the number of trials in such problem and realize the flexibility in feasible trials.

The remainder of this paper is organized as follows. In Section 2, we describe the concept of thinning-out and propose two inferring methods for it. In

Section 3, we evaluate our method using the minimization problem of mathematical test functions. In Section 4, we apply our method to discovery of good shot motions by legged robots in a simulation environment. Finally, Section 5 presents our conclusions.

2   Thinning-Out

In this section, we treat the maximization problem of unknown score function. We assume that the score function is continuous and, to some extent, smooth over the search space. Our assumption seems reasonable because each robot movement is continuous and thus small changes of parameters will not affect the score significantly. Based on this assumption, we infer local shapes of the score function. Given a candidate point to try, we estimate the score of the candidate point by using the distance from the nearest neighborhood whose score is known. We thin-out the candidate point if the estimated upperbound of the score is lower than the current highest score. Note that our method does not take into account the distance from the current highest point. In other words, we do not assume that the true highest point lies near the current highest one. Therefore an expected point with high score, even it is far from the current highest point, has a chance to be tried. In summary, our method is robust and it is unlikely to get stuck in local maxima.

Now we define the local smoothness of the score function in terms of Lipschitz condition, which is found in standard textbooks on calculus. We use g-Lipschitz continuous for some function g, as natural extension of c-Lipschitz for some constant c in the textbooks.

Definition 1 (Lipschitz condition). *Let R be the set of real numbers, X be a metric space with metric d, and $f : X \to R$ be a score function on it. Given a function $g : R \to R$, f is said to be g-Lipschitz continuous, if it holds for any $x_1, x_2 \in R$ that*

$$|f(x_1) - f(x_2)| \le g(d(x_1, x_2)).$$

The function g is called Lipschitz function.

Suppose that a score function f is g-Lipschitz continuous. Then, for any points x_1 and x_2, an upperbound of $f(x_1)$ is obtained by

$$f(x_1) \le f(x_2) + g(d(x_1, x_2)).$$

Our thinning-out strategy is to infer a proper Lipschitz function which characterizes the score function f, so as to obtain an upperbound of the score of a candidate point. If the upperbound is smaller than the current best score, we do not have to try the candidate point. We will explain the details of our methods to infer Lipschitz functions soon. Our thinning-out condition is formally described as follows:

Definition 2 (Thinning-out condition). *Let x_b be the point whose score $f(x_b)$ is the current best. Let x_c be a candidate to try. Let x_n be the nearest neighbor of the candidate. Given an inferred function $\hat{g} : \mathbf{R} \to \mathbf{R}$, x_c is said to satisfy the thinning-out condition with respect to $\hat{g}$, if it holds that*

$$f(x_n) + \hat{g}(d(x_c, x_n)) \leq f(x_b)$$

Now we propose two methods to infer Lipschitz functions.

Max Gradient Method

Suppose that we know the maximum gradient

$$c = \max_{x_1, x_2 \in X, \ x_1 \neq x_2} \frac{|f(x_1) - f(x_2)|}{d(x_1, x_2)}$$

of the score function f, over any two different points x_1 and x_2 in X. Then for the function defined by $g(d) = c \cdot d$, it is easy to verify that f is g-Lipschitz continuous. Thus g can be used to thin-out candidates without errors. Since c itself is unavailable in practice, we substitute the maximum gradient from every two points in past trials so far, which will become a good approximation of c after enough trials. We call it *Max Gradient* (MG) method. This method may have small error rate, because it deals with the worst case scenario. However, it can hardly thin-out candidates in rough score functions obviously, since the estimated value of the Lipschitz function is too conservative in many cases.

Gathering Differences Method

Meta-heuristics picks up many samples from an interesting region expected to have the best score. We can get the shape of the interesting region by using information of points which are densely packed in past trials so far. Thus we infer Lipschitz functions by gathering the differences of the scores, from the smallest one in ascending order of the distance between the points, until the summation of the distances become greater than the distance between x_c and x_n, as shown in Algorithm 1. It will become a good approximation after enough trials, since a line connecting fairly close two points can approximate the gradient of the function nearby the points. We call it *Gathering Differences* (GD) method. This method can thin-out many candidates, since it may infer the local shape of the score function in the interesting region. However, it may wrongly thin-out them, because it is just heuristics and does not have any theoretical propriety.

3 Performance Evaluation of Thinning-Out

We need efficient sampling methods for picking up candidates, since our thinning-out method in the previous section just skips over the candidates. In this paper, we utilize Genetic Algorithm (GA), which is one of the *meta-heuristics* methods, because we intend to address discovery problems in which the score function is

---

Algorithm 1. Gathering differences method

---

input : distance and the set Hist of pairs $(x, f(x))$ observed so far
output : An inferred value of the Lipschitz function $g(\text{distance})$

initialize Diff as a map from $\mathcal{R}$ to $\mathcal{R}$;
foreach x_1, $f(x_1)$ *in* Hist **do**
    foreach x_2, $f(x_2)$ *in* Hist **do**
        Diff $[d(x_1, x_2)] \leftarrow |f(x_1) - f(x_2)|$;
    end
end
sum_diff_point $\leftarrow$ 0;
sum_diff_score $\leftarrow$ 0;
foreach diff_point, diff_score *in* Diff *in ascending order w.r.t.* diff_point **do**
    sum_diff_point $\leftarrow$ sum_diff_point + diff_point ;
    sum_diff_score $\leftarrow$ sum_diff_score + diff_score ;
    if sum_diff_point $\geq$ distance **then**
        return sum_diff_score ;
    end
end
return ∞ ;

Algorithm 2. Evaluation of a candidate with thinning-out.

---

input : candidate
output : score

while candidate *satisfies the thinning-out condition* **do**
    candidate $\leftarrow$ a random perturbation of candidate ;
end
score $\leftarrow$ Evaluate(candidate);
return score ;

---

unknown. We can combine GA and thinning-out by designing the evaluation function of a candidate as shown in Algorithm 2. Since thinning-out is meta strategy, we can easily utilize other meta-heuristics methods in the same way as GA. Although we have experimented other methods such as hill climbing, simulated annealing, and policy gradient and verified that they worked well with thinning-out, we omitted them here because of space limitations.

We use the minimization problem of mathematical test functions for verifying the performance of our thinning-out method. The evaluation by test functions is commonly performed for verifying the performance of meta-heuristics. In this paper, we use Rastrigin, Schwefel, Griewank, Rosenbrock, and Ridge functions, which are used by Hiroyasu et al. [9]. In addition, we add Ackley function [10], because we think our thinning-out method is not good at the function with deep rapid valleys. Characteristics of these functions are as follows. Rastrigin, Schwefel, Griewank, and Ackley functions have multiple peaks, although Griewank and Ackley functions have a single peak with a global view. Griewank, Rosenbrock

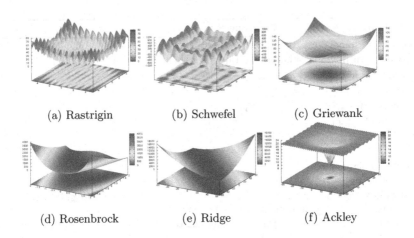

(a) Rastrigin   (b) Schwefel   (c) Griewank

(d) Rosenbrock   (e) Ridge   (f) Ackley

Fig. 1. The shape of test functions in 2 dimensions

Table 1. Thinning-out rate and error rate of GA+MG and GA+GD in 6 test functions in 2 dimensions. Each rate is the average over 100 experiments using 100 candidates.

function	GA+MG		GA+GD	
	thinning (%)	error (%)	thinning (%)	error (%)
Rastrigin	81.47	0.23	78.60	1.99
Schwefel	82.98	0.18	78.97	1.42
Griewank	82.82	0.24	78.98	1.01
Rosenbrock	82.23	0.05	78.71	0.69
Ridge	81.89	0.00	80.40	0.75
Ackley	79.91	2.77	71.48	2.94

and Ridge functions have the design variables' dependency. Fig. 1 shows the shapes of these functions in 2 dimensions.

We compared the performance of our methods by three different viewpoints: the kind of test functions, the number of candidates, and the dimension of test functions. In each experiment, the step size parameter (for mutation and perturbation) of GA is 1% of the domain size of each dimension. We used ϵ-*thinning-out* in the same way as ϵ-*greedy* in reinforcement learning [11], because our methods can not always thin-out candidates safely. ϵ-thinning-out evaluates a candidate with probability ϵ, and otherwise, it skips over the candidate. Consequently, ϵ-thinning-out can hold out the possibility for evaluating candidates that are wrongly thinned-out once and avoid never halting by thinning-out all candidates. We set $\epsilon = 0.01$.

Firstly, we compared the performance by the kind of test functions. Table 1 shows thinning-out rate and error rate of GA+MG and GA+GD in each test function. The thinning-out rate means the rate of thinned-out candidates in all

Table 2. Results of minimization by GA, GA+MG, and GA+GD in 6 test functions in 2 dimensions. Each result means the minimum score of 50 actual trials and is the average over 100 experiments.

function	GA	GA+MG	GA+GD
Rastrigin	24	13	19
Schwefel	712	435	439
Griewank	43	32	33
Rosenbrock	418	330	296
Ridge	11542427	8233764	8878178
Ackley	19	18	18

candidates, and the error rate means the rate of wrongly thinned-out candidates in thinned-out candidates. Both MG and GD totally reduced the number of trials by more than 70 % with low error rates. As anticipated, both MG and GD have slightly higher error rates in Ackley function. Table 2 shows the results of minimization by GA, GA+MG, and GA+GD. Both GA+MG and GA+GD always got better results than GA in all test functions. This is because a small number of errors (i.e., wrongly thinned-out candidates) will not affect final results as shown in Fig. 2. Contrary to our expectation, these tables indicate that

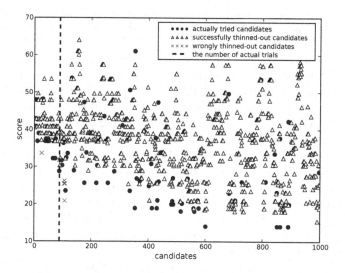

Fig. 2. This graph shows an example of minimization process with thinning-out. The example is in an experiment using 1000 candidates in 2 dimensional Rastrigin function. Circles, triangles, and crosses represent actually evaluated candidates, successfully thinned-out candidates, and wrongly thinned-out candidates, respectively. The dashed line indicates the same number of candidates as actual trials with thinning-out. The minimum score in the left side of the line means the minimization result without thinning-out.

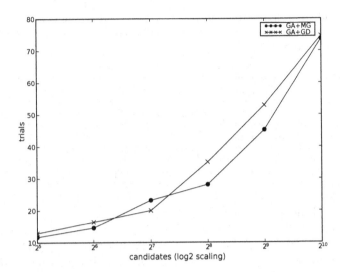

Fig. 3. This graph shows the number of actual trials over the number of candidates by GA+MG and GA+GD in Rastrigin function in 2 dimensions. Each result is the average over 10 experiments

the thinning-out rate of MG is higher than that of GD. The later experiment, however, finds that the thinning-out rate of GD is higher than that of MG in higher dimensions.

Secondly, we compared the performance by the number of candidates. Fig. 3 shows the relationship between the number of trials and that of candidates. The graph indicates that the thinning-out rate gets higher as the number of candidates gets larger. For example, the results of 80 trials by both GA+MG and GA+GD is almost the same as that of $2^{10} = 1024$ trials by GA, if there are no critical errors. We theoretically analyzed the number of trials with respect to the number of candidates for a simplified case. For a score function $f(x) = x$, let n_c be the number of candidates by random search. Then the number of trials is reduced to $O(\log(n_c))$ by our thinning-out method. If we can prove it in more practical functions (e.g., $f(x) = x^n$), the result has practical significance, because random sampling with thinning-out worked better than GA in low dimensions. In this paper, we picked up GA since random search can hardly suggest good candidates in high dimensions.

Finally, we compared the performance by the dimension of test functions. Table 3 shows thinning-out rate and error rate of GA+MG and GA+GD in various dimensions. The table indicates that MG can hardly thin-out the large number of candidates in high dimensions. On the other hand, GD can also thin-out almost the same number of candidates in high dimensions as that in low dimensions. This result indicates that GD is better than MG in high dimensions, if there are no critical errors. In the next section, we show that there are no critical errors in our intended problem, skill discovery.

Table 3. Thinning-out rate and error rate of GA+MG and GA+GD in Rastrigin function in 2, 5, 10, 50, and 100 dimensions. Each rate is the average over 100 experiments using 100 candidates

dimension	GA+MG		GA+GD	
	thinning (%)	error (%)	thinning (%)	error (%)
2	82.07	0.41	77.44	2.44
5	59.83	0.16	77.26	4.08
10	45.52	0.12	75.08	5.09
50	35.23	0.19	70.84	6.89
100	35.70	0.13	69.89	6.12

4   Discovery of Good Shot Motions

4.1   Creation of Initial Motions

In this paper, we use AIBO, which was developed by Sony Corporation, as a legged robot. Physical motions of AIBO are realized by sending *frames*, consisting of the 15 joint angles for its head and legs, to OVirtualRobot every 8 ms. OVirtualRobot is a kind of proxy object that is defined in the software development kit OPEN-R for AIBO. In our framework, these frames are generated from *key-frames*. The key-frames are the characteristic frames shaping the skeleton of each motion. For example, a kick motion needs at least two key-frames, since robots must pull and push its leg when executing it. We indicate the number of interpolations for each key-frame, so that whole frames can be generated by using a linear interpolation method. Thus, our motion takes $8n_i$ ms, where n_i is the total number of interpolations.

4.2   Discovery Process

We directly utilize the key-frames for discovering good shot motions. All we do is to create sketchy motions, that is to indicate the key frames for the motion, and it is possible to realize flexible search in the neighborhood of the skeleton without modeling the movement and setting extra-parameterization. We fix the number of key-frames and interpolations. In other words, the search space of our discovery process has $15n_k$-dimensions, where n_k means the number of key-frames. In the process, we sample $x \in \mathbf{R}^{15n_k}$, make our robot perform a shot motion generated from x, and calculate scores as the formula $r_b \cdot (1 - |\theta_t - \theta_b| / \theta_c)$, where r_b and θ_b mean the distance and angle to the kicked ball, and θ_t means the target direction for shots. The formula linearly reduces the value in inverse proportion to the difference between θ_b and θ_t. θ_c is a constant for the degree of reducing, and we set $\theta_c = \pi/4$.

4.3   Experiments and Results

We applied our methods to discovery of good shot motions by legged robots in a simulation environment. We slightly extended the 3D simulator developed

Table 4. Results of discovery of good motions by GA, GA+MG, and GA+GD. Each result means the maximum score of 50 actual trials and is the average of 10 experiments.

GA	GA+MG	GA+GD
936	940	1058

by Zaratti et al. [12] and used it. Although this simulator can absolutely not produce complete, real environments, it is suitable for verifying the performance of such new methods, because we can perform reproducible measurement without annoying real noise, as well as without damaging our robots.

Our experiments require much more time since the simulation of physical motions itself requires complex computation, even though our discovery processes lies in the simulator. Actually, each experiment in this section took a couple dozens hours. Therefore, thinning-out can make discovery processes more efficient in a simulation environment as well as real environments, because it can reduce time-consuming trials themselves.

We experimented using the motion for shooting a ball to a left oblique direction with its right leg, as an initial motion. The search space is 75(= 15 joint angles * 5 key frames)-dimensions. The step size parameter of GA is $\pi/36$ in each dimension. Table 4 shows the results using GA, GA+MG and GA+GD. The table indicates that GA+GD get better results than GA and GA+MG. Although the difference is small, it should be noted that GA+GD used several hundred candidates. In other words, the result of GA+GD is almost the same result of several hundred trials in GA. The result ought to be improved by using a better sampling method. Fig. 4 shows the initial motion and two better motions which were discovered by using our methods. The motion (b) uses its whole body, although the initial motion (a) uses almost only

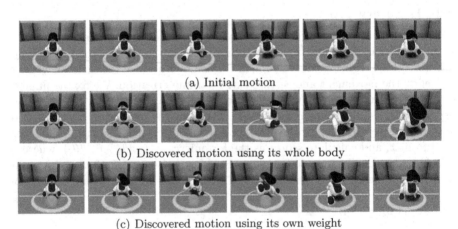

(a) Initial motion

(b) Discovered motion using its whole body

(c) Discovered motion using its own weight

Fig. 4. Initial motion and discovered motions

its right leg. The motion (c) uses its own weight, swinging down its right leg. It should be noted that the motion (c) is much different from the initial motion (a). This result can not be achieved by modeling the initial motion and adjusting the parameters of the model. These results, especially the motion (c), indicate that our skill discovery method with key-frames has flexibility suitable for practical use. The movies of these discovered motions are available online (http://www.shino.ecei.tohoku.ac.jp/~{}kobayashi/movies.html).

5 Conclusions and Future Work

In this paper, we proposed the concept "thinning-out", which is effective for the problems that take much more evaluation time in each trial. We proposed two methods (MG and GD), which infer Lipschitz functions for thinning-out. By the experiments on the minimization problem of several test functions, we verified that MG can safely thin-out few candidates, and conversely GD can fearlessly thin-out many candidates, especially in high dimensions. The results of test functions also suggests that thinning-out can be utilized widely in other different problems. In addition, we applied our methods to discovery of good shot motions by legged robots in a simulation environment. Our virtual robots discovered sophisticated motions that is much different from the initial motion in a feasible number of trials.

From now on, the experiments using real robots will be needed to verify that thinning-out can treat real noise. Discovery of good shot motions in real environments, however, will be unrealistic, because we must estimate the distance to the kicked ball and restore the ball to the initial point carefully with each trial for themselves. Therefore, we plan to make our robots perform *autonomous learning* in the same way as Kobayashi et al. [6]. Autonomous learning of forward shots is readily achievable in much the same way as the method of them, and that of other shots may be possible by utilizing ceiling cameras.

We also need to more accurate inference methods for Lipschitz functions, because the two methods proposed in this paper have both merits and demerits. Although we came up with several ideas, which include a method using the average, median, and weighted average of gradients, other than the two methods, they did not work well. For example, we may be able to infer more proper values by utilizing heuristics depending on each problem.

References

1. Ratle, A.: Accelerating the Convergence of Evolutionary Algorithms by Fitness Landscape Approximation. In: Eiben, A.E., Bäck, T., Schoenauer, M., Schwefel, H-P. (eds.) Proceedings of the Fifth International Conference on Parallel Problem Solving from Nature (PPSN V). LNCS, vol. 1498, pp. 87–96. Springer, Heidelberg (1998)

2. Sano, Y., Kita, H., Kamihira, I., Yamaguchi, M.: Online Optimization of an Engine Controller by means of a Genetic Algorithm using History of Search. In: Proceedings of the 3rd Asia-Pacific Conference on Simulated Evolution and Learning (SEAL), pp. 2929–2934 (2000)
3. Kim, M.S., Uther, W.: Automatic Gait Optimisation for Quadruped Robots. In: Proceedings of 2003 Australasian Conference on Robotics and Automation, pp. 1–9 (2003)
4. Kohl, N., Stone, P.: Machine Learning for Fast Quadrupedal Locomotion. In: The Nineteenth National Conference on Artificial Intelligence (AAAI2004), pp. 611–616 (2004)
5. Fidelman, P., Stone, P.: The Chin Pinch: A Case Study in Skill Learning on a Legged Robot. In: Lakemeyer, G., Sklar, E., Sorrenti, D.G., Takahashi, T. (eds.) RoboCup 2006. LNCS (LNAI), vol. 4434, pp. 59–71. Springer, Heidelberg (2007)
6. Kobayashi, H., Osaki, T., Williams, E., Ishino, A., Shinohara, A.: Autonomous Learning of Ball Trapping in the Four-legged Robot League. In: Lakemeyer, G., Sklar, E., Sorrenti, D.G., Takahashi, T. (eds.) RoboCup 2006. LNCS (LNAI), vol. 4434, pp. 86–97. Springer, Heidelberg (2007)
7. Zagal, J.C., del Solar, J.R.: Learning to Kick the Ball Using Back to Reality. In: Nardi, D., Riedmiller, M., Sammut, C., Santos-Victor, J. (eds.) RoboCup 2004. LNCS (LNAI), vol. 3276, pp. 335–347. Springer, Heidelberg (2005)
8. Lee, H., Shen, Y., Yu, C.H., Singh, G., Ng, A.Y.: Quadruped Robot Obstacle Negotiation via Reinforcement Learning. In: Proceedings of the 2006 IEEE International conference on robotics and Automation (ICRA2006) (2006)
9. Hiroyasu, T., Miki, M., Sano, M., Shimosaka, H., Tsutsui, S., Dongarra, J.: Distributed Probabilistic Model-Building Genetic Algorithm. In: Cantú-Paz, E., Foster, J.A., Deb, K., Davis, L., Roy, R., O'Reilly, U-M., Beyer, H-G., Kendall, G., Wilson, S.W., Harman, M., Wegener, J., Dasgupta, D., Potter, M.A., Schultz, A., Dowsland, K.A., Jonoska, N., Miller, J., Standish, R.K. (eds.) GECCO 2003. LNCS, vol. 2723, pp. 1015–1028. Springer, Heidelberg (2003)
10. Bäck, T.: Evolutionary Algorithms in Theory and Practice. Oxford University Press (1996)
11. Sutton, R.S., Barto, A.G.: Reinforcement Learning: An Introduction. MIT Press, Cambridge (1998)
12. Zaratti, M., Fratarcangeli, M., Iocchi, L.: A 3D Simulator of Multiple Legged Robots based on USARSim. In: Lakemeyer, G., et al. (eds.) RoboCup 2006. LNCS (LNAI), vol. 4434, pp. 13–24. Springer, Heidelberg (2007)

A Theoretical Study on
Variable Ordering of Zero-Suppressed BDDs
for Representing Frequent Itemsets

Shin-ichi Minato

Graduate School of Information Science and Technology,
Hokkaido University
Sapporo, 060-0814 Japan

Abstract. Recently, an efficient method of database analysis using Zero-suppressed Binary Decision Diagrams (ZBDDs) has been proposed. BDDs are a graph-based representation of Boolean functions, now widely used in system design and verification. Here we focus on ZBDDs, a special type of BDDs, which are suitable for handling large-scale combinatorial itemsets in frequent itemset mining. In general, it is well-known that the size of ZBDDs greatly depends on variable ordering; however, in the specific cases of applying ZBDDs to data mining, the effect of variable ordering has not been studied well. In this paper, we present a theoretical study on ZBDD variable ordering for representing frequent itemsets. We show two instances of databases we composed, where the ZBDD sizes are exponentially sensitive to the variable ordering. We also show that there is a case where the ZBDD size must be exponential in any variable ordering. Our theoretical results are helpful for developing a good heuristic method of variable ordering.

1   Introduction

Discovering useful knowledge from large-scale databases has attracted a considerable attention during the last decade. Frequent pattern mining is one of the fundamental problems for knowledge discovery. Since the pioneering paper by Agrawal *et al.*[1], various algorithms have been proposed to solve the frequent pattern mining problem (cf., e.g., [11,5]).

Recently, we have attacked the problem of efficiently generating the frequent patterns in a transaction database by using a data structure called *Zero-suppressed Binary Decision Diagrams* (abbr. ZBDDs), see [7,8]. ZBDDs are a special case of Binary Decision Diagrams (abbr. BDDs)[2]. Using ZBDDs one can implicitly enumerate sets of combinations. Moreover, one can then perform efficiently various operations including the discovery and analysis of frequent patterns.

In general, it is well-known that the size of ZBDDs greatly depends on variable ordering, however, in the specific cases of applying ZBDDs to data mining, the effect of variable ordering has not been studied well. In this paper, we present a theoretical study on ZBDD variable ordering for representing frequent itemsets.

V. Corruble, M. Takeda, and E. Suzuki (Eds.): DS 2007, LNAI 4755, pp. 139–150, 2007.
© Springer-Verlag Berlin Heidelberg 2007

We show two instances of databases we composed, where the ZBDD sizes are exponentially sensitive to the variable ordering. We also show that there is a case where the ZBDD size must be exponential in any variable ordering.

2   Database Representation Using ZBDDs

In this section, we first describe the database representation to be discussed. Here we consider databases of the following type. Let $M \neq \emptyset$ be any set. We refer to the elements of M as to *items*. In our examples below, we use $M = \{a, b, c\}$. Then the set of all possible combinations is the power set $\wp(M)$ of M. Any subset $\mathcal{C} \subseteq \wp(M)$ is said to be a set of combinations. The elements of a set of combinations are sets of items, e.g., $\{a, c\}$. To simplify notation, we write ac instead of $\{a, c\}$ and we refer to the elements of a set of combinations as to *tuples*. A transaction database is just a list of tuples.

2.1   BDDs and ZBDDs

A *Binary Decision Diagram (BDD)* is a graph representation for a Boolean function. An Example is shown in Fig. 1 for $F(a, b, c) = a\bar{b}c \vee \bar{a}b\bar{c}$.

Given a variable ordering (in our example a, b, c), one can use Bryant's algorithm[2] to construct the BDD for any given Boolean function. For many Boolean functions appearing in practice this algorithm is quite efficient and the resulting BDDs are much more efficient representations than binary decision trees.

BDDs were originally invented to represent Boolean functions. But we can also map a set of combinations into Boolean space of n variables, where n is the cardinality of the item set M (see Fig. 2). So, one could also use BDDs to represent sets of combinations. However, one can even obtain a more efficient representation by using *Zero-suppressed* BDDs (ZBDDs)[7].

If there are many similar combinations then the subgraphs are shared resulting in a smaller representation. In addition, ZBDDs have a special type of node deletion rule. As shown in Fig. 3, All nodes whose 1-edge directly points to the

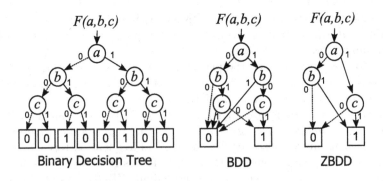

Fig. 1. Binary Decision Tree, BDDs and ZBDDs

a	b	c		F	
0	0	0		0	$\to S$
0	0	1		0	
0	1	0		1	$\to b$
0	1	1		0	
1	0	0		0	
1	0	1		1	$\to ac$
1	1	0		0	
1	1	1		0	

As a Boolean function:
$$F = a\bar{b}c \vee \bar{a}b\bar{c}$$

As a set of combinations:
$$S = \{ac, b\}$$

Fig. 2. Correspondence of Boolean functions and sets of combinations

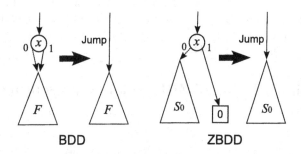

Fig. 3. BDD and ZBDD reduction rule

0-terminal node are deleted. Because of this, the nodes of items that do not appear in any sets of combinations are automatically deleted as shown in Fig.1. This ZBDD reduction rule is extremely effective if we handle a set of sparse combinations. If the average appearance ratio of each item is 1%, ZBDDs are possibly more compact than ordinary BDDs, up to 100 times.

ZBDD representation has another good property that each path from the root node to the 1-terminal node corresponds to each combination in the set. Namely, the number of such paths in the ZBDD equals to the number of combinations in the set. This beautiful property indicates that, even if there are no equivalent nodes to be shared, the ZBDD structure explicitly stores all items of each combination, as well as using an explicit linear linked list data structure. In other words, (the order of) ZBDD size never exceeds the explicit representation. If more nodes are shared, the ZBDD is more compact than linear list.

2.2 ZBDD-Based Representation for Frequent Itemsets

Frequent itemset mining (or frequent pattern mining) is the problem of enumerating all possible subsets of itemset M (also called patterns) which appear more than or equal to α times in the database, for given α. Since their introduction by Agrawal et al.[1], many papers have been published about new algorithms and improvements for solving such mining problems[4,5,11]. Recently, graph-based methods, such as FP-growth[5], have received a great deal of attention, because they can quickly manipulate large-scale itemset data by constructing compact graph structures in main memory.

The ZBDD-based method is a similar approach to handling sets of combinations in main memory but is more efficient because ZBDD is a kind of DAG for representing itemsets, while FP-growth uses a tree representation for the same objects. In general, DAGs can be more compact than trees.

Recently, our research group has developed an efficient algorithm *ZBDD-growth*[9] to generate ZBDDs compactly representing all frequent itemsets for given databases. Our method is not only enumerating/listing the frequent patterns but also efficiently analyzing the huge size of mining results by using ZBDD operations. For example, extracting all patterns including a certain items, or computing the intersection/union/difference set for given two sets of patterns. The computation time of those operations does not directly depend on the number of patterns but almost linear to the (compressed) ZBDD size. It is an important advantage of using ZBDDs.

3   Variable Ordering of ZBDDs for Representing Frequent Itemsets

As described above, it is possible for us to represent histograms of frequent pattern sets compactly by using the ZBDD data structure. However, the ZBDD size is quite sensitive with respect to the underlying ordering of the variables. So, it is important to find a good variable ordering such that the resulting ZBDD size is close to the smallest size possible.

3.1   Properties of Variable Ordering in Ordinary BDDs

In the field of logic VLSI circuit design, many researchers have dealt with the problem of finding good variable orderings for BDDs. For ordinary BDDs, two features of the variable ordering are known that affect the size of the resulting BDDs [3].

(1) Pairs of inputs having the local computability property had better be kept close to one another in the ordering.
(2) Inputs having a strong controllability to the output had better be located at higher order.

As a typical example where the local computability dominates the BDD size, the following AND-OR two-level logic function is known.

$$x_1 x_2 \vee x_3 x_4 \vee x_5 x_6 \vee \cdots x_{2n-1} x_{2n}$$

This function can be represented by only $2n$ BDD nodes with the variable order: $x_1, x_2, x_3, x_4, \ldots, x_{2n}$, where each pair of variables in the same product term are kept together, while we need $(2^{n+1} - 2)$ BDD nodes with the order: $x_1, x_3, \ldots, x_{2n-1}, x_2, x_4, \ldots, x_{2n}$, where those pairs are kept away from each other. (see Fig. 4.)

$$x_1\, x_2 + x_3\, x_4 + x_5\, x_6 + x_7\, x_8 \qquad\qquad x_1\, x_5 + x_2\, x_6 + x_3\, x_7 + x_4\, x_8$$

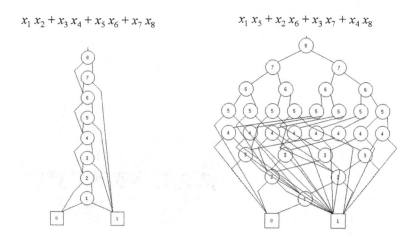

Fig. 4. Effect of BDD variable ordering for the AND-OR two-level logic function

On the other hand, the data-selector function is known as another example where the output controllability dominates the BDD size. For example, 8-bit data-selector with three control inputs and eight data inputs has the function that one of the data input is just selected by the 3-bit binary code of control inputs. In this case, the function can be represented by a linear size of BDDs when the control inputs are higher than data inputs, but the BDD becomes an exponential size in the reversal order. (see Fig. 5.)

As shown the above, each of the both properties may have an exponential impact to the BDD size. Although one should try to come up with orderings obeying these two features, it may be difficult to do so, since these requirements are sometimes contradictory. The problem of finding the optimal variable ordering for BDDs is known to be NP-complete [10]. In addition, there exist functions which always require an exponential number of BDD nodes for any variable ordering, so in such cases, variable ordering will be useless.

3.2 Consideration on ZBDDs Representing Frequent Patterns

Now we consider the effect of variable ordering for the ZBDDs representing frequent itemsets. For the sake of simple discussion, first we assume the minimum frequency $\alpha = 1$, namely, we consider the ZBDD enumerating all possible patterns which appear at least once in the database. In this case, each tuple with k items generates 2^k patterns, and the total number of patterns may become $O(2^n)$ when n is the size of database description. Thus, the ZBDD size may become exponential in the worst case.

The ZBDD of all patterns in the given database can be generated by computing the union of $P(T_k)$ for all tuples $T_1, T_2, \ldots, T_m$ in the database, where $P(T_k)$ means the sets of all patterns included in a tuple T_k. Here we can observe the following property. If an item x appears in a tuple T_k, the patterns in $P(T_k)$ may

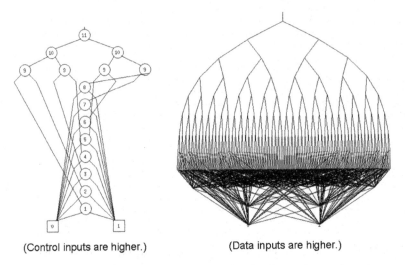

(Control inputs are higher.)   (Data inputs are higher.)

Fig. 5. Effect of BDD variable ordering for the data-selector function

or may not include x, therefore, the item x contributes to double the number of patterns in $P(T_k)$. On the other hand, if x does not appear in T_k, the patterns in $P(T_k)$ never include x, and thus the item x does not contribute to increase the patterns.

Let us consider the analogy with the process of generating an ordinary BDD for a given logic expression. When an input variable x appears in a product term T_k in the logic expression, the variable x should be true to satisfy T_k, so the variable x does not contribute to increase the satisfiable solutions for the logic of T_k. On the other hand, if x does not appear in T_k, the value of x may be false or true, both possible to satisfy T_k, so the number of solutions becomes twice.

Consequently, we can observe the completely opposite effects between the two facts that a variable appears in a term of the logic expression, and that an item appears in a tuple of the database. This observation indicates that we may discuss the effect of ZBDD variable ordering by looking the missing items in a tuple as well as the variables appearing in the logic expressions.

3.3 An Instance Dominated by Local Computability

Based on the above consideration, we made an artificial database where the property of local computability dominates the ZBDD size, as follows.

T_1	a_2 a_3 a_4 $\cdots$	a_n	b_2 b_3 b_4 $\cdots$	b_n
T_2	a_1 a_3 a_4 $\cdots$	a_n	b_1 b_3 b_4 $\cdots$	b_n
T_3	a_1 a_2 a_4 $\cdots$	a_n	b_1 b_2 b_4 $\cdots$	b_n
$\vdots$	$\vdots$ $\vdots$ $\vdots$	$\vdots$	$\vdots$ $\vdots$ $\vdots$	$\vdots$
T_n	a_1 a_2 a_3 $\cdots$ a_{n-1}		b_1 b_2 b_3 $\cdots$ b_{n-1}	

Table 1. Experimental results for the databases dominated by local computability

Order1: $a_1\ b_1\ a_2\ b_2\ \ldots\ a_n\ b_n$
Order2: $a_1\ a_2\ \ldots\ a_n\ b_1\ b_2\ \ldots\ b_n$

n	ZBDD size (order1)	ZBDD size (order2)	Total patterns
3	10	16	37
4	15	39	175
5	20	86	781
6	25	181	3,367
7	25	372	14,197
8	30	755	58,975
9	35	1,522	242,461
10	40	3,057	989,527
11	45	6,128	4,017,157
12	50	12,271	16,245,775

This database consists of the tuples T_k ($k = 1, ..., n$) each of which has almost all items $a_1\ a_2\ \ldots\ a_n\ b_1\ b_2\ \ldots\ b_n$ but only one pair of items a_k and b_k are missing. If we consider the opposite property of item appearance, this database corresponds to the AND-OR two-level logic expression shown in Section 3.1, and we can expect that the pairs of two items missing from the same tuple have the property of local computability.

To confirm our consideration, we generated the ZBDDs representing all patterns included in the database. Table 1 shows the ZBDD size of the two different variable orders, and the two ZBDDs for $n = 8$ are shown in Fig. 6. The result obviously shows an exponential difference between the two ordering. It is

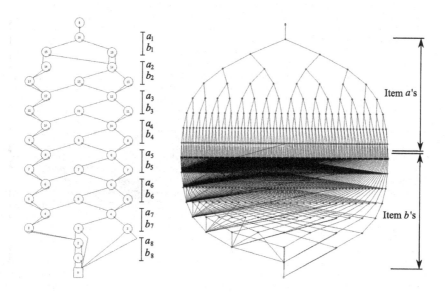

Fig. 6. ZBDDs for the instance dominated by local computability ($n = 8$)

intuitively explained as follows. In the order2, the items a_1 to a_k are ordered in the higher positions, but there is no opportunity to share the ZBDD nodes only by the item a's information without b's, so just a binary tree are generated for the first n stages and thus at least $(2^n - 1)$ nodes are generated. On the other hand, when we use the other order such that a_k and b_k are kept closer, the ZBDD nodes can be shared by using combinatorial information of a_k and b_k on each stage, thus the ZBDD size becomes $O(n)$.

3.4   An Instance Dominated by Output Controllability

Based on the above consideration, we made an artificial database where the property of output controllability dominates the ZBDD size. The following database corresponds to the 8-bit data-selector function, shown in Section 3.1.

T_0	x_1 x_2 x_3 x_4 x_5 x_6 x_7 y_0 y_2 y_4
T_1	x_0 x_2 x_3 x_4 x_5 x_6 x_7 y_0 y_2 y_5
T_2	x_0 x_1 x_3 x_4 x_5 x_6 x_7 y_0 y_3 y_4
T_3	x_0 x_1 x_2 x_4 x_5 x_6 x_7 y_0 y_3 y_5
T_4	x_0 x_1 x_2 x_3 x_5 x_6 x_7 y_1 y_2 y_4
T_5	x_0 x_1 x_2 x_3 x_4 x_6 x_7 y_1 y_2 y_5
T_6	x_0 x_1 x_2 x_3 x_4 x_5 x_7 y_1 y_3 y_4
T_7	x_0 x_1 x_2 x_3 x_4 x_5 x_6 y_1 y_3 y_5

This database has the items x_0 x_1 $\ldots$ x_{n-1} as the data inputs and y_0 y_1 $\ldots$ y_m ($m = 2\lceil \log_2 n \rceil$) as the control inputs. The control inputs have

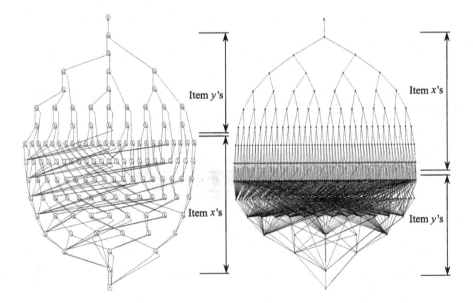

Fig. 7. ZBDDs for the instance dominated by output controllability ($n = 8$)

Table 2. Experimental results for the databases dominated by output controllability

n (number of x's)	ZBDD size(y's higher)	ZBDD size(x's higher)	Total patterns
8	126	579	5,023
12	339	5,117	227,295
16	650	137,444	4,159,487
20	1,151	2,435,284	161,496,559

the pair-wise structure as $(y_0, y_1), (y_2, y_3), \ldots$, each of which represents a digit of binary coded number. Either of odd or even numbered y appears on each tuple T_k in the database, to represent the value $0/1$ of the binary code for k. The tuple T_k also includes all items x_0 to x_n except x_k. With the opposite property of item appearance, we can consider that this database selects one of the data input x_k according to the binary coded number specified by the control inputs, as well as the data-selector function shown in Section 3.1.

Figure 7 shows ZBDDs for the instance of $n = 8$ with two different variable orders. We can observe that the ZBDD will become a polynomial size when the control inputs y's are higher than data inputs x's, but it will become an exponential size in the reversal order. It is explained as follows. If we first assign a set of values into all the items y's, the rest of patterns of x's are related only to one or two tuples in the database, so each ZBDD subgraph for items x's becomes a beautiful array structure with n nodes. A pair of (y_i, y_{i+1}) may cause three patterns: only y_i appears, only y_{i+1} appears, or both absent, so, the upper part of ZBDD for y's become a ternary tree for the $\lceil \log_2 n \rceil$-bit of binary code. The total ZBDD nodes are bounded by $O(n \cdot 3^{\log_2 n}) \approx O(n^{2.7})$.

On the other hand, when we use the reversal order such that x's are higher than y's, there is no opportunity to share the ZBDD nodes only by x's information without y's, so just a binary tree are generated for the first n stages and thus at least $(2^n - 1)$ nodes are required.

To confirm our consideration, we generated the ZBDDs representing all patterns included in the database. Table 2 shows the ZBDD size of the two different variable orders. The result obviously shows an exponential difference between the two orders. Thus, we can see that there exists an example where the property of output controllability has an exponential impact for the ZBDD size.

3.5   A Case of Generating Exponential ZBDD in Any Variable Ordering

Based on the above observation, we can show that there exists a database where the ZBDD representing the set of patterns must be an exponential size in any variable ordering. The "data-selector" database, shown in the last section, consists of the two sorts of items x's and y's. We must put the y's higher than x's to avoid exponential explosion of ZBDD size. Now, let us define N as the total number of items ($= n + 2\lceil \log_2 n \rceil$), then we make N copies of the databases with the N different variable orders by rotating the items one by one, as shown in Fig. 8, and finally we merge the all N blocks into one database. If we generate a ZBDD

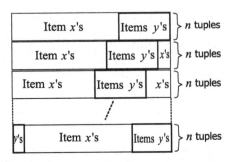

Fig. 8. N copies of "data-selector" databases with the rotated variable orders

for the sets of all patterns in this database, any variable ordering cannot avoid a bad variable order, because, at least one of the blocks, more than $\frac{n}{2\lceil \log_2 n \rceil}$ items of x's become higher than y's. So, we need at least $O(2^{\frac{n}{2\lceil \log_2 n \rceil}})$ ZBDD nodes for representing the patterns for the block of the bad variable order. Since the given database has a polynomial size $O(N^3)$, we can conclude that this database requires an exponential size of ZBDD in any variable ordering.

3.6 Effect of the Minimum Frequency Threshold

In the above discussions, we assume the minimum frequency threshold $\alpha = 1$, it means that the ZBDDs represent the sets of all possible patterns included in the databases. However, in the real applications, we specify a larger α to reduce the number of frequent patterns into a feasible amount, and the size of the ZBDD is also reduced. If the number of frequent patterns are not exponential for a given large α, then the ZBDD size never become exponential, and in such cases, the variable ordering does not have an exponential effect to the ZBDD size.

Table 3 is the experimental result for the same database as shown in Section3.3 with a different threshold $\alpha = n/2$, namely, representing the set of frequent

Table 3. Same database as Table 1 with the different threshold $\alpha = n/2$

Order1: $a_1\ b_1\ a_2\ b_2\ \ldots\ a_n\ b_n$
Order2: $a_1\ a_2\ \ldots\ a_n\ b_1\ b_2\ \ldots\ b_n$

n	ZBDD size (order1)	ZBDD size (order2)	Total patterns
3	8	8	67
4	16	30	106
5	22	47	781
6	33	142	694
7	42	222	1,156
8	56	616	7,459
9	68	969	12,896
10	85	2,564	81,922
11	100	4,074	143,980
12	120	10,503	912,718

patterns which appears at more than half tuples in the database. The result shows that the variable ordering still has an exponential impact to the ZBDD size when $\alpha = n/2$. In this example, we expect that as long as we specify an α proportional to n (i.e. $\alpha = c \cdot n, \ \ 0 < c < 1$), the similar exponential impact will be observed.

4   Conclusion

In this paper, we presented a theoretical study on ZBDD variable ordering for representing frequent itemsets. We composed two instances of databases where the ZBDD sizes are exponentially sensitive to the variable ordering, and we discussed why such a remarkable difference occurs. In addition, we also showed that there is a case where the ZBDD size must be exponential in any variable ordering. These discussions clarify the property of variable ordering when we apply the ZBDD-based data structure to data mining problems. Recently, we proposed a heuristic variable ordering method[6] that finds a good variable order before generating ZBDDs, by using the structural information of the given database. Such a heuristic method is developed based on the theoretical results discussed in this paper.

Acknowledgment

The author would like to thank Thomas Zeugmann, Hiroki Arimura, and Takuya Kida of Hokkaido University for their valuable discussions and comments. This research was partially supported by Japan Society for the Promotion of Science, Grant-in-Aid for Scientific Research (B) on "ZBDD-based Database Analysis," 17300041, 2007, and Grant-in-Aid for Specially Promoted Research on "Semi-Structured Data Mining," 17002008.

References

1. Agrawal, R., Imielinski, T., Swami, A.N.: Mining association rules between sets of items in large databases. In: Buneman, P., Jajodia, S. (eds.) Proc. of the 1993 ACM SIGMOD International Conference on Management of Data. SIGMOD Record, 22(2), 207–216 (1993)
2. Bryant, R.E.: Graph-based algorithms for Boolean function manipulation. IEEE Transactions on Computers C-35(8), 677–691 (1986)
3. Fujita, M., Fujisawa, H., Kawato, N.: Evaluation and implementation of boolean comparison method based on binary decision diagrams. In: Proc. of ACM/IEEE International Conf. on Computer-Aided Design (ICCAD-88), pp. 2–5 (1988)
4. Goethals, B.: Survey on frequent pattern mining (2003), http://www.cs.helsinki.fi/u/goethals/publications/survey.ps
5. Han, J., Pei, J., Yin, Y., Mao, R.: Mining frequent patterns without candidate generation: a frequent-pattern tree approach. Data Mining and Knowledge Discovery 8(1), 53–87 (2004)

6. Iwasaki, H., Minato, S., Zeugmann, T.: A method of variable ordering for zero-suppressed binary decision diagrams in data mining applications. In: Proc. of The Third IEEE International Workshop on Databases for Next-Generation Researchers (SWOD2007) (to appear, 2007)
7. Minato, S.: Zero-suppressed BDDs for set manipulation in combinatorial problems. In: Proc. of 30th ACM/IEEE Design Automation Conference, pp. 272–277 (1993)
8. Minato, S.: Zero-suppressed BDDs and their applications. International Journal on Software Tools for Technology Transfer (STTT, 3(2), 156–170 (2001)
9. Minato, S., Arimura, H.: Frequent pattern mining and knowledge indexing based on zero-suppressed BDDs. In: Bonchi, F., Boulicaut, J.-F. (eds.) KDID 2006. LNCS, vol. 3933, pp. 83–94. Springer, Heidelberg (2006)
10. Tani, S., Hamaguchi, K., Yajima, S.: The complexity of the optimal variable ordering problems of shared binary decision diagrams. In: Ng, K.W., Balasubramanian, N.V., Raghavan, P., Chin, F.Y.L. (eds.) ISAAC 1993. LNCS, vol. 762, pp. 389–398. Springer, Heidelberg (1993)
11. Zaki, M.J.: Scalable algorithms for association mining. IEEE Trans. Knowl. Data Eng. 12(2), 372–390 (2000)

Fast NML Computation for Naive Bayes Models

Tommi Mononen and Petri Myllymäki

Complex Systems Computation Group (CoSCo)
Helsinki Institute for Information Technology (HIIT)
University of Helsinki & Helsinki University of Technology
P.O.Box 68 (Department of Computer Science)
FIN-00014 University of Helsinki, Finland
{Firstname.Lastname}@hiit.fi

Abstract. The *Minimum Description Length* (MDL) is an information-theoretic principle that can be used for model selection and other statistical inference tasks. One way to implement this principle in practice is to compute the *Normalized Maximum Likelihood* (NML) distribution for a given parametric model class. Unfortunately this is a computationally infeasible task for many model classes of practical importance. In this paper we present a fast algorithm for computing the NML for the Naive Bayes model class, which is frequently used in classification and clustering tasks. The algorithm is based on a relationship between powers of generating functions and discrete convolution. The resulting algorithm has the time complexity of $\mathcal{O}(n^2)$, where n is the size of the data.

1 Introduction

The information-theoretical *Minimum Description Length* (MDL) principle [15, 4, 17, 3] for model selection is based on the conceptually simple idea that given a data set, the best model for the data is the one which results in the shortest description for the data together with the model. Hence, we wish to select a model representing a balance between too simple models (in which case the code length for the data is large) and too complex models (in which case the code length for the data is small, but for the model itself large).

Consider a parametric probabilistic model class, i.e., a set of models each defining a probability distribution over all possible data sets. Let us call the shortest possible code length obtainable with the given set of models *stochastic complexity*. Consequently, given a data set, we can choose between alternative parametric models (model classes) with different number of parameters by comparing the corresponding stochastic complexities for the given data.

However, there remains the question of how to formally define the stochastic complexity for a model class. As each of the probability distributions in a probabilistic model class corresponds to a code length, it is obvious that no code can be shorter than all the other codes in the model class for all data sets, because no probability distribution can dominate another probability distribution over all data sets. A *universal model* is a model (code) which can imitate any model

V. Corruble, M. Takeda, and E. Suzuki (Eds.): DS 2007, LNAI 4755, pp. 151–160, 2007.
© Springer-Verlag Berlin Heidelberg 2007

in a given parametric model class. The *normalized maximum likelihood (NML)* distribution [16, 18] is the *worst-case optimal* universal model giving a desired formal definition for the stochastic complexity (see the next Section).

Unfortunately, computing the NML is very difficult for many model classes of practical interest. In this paper we consider Bayesian networks, probabilistic model classes defined by acyclic directed graphs [14, 5]. The Naive Bayes model is a simple Bayesian network, which is continuously used with success in areas such as clustering and classification. It has been earlier shown [11] how to compute the NML for the Naive Bayes model family in $\mathcal{O}(n^2 \log L)$ time, where L denotes the number of values of the class variable of the Naive Bayes model. In this paper we introduce a faster $\mathcal{O}(n^2)$ algorithm for this task, based on generating functions.

2   The Problem

The normalized maximum likelihood (NML) distribution [16, 18] is defined by

$$P_{NML}(\mathbf{x}^n|\mathcal{M}) = \frac{P(\mathbf{x}^n|\hat{\theta}(\mathbf{x}^n, \mathcal{M}))}{\sum_{\mathbf{y}^n} P(\mathbf{y}^n|\hat{\theta}(\mathbf{y}^n, \mathcal{M}))}, \tag{1}$$

where the numerator is the maximum likelihood for the observed data $\mathbf{x}^n$ within the model class $\mathcal{M}$. The *normalizing term* in the denominator is the sum over maximum likelihoods of all possible data sets of size n, with respect to the model class. As shown in [18], this yields the worst case universal distribution with respect to the model class $\mathcal{M}$.

Although NML was defined as the worst-case optimal universal model, without considering model complexity regularization, it is interesting to note how it behaves as a model class selection criterion. Namely, if the model class is very complex, then the maximum likelihood for the given data (the numerator in (1)) is large, but so is also the denominator as a complex model gives a high maximum likelihood for many data sets. For simple model classes the sum in the denominator is small, but so is the numerator. Consequently, the denominator behaves a a regularization term, and the model class optimizing the stochastic complexity $-\log(P_{NML}(\mathbf{x}^n|\mathcal{M}))$ has to balance between model complexity and fit to the given data.

Let us now consider Naive Bayes models. The *Naive Bayes* is a Bayesian network with one root node and m leaf nodes attached to the root node. Variables related to nodes are multinomially distributed. The joint distribution corresponding to the Naive Bayes is defined by

$$P(\mathbf{x}) = P(x_0) \prod_{i=1}^{m} P(x_i|x_0), \tag{2}$$

where $\mathbf{x} = (x_0, x_1, \ldots, x_m)$ is a vector of variable value assignments and x_0 is the value in the root.

For Naive Bayes models, computing the numerator of (1) is trivial, but this is not the case with the denominator. In this paper we derive an efficient algorithm for computing the normalizing term for this model family and call it the *Naive Bayes normalizing term*.

3 Generating Functions and the Naive Bayes

The normalizing term for a single multinomial variable is called *multinomial normalizing term*. We can compute the multinomial normalizing term efficiently: the most efficient known method is proved using generating functions [9, 10]. We now use this same methodology in the Naive Bayes case. First we have to define the needed operations, which we use with generating functions, and then take a closer look at the Naive Bayes normalizing term.

3.1 Generating Functions

An *ordinary generating function* (OGF) of a sequence a_n is

$$F(z) = \sum_{n=0}^{\infty} a_n z^n = a_0 + a_1 z + a_2 z^2 + \cdots, \tag{3}$$

where $z \in \mathbb{C}$ [2]. We are only interested in coefficients a_n, not the value of the function $F(z)$ itself. The function $F(z)$ is only used for computation of some a_n coefficients or in derivation of recurrence formulas. With a *recurrence formula* we can compute the coefficient a_{n+1} with the help of the fixed and finite set of previous coefficients. A generating function may have a closed form, in which case manipulation is easier.

As a generating function is also a formal power series, all general formal power series operations are applicable. In the case of the multinomial normalizing term, however, we need the *exponential generating function* (EGF), which is of form

$$G(z) = \sum_{n=0}^{\infty} b_n \frac{z^n}{n!}. \tag{4}$$

We need to define for later use also two operations: a *coefficient extraction* from a formal power series and taking the power of the exponential generating function. The first operation is defined by

$$[z^n]G(z) = \frac{b_n}{n!}, \tag{5}$$

which means that $[z^n]$ gives us the coefficient of term z^n. The second operation defines what happens to coefficients when we exponentiate the generating function. Rising the generating function $G(z)$ to the power of two, denoted by

$$G^2(z) = \left(\sum_{n=0}^{\infty} b_n \frac{z^n}{n!} \right)^2 = \sum_{n=0}^{\infty} c_n \frac{z^n}{n!}, \tag{6}$$

corresponds to the binomial convolution

$$c_n = \sum_{h=0}^{n} \binom{n}{h} b_h b_{n-h} \tag{7}$$

in the level of coefficients. Similarly, the power of L gives

$$d_n = \sum_{h_1+\cdots+h_L=n} \left(\frac{n!}{h_1! h_2! \cdots h_L!} \right) b_{h_1} b_{h_2} \cdots b_{h_L}, \tag{8}$$

which is the multinomial convolution [2, 8]. This relation between the *expanded form* and the *power form* is the key feature for achieving a new, more efficient algorithm for computing the Naive Bayes normalizing term.

3.2   Naive Bayes Generating Function in the Power Form

First we have to define the generating function for the multinomial normalizing term. We do not give this function in the expanded form, but use a more compact notation: Lth power of a generating function [9, 10]. The power form is

$$\mathcal{B}^L(z) = \left(\sum_{n=0}^{\infty} n^n \frac{z^n}{n!} \right)^L. \tag{9}$$

We call the series inside the parentheses a *basic series*. The basic series here is of exponential type and formal power series coefficients are now $\frac{n^n}{n!}$. Coefficients of the exponential generating function are n^n. When we expand power L, we get an exponential generating function

$$\mathcal{B}^L(z) = \sum_{n=0}^{\infty} \mathcal{C}_{MN}(L, n) n^n \frac{z^n}{n!}, \tag{10}$$

where $\mathcal{C}_{MN}(L, n)$ is the multinomial normalizing term with L values and n data vectors [9, 10]. By a strict definition of generating functions this is not such a function, as it is not explicitly defined. However, we misuse the definition here slightly and in same way also later in the Naive Bayes case, because the implicit form is sufficient for our purposes. There are efficient ways to compute the term $\mathcal{C}_{MN}(L, n)$, and we will show one of them later.

Now we focus on the Naive Bayes normalizing term. The normalizing term is represented in the previous papers only using the expanded form. We denote the Naive Bayes normalizing term by $\mathcal{C}_{NB}(L, K_1, \ldots, K_m, n)$, where L is the number of values of the root variable and K_i is the number of values in leaf variable i. The following theorem shows the simple power form of the normalizing term.

Theorem 1

$$\frac{n^n}{n!} \mathcal{C}_{NB}(L, K_1, \ldots, K_m, n) = [z^n] \left(\sum_{n=0}^{\infty} n^n \left(\prod_{i=1}^{m} \mathcal{C}_{MN}(K_i, n) \right) \frac{z^n}{n!} \right)^L.$$

Proof. A vector $(h_1, \ldots, h_L)$ is a sufficient statistics i.e. data counts of the root variable. The used formula for the Naive Bayes normalizing term is from the paper [11]. With standard manipulation we get

$$\frac{n^n}{n!} \mathcal{C}_{NB}(L, K_1, \ldots, K_m, n) \tag{11}$$

$$= \frac{n^n}{n!} \sum_{h_1 + \cdots + h_L = n} \frac{n!}{h_1! \cdots h_L!} \left(\prod_{k=1}^{L} \left(\frac{h_k}{n} \right)^{h_k} \right) \prod_{i=1}^{m} \prod_{k=1}^{L} \mathcal{C}_{MN}(K_i, h_k) \tag{12}$$

$$= \frac{n^n}{n!} \sum_{h_1 + \cdots + h_L = n} \frac{n!}{h_1! \cdots h_L!} \left(\frac{1}{n^n} \prod_{k=1}^{L} h_k^{h_k} \right) \prod_{k=1}^{L} \left(\prod_{i=1}^{m} \mathcal{C}_{MN}(K_i, h_k) \right) \tag{13}$$

$$= \frac{1}{n!} \sum_{h_1 + \cdots + h_L = n} \frac{n!}{h_1! \cdots h_L!} \prod_{k=1}^{L} \left(h_k^{h_k} \prod_{i=1}^{m} \mathcal{C}_{MN}(K_i, h_k) \right) \tag{14}$$

$$= [z^n] \left(\sum_{n=0}^{\infty} n^n \left(\prod_{i=1}^{m} \mathcal{C}_{MN}(K_i, n) \right) \frac{z^n}{n!} \right)^L. \tag{15}$$

The last form is the power form, from where we can easily extract the basic series. We started from the expanded form and ended up with the power form. □

Let us compare the generating functions of the multinomial and the Naive Bayes normalizing terms. In the multinomial case we have

$$\left(\sum_{n=0}^{\infty} n^n \frac{z^n}{n!} \right)^L \tag{16}$$

and in the Naive Bayes case we have

$$\mathcal{E}^L = \left(\sum_{n=0}^{\infty} \mathcal{C}_{MN}(K_1, n) \cdots \mathcal{C}_{MN}(K_m, n) n^n \frac{z^n}{n!} \right)^L. \tag{17}$$

The two forms seem to be quite similar, except that in the Naive Bayes case we have additional multinomial normalizing terms inside the basic series terms. These extra multinomial normalizing terms makes the expanded form look quite ugly. However, despite of the complex terms, there exists an $\mathcal{O}(n^2 \log L)$ algorithm for computing the Naive Bayes normalizing term [11]. The basic idea is very simple: we can split the exponent L into two parts. Let's call these parts L^* and $L - L^*$. Then we get $\mathcal{E}^L = \mathcal{E}^{L^*} \mathcal{E}^{L - L^*}$. Now we can simply take the normal discrete convolution in the right hand side to get one term of the series in the left hand side. If we require that the result is also a normalizing term, we get the known recurrence formula

$$\mathcal{C}_{NB}(L, K_1, \ldots, K_m, n) = \sum_{k=0}^{n} \binom{n}{k} \left(\frac{k}{n} \right)^k \left(\frac{n-k}{n} \right)^{n-k}$$

$$\cdot \mathcal{C}_{NB}(L^*, K_1, \ldots, K_m, k) \, \mathcal{C}_{NB}(L - L^*, K_1, \ldots, K_m, n-k). \tag{18}$$

So two lower exponents produce a higher one. To achieve the $\log L$ -term in the time complexity, we have to merge exponents wisely, so that we do not make any unnecessary steps. For example, if we want to compute $\mathcal{E}^L$ with $L = 16$, we first compute $\mathcal{E}^2$ and then compute $\mathcal{E}^2\mathcal{E}^2$ to get $\mathcal{E}^4$. In the same way we get the series $\mathcal{E}^8$ and finally the series $\mathcal{E}^{16}$. If the target value is not two to some power, then we have to do more complicated multiplications based on same idea. However, in the next section we present a novel, even more efficient way for computing the Naive Bayes normalizing term.

4   Powers of Formal Power Series

As basic series are formal power series, we can use some known powers of formal power series formula. One of these formulas is the *Miller formula* [6]. It is originally a result of Euler and it has time complexity of $\mathcal{O}(n^2)$ for any real number exponent, but of course only natural numbers are meaningful in our case.

The proof of the Miller formula has been sketched many times in history [7, 13, 6], but we were not able to find a detailed proof in the literature. The detailed proof is relatively straightforward and uses only standard manipulation. We complete below the missing parts of Knuth's proof for sake of clarity.

Theorem 2 (The Miller formula). *If two formal power series are* $V(z) = 1 + \sum_{k=1}^{\infty} v_k z^k$ *and* $W(z) = \sum_{k=0}^{\infty} w_k z^k$ *and* $W(z) = (V(z))^{\alpha}$, $\alpha \in \mathbb{R}$, *then* $w_0 = 1$ *and* $w_n = \sum_{k=1}^{n}\left(\left(\frac{\alpha+1}{n}\right)k - 1\right)v_k w_{n-k}$.

Proof. It is evident that $w_0 = 1$, since $v_0 = 1$ and $1 = 1^{\alpha}$. Next we derivate the basic equation of the theorem and get

$$W'(z) = \alpha V(z)^{\alpha-1}V'(z).$$

Then we multiply both sides with $V(z)$ and substitute $V(z)^{\alpha} = W(z)$, which gives us the equation

$$W'(z)V(z) = \alpha W(z)V'(z).$$

Let us now look at the z^{n-1}-coefficients of both sides:

$$[z^{n-1}]W'(z)V(z) = \alpha[z^{n-1}]W(z)V'(z) \tag{19}$$

$$\sum_{k=0}^{n} k w_k v_{n-k} = \alpha \sum_{k=0}^{n}(n-k)w_k v_{n-k} \tag{20}$$

$$\sum_{k=0}^{n-1} k w_k v_{n-k} + n w_n v_0 = \alpha \sum_{k=0}^{n-1}(n-k)w_k v_{n-k} + 0 \tag{21}$$

$$n w_n v_0 = \sum_{k=0}^{n-1}\left(\alpha(n-k)w_k v_{n-k} - k w_k v_{n-k}\right) \tag{22}$$

$$w_n = \frac{1}{n v_0}\sum_{k=0}^{n-1}\left((\alpha(n-k) - k)w_k v_{n-k}\right) \tag{23}$$

$$w_n = \sum_{k=0}^{n-1} \left(\frac{\alpha(n-k) - k}{n} \right) w_k v_{n-k} \tag{24}$$

$$w_n = \sum_{k=1}^{n} \left(\frac{\alpha(n-n+k) - n + k}{n} \right) w_{n-k} v_{n-n+k} \tag{25}$$

$$w_n = \sum_{k=1}^{n} \left(\left(\frac{\alpha+1}{n} \right) k - 1 \right) w_{n-k} v_k. \tag{26}$$

After some straightforward manipulation we get the result. □

We can obviously use this method for computing the normalizing term of the Naive Bayes model. It should be noted that while this result is elegant, if we use the discrete Fourier transform, we can achieve the time complexity $\mathcal{O}(n \log n)$ by using the basic identity $(V(z))^\alpha = \exp(\alpha \log(V(z)))$ and the *fast Fourier transform* (FFT). The FFT method involves utilization of Newton's method and is explained in the paper [1]. However, the usefulness of this approach is unclear as some earlier tests with the multinomial normalizing term [12] show that the used floating point numbers must have very high precision in practical cases. This is due to the fact that the values of the normalizing terms can be quite large, and consequently, as the data size increases, the precision of the floating point numbers must also increase. This means that increasing the precision will affect the efficiency of the algorithm, although the number of operations remains in principle the same.

5   Computation of the Naive Bayes Normalizing Term

The computation of the Naive Bayes normalizing term is quite straightforward given the results derived above. Now we collect these results in a form of an algorithm. As we did not describe earlier how to compute efficiently the multinomial normalizing term, we start by defining that.

5.1   Recurrence Formula for the Multinomial Normalizing Term

The multinomial normalizing term can be computed by using a recurrence formula [9, 10]. Initial values for this formula are

$$\mathcal{C}_{MN}(1, n) = 1 \quad \text{and} \tag{27}$$

$$\mathcal{C}_{MN}(2, n) = \sum_{k=0}^{n} \binom{n}{k} \left(\frac{k}{n} \right)^k \left(\frac{n-k}{n} \right)^{n-k}, \tag{28}$$

where $\mathcal{C}_{MN}(2, n)$ is a *binomial normalizing term*. After this we use the recurrence formula

$$\mathcal{C}_{MN}(L+2, n) = \mathcal{C}_{MN}(L+1, n) + \frac{n}{L} \mathcal{C}_{MN}(L, n) \tag{29}$$

to get the normalizing term with the wanted number of values in a multinomial variable. Time complexity of this whole method is $\mathcal{O}(n)$, as the number of values in a variable is usually much smaller than the number of the data points n. If we have to compute all normalizing terms between $[0, n]$ and we choose to use FFT, then binomial normalizing terms should be computed using (16). For the multiplication we can apply FFT.

5.2   The Algorithm

Now we have all the components we need for our algorithm. As said before, our main theorem is quite obvious given the earlier results (Theorems 1 and 2) and needs no proof.

Theorem 3. *The Naive Bayes normalizing term can be efficiently calculated in following way:*

1. *Compute first $n + 1$ binomial normalizing terms.*
2. *Use the recurrence formula to get the needed multinomial normalizing terms.*
3. *Compute the basic series $\sum_{k=0}^{n} \mathcal{C}_{MN}(K_1, k) \cdots \mathcal{C}_{MN}(K_m, k) k^k \frac{z^k}{k!}$.*
4. *Use the Miller formula to compute a new series, which is the basic series to the power of L.*
5. *Extract the Naive Bayes normalizing terms from the computed series by extracting coefficients and multiplying every coefficient so that the kth coefficient is multiplied by $\frac{k!}{k^k}$.*

Time complexity is $\mathcal{O}(n^2)$ for any exponent, because complexities of the steps are $\mathcal{O}(n^2)$, $\mathcal{O}(n \cdot \max(K_i))$, $\mathcal{O}(n \cdot m)$, $\mathcal{O}(n^2)$ and $\mathcal{O}(n)$, respectively. This way we get all the Naive Bayes normalizing terms between $[0, n]$ in the given time, not just the nth of them. Notice that if the FFT approach could be used, time complexities of the first (explained in the Sect. 5.1) and the fourth steps would become $\mathcal{O}(n \log n)$. In this case the Miller formula in the fourth step is replaced with the algorithm mentioned in Section 4. Theorem 3 gives us actually a general framework for designing this kind of algorithms, as step 4 can be replaced with any exponentiation algorithm.

The method given in Theorem 3 is more efficient than the $\mathcal{O}(n^2 \log L)$-algorithm presented in [11]. This is easy to see, as the previous algorithm essentially performs in the fourth step at minimum $\log L$- power series multiplications instead of something which corresponds just one power series multiplication. In fact even when using the previous method, in some case it can be wise to compute series coefficients and not to require that all sub-results has to be normalizing terms. This way we can replace (18) just with normal convolution and achieve some more efficiency by omitting unnecessary multipliers present in the old formula. Furthermore, the old formula is applicable for values L greater than 2, but we can use normal convolution for values L greater than 1. In the fifth step we then convert wanted series coefficients into normalizing terms.

The new Miller method algorithm works perfectly fine with exact rational numbers. However our preliminary implementations show that in practice this is not necessarily the case with fixed precision floating point numbers and all formal power series: for some tested basic series, small errors in elementary operations tend to corrupt the normalizing terms very fast as n grows (because the algorithm uses iteratively previous values). Therefore with finite precision floating point numbers, using the previous, slower algorithm may be more advisable.

We have derived an efficient algorithm for computing the Naive Bayes normalizing term exactly. The computational complexity of computing the NML criterion for a Naive Bayes model is the same as for this algorithm, as the numerator of (1) is trivial to compute. Further information on computing the stochastic complexity for Naive Bayes models can be found in papers [11, 12].

6   Concluding Remarks

We presented an $\mathcal{O}(n^2)$ time algorithm for computing the normalizing term of the NML distribution exactly in the case of the Naive Bayes model. As the remaining term of the NML distribution is trivial to compute in this case, this result leads to a computationally efficient algorithm for computing the NML exactly for Naive Bayes models. We also defined a general framework for developing efficient algorithms for the NML computation in the Naive Bayes case and showed how the old $\mathcal{O}(n^2 \log L)$-algorithm can be seen as an special case of the framework, and how to make the algorithm more efficient.

We believe that it is not possible to do formal power series exponentiation in this case faster than $\mathcal{O}(n^2)$ without resorting to the Fast Fourier transform, which would easily lead to numerical problems, as discussed earlier. So unless the basic series for the Naive Bayes model reveals new hidden regularities with respect to exponentiation, our algorithm meets the lower limit of the time complexity for computing the NML exactly for Naive Bayes models.

Acknowledgements

The authors would like thank the anonymous reviewers for constructive comments. This work was supported in part by the Academy of Finland under the project Civi and by the Finnish Funding Agency for Technology and Innovation under the projects Kukot and PMMA. In addition, this work was supported in part by the IST Programme of the European Community, under the PASCAL Network of Excellence, IST-2002-506778. This publication only reflects the authors' views.

References

[1] Brent, R.P.: Multiple-precision zero-finding methods and the complexity of elementary function evaluation. In: Traub, J.F. (ed.) Analytic Computational Complexity, Academic Press, New York (1976)

[2] Flajolet, P., Sedgewick, R.: Analytic Combinatorics (in preparation)

[3] Grünwald, P.: The Minimum Description Length Principle. MIT Press, Cambridge (2007)

[4] Grünwald, P., Myung, J., Pitt, M. (eds.): Advances in Minimum Description Length: Theory and Applications. MIT Press, Cambridge (2005)

[5] Heckerman, D., Geiger, D., Chickering, D.M.: Learning Bayesian networks: The combination of knowledge and statistical data. Machine Learning 20(3), 197–243 (1995)

[6] Henrici, P.: Automatic computations with power series. Journal of the ACM 3(1), 11–15 (1956)

[7] Knuth, D.E.: The Art of Computer Programming, volume. 2: Seminumerical Algorithms, 3rd edn. Addison-Wesley, Reading (1998), ISBN: 0201896842

[8] Knuth, D.E., Pittel, B.: A recurrence related to trees. Proceedings of the American Mathematical Society 105(2), 335–349 (1989)

[9] Kontkanen, P., Myllymäki, P.: Analyzing the stochastic complexity via tree polynomials. Technical Report 2005-4, Helsinki Institute for Information Technology (HIIT) (2005)

[10] Kontkanen, P., Myllymäki, P.: A linear-time algorithm for computing the multinomial stochastic complexity. Information Processing Letters 103(6), 227–233 (2007)

[11] Kontkanen, P., Myllymäki, P., Buntine, W., Rissanen, J., Tirri, H.: An MDL framework for data clustering. In: Grünwald, P., Myung, I.J., Pitt, M. (eds.) Advances in Minimum Description Length: Theory and Applications, MIT Press, Cambridge (2006)

[12] Kontkanen, P., Wettig, H., Myllymäki, P.: NML computation algorithms for tree-structured multinomial Bayesian networks. EURASIP Journal on Bioinformatics and Systems Biology,(to appear)

[13] Nakos, G.: Expansions of powers of multivariate formal power series. Mathematica Journal 3(1), 45–47 (1993)

[14] Pearl, J.: Probabilistic Reasoning in Intelligent Systems: Networks of Plausible Inference. Morgan Kaufmann, San Mateo (1988)

[15] Rissanen, J.: Stochastic Complexity in Statistical Inquiry. World Scientific, New Jersey (1989)

[16] Rissanen, J.: Fisher information and stochastic complexity. IEEE Transactions on Information Theory 42(1), 40–47 (1996)

[17] Rissanen, J.: Information and Complexity in Statistical Modeling. Springer, Heidelberg (2007)

[18] Shtarkov, Y.M.: Universal sequential coding of single messages. Problems of Information Transmission 23, 3–17 (1987)

Unsupervised Spam Detection Based on String Alienness Measures

Kazuyuki Narisawa, Hideo Bannai, Kohei Hatano, and Masayuki Takeda

Department of Informatics, Kyushu University, Fukuoka 819-0395, Japan
{k-nari, bannai, hatano, takeda}@i.kyushu-u.ac.jp

Abstract. We propose an unsupervised method for detecting spam documents from a given set of documents, based on *equivalence relations* on strings. We give three measures for quantifying the *alienness* (i.e. how different they are from others) of substrings within the documents. A document is then classified as spam if it contains a substring that is in an equivalence class with a high degree of alienness. The proposed method is unsupervised, language independent, and scalable. Computational experiments conducted on data collected from Japanese web forums show that the method successfully discovers spams.

1 Introduction

Due to its remarkable development, the Web has become a major means of advertisement [1]. Not only normal websites, but CGM (Consumer Generated Media), such as Weblogs, forums and SNS, made and written by the casual user, is also exploited as an advertisement media. *Spam* messages, which are unsolicited, unwanted advertisement messages sent or posted by spammers, is becoming a huge issue on this media, because in general, any user can freely and easily post messages.

There exist various types of spam: webspam (spam in web sites), linkspam (spam used linkfarm), wikispam (spam in Wikis), splog (spam in Weblogs) [2], commentspam (spam in forums), spam mail (spam in email), and more recently, spim (spam over Instant Messaging) [3], and spit (spam over IP Telephony). These spams advertise their goods and websites, mislead users to access other websites, manipulate the PageRank [4] of their sites and so on. Not only do these messages interfere with the user trying to obtain useful information, but they can overload the servers which provide various services to the users. Hence, developing methods to detect such spams automatically is an important problem.

In this paper, we consider an unsupervised and language independent method for the detection of spam in document sets, based on the *alienness* of the substrings contained in each document. In order to effectively transmit their advertisement message to their potential customers (victims), spammers send many identical, or nearly identical spam messages. We assume that such redundancies in spam causes their substring frequencies distribution to deviate from that of other normal messages, and quantify this amount using several measures based

V. Corruble, M. Takeda, and E. Suzuki (Eds.): DS 2007, LNAI 4755, pp. 161–172, 2007.

on the substring equivalence relation defined in [5]. A document is then classified as spam if it contains a substring that is in an equivalence class with a high degree of alienness.

In Section 2, we review related work. In Section 3, we introduce some notations, as well as the substring amplification method [6] which is our previous unsupervised method for detecting spam. In Section 4, we describe our new spam detection method. We show results of computational experiments conducted on Japanese web forum postings in Section 5. In Section 6, we conclude the paper.

2   Related Work

There are roughly three strategies for detecting spam.

Link Analysis: This detects malicious link sets called linkfarms, by analyzing link structures [7,8,9]. It can detect linkspams with high accuracy, and does not depend on languages. However, it suffers from the drawback that it generally has a high computational cost, and that it can only be used for spam messages that contain links.

Machine Learning: There are various machine learning based filters such as Bayesian filters [10], which are fairly effective for spam mails using header information in addition to contents. However, such supervised methods must first be fed with a large amount of training message data marked as spam or nonspam, which may be costly to generate.

Statistical Analysis: This approach detects spams by considering various statistics of words or n-grams in documents [11]. However, word statistics requires word segmentation for languages that do not have word boundaries, such as Japanese or Chinese. Concerning n-gram statistics, a good n must somehow be chosen.

Our proposed method can be classified as a Statistical Analysis strategy, and uses the entire set of substrings instead of words or n-grams. Although the number of substrings in a document is quadratic in its length, our method runs in linear time by grouping the substrings into equivalence classes.

3   Preliminaries

Let Σ be a finite alphabet. An element of Σ^* is called a *string*. Strings x, y and z are said to be a *prefix*, *substring*, and *suffix* of the string $u = xyz$, respectively, and the string u is said to be a *superstring* of y. The length of a string u is denoted by $|u|$. The empty string is denoted by ε, that is, $|\varepsilon| = 0$. Let $\Sigma^+ = \Sigma^* - \{\varepsilon\}$. The i-th character of a string u is denoted by $u[i]$ for $1 \leq i \leq |u|$, and the substring of u that begins at position i and ends at position j is denoted by $u[i : j]$ for $1 \leq i \leq j \leq |u|$. For convenience, let $u[i : j] = \varepsilon$ for $j < i$. The set of substrings of a string w is denoted by $Sub(w)$, and let $Sub(S) = \bigcup_{w \in S} Sub(w)$ for a set S of strings. The elements of $Sub(S)$ are called *substrings* of S. Let $Sub_f(S)$ denote the set of substrings appearing f times in S. Let $|S|$ denote the cardinality of S.

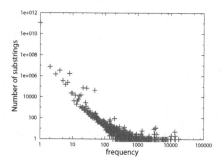

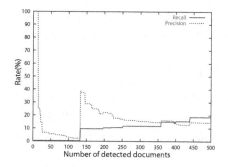

Fig. 1. An f-$|Sub_f(S)|$ plot. Outliers tend to correspond to spams.

Fig. 2. The performance of the Substring Amplification Method

3.1 Our Previous Method: Substring Amplification

We describe the Substring Amplification Method presented in [6], which is an unsupervised spam detection method. It is conceptually similar to the method in this paper in that it tries to detect spams by finding deviations in occurrence frequencies of substrings in documents. Unlike n-gram analysis and word analysis, it uses the frequency distribution of all substrings of the input documents. We assume that the Zipf's law [12,13] holds between the frequencies f and the number $|Sub_f(S)|$ of distinct substrings with frequency f, and look for outliers. Figure 1 is an example plot for the web data described in Section 4 (forum 4314). Looking more closely at this graph, outliers from the distribution with unexpectedly large $|Sub_f(S)|$ are observed to be due to substrings from spam documents. Figure 2 shows the performance of the Substring Amplification Method run on the same data.

The Substring Amplification Method finds suspicious frequencies f and outputs the set of substrings with frequency f. However, this set is comprised of substrings of essentially identical occurrences, as well as substrings that just happened to have the same frequency. Moreover, it was observed that usually, only a single group of substrings having essentially identical occurrences correspond to spam, and is responsible for large $|Sub_f(S)|$ values. In order to improve the accuracy of the Substring Amplification Method, we formalize this observation and propose a new method using the equivalence relation on substrings defined by [14]. In the next section, we describe our method in detail.

4 New Method

We consider the equivalence relation over substrings, introduced by Blumer et al. [14] based on their occurrences. Intuitively, each equivalence class gathers the substrings whose "occurrences" are the same.

We note that by using the suffix array data structure [15] together with its lcp array, we can enumerate in linear time, the equivalence classes, as well as their values for each of the measures that will be used in this paper as shown in [16]. The algorithm is a non-trivial extension of the algorithm of [17], but it is beyond the scope of this paper.

4.1 Equivalence Relations on Substrings

In this subsection, we give definitions of the equivalence relations of Blumer et al. [14], and then state some properties.

Definition 1. *Let S be a non-empty finite subset of Σ^+. For any x in $Sub(S)$, let*

$$BegPos_S(x) = \left\{ \langle w, j \rangle \mid w \in S, 0 \le j \le |w|, x = w[j+1 : j+|x|] \right\},$$
$$EndPos_S(x) = \left\{ \langle w, j \rangle \mid w \in S, 0 \le j \le |w|, x = w[j-|x|+1 : j] \right\}.$$

For any $x \notin Sub(S)$, let $BegPos_S(x) = EndPos_S(x) = \emptyset$. In this paper, we omit the set S, and write simply $BegPos$ and $EndPos$.

For example, if $S = \{\texttt{discover}, \texttt{cover}, \texttt{November}, \texttt{vertical}\}$, then the sets $BegPos$ and $EndPos$ for their substrings are as follows. $BegPos(\texttt{o}) = BegPos(\texttt{ov}) = BegPos(\texttt{ove}) = \{\langle \texttt{discover}, 4 \rangle, \langle \texttt{cover}, 1 \rangle, \langle \texttt{November}, 1 \rangle\}$, $BegPos(\texttt{c}) = \{\langle \texttt{discover}, 3 \rangle, \langle \texttt{cover}, 0 \rangle, \langle \texttt{vertical}, 5 \rangle\}$, $BegPos(\texttt{co}) = BegPos(\texttt{cov}) = BegPos(\texttt{cove}) = BegPos(\texttt{cover}) = \{\langle \texttt{discover}, 3 \rangle, \langle \texttt{cover}, 0 \rangle\}$, and $EndPos(\texttt{r}) = EndPos(\texttt{er}) = \{\langle \texttt{discover}, 8 \rangle, \langle \texttt{cover}, 5 \rangle, \langle \texttt{November}, 8 \rangle, \langle \texttt{vertical}, 3 \rangle\}$, $EndPos(\texttt{o}) = \{\langle \texttt{discover}, 5 \rangle, \langle \texttt{cover}, 2 \rangle, \langle \texttt{November}, 2 \rangle\}$, $EndPos(\texttt{over}) = EndPos(\texttt{cover}) = \{\langle \texttt{discover}, 5 \rangle, \langle \texttt{cover}, 2 \rangle\}$.

Definition 2. *Let x and y be arbitrary strings in Σ^*. The equivalence relations $\equiv_L$ and $\equiv_R$ are defined by*

$$x \equiv_L y \Leftrightarrow BegPos(x) = BegPos(y),$$
$$x \equiv_R y \Leftrightarrow EndPos(x) = EndPos(y).$$

The equivalence class of a string x in Σ^* with respect to $\equiv_L$ and $\equiv_R$ is denoted by $[x]_{\equiv_L}$ and $[x]_{\equiv_R}$, respectively.

For example, if $S = \{\texttt{discover}, \texttt{cover}, \texttt{November}, \texttt{vertical}\}$, then $[\varepsilon]_{\equiv_L} = [\varepsilon]_{\equiv_R} = \{\varepsilon\}$, $[\texttt{o}]_{\equiv_L} = [\texttt{ov}]_{\equiv_L} = [\texttt{ove}]_{\equiv_L} = \{\texttt{o}, \texttt{ov}, \texttt{ove}\}$, $[\texttt{c}]_{\equiv_L} = \{\texttt{c}\}$, $[\texttt{co}]_{\equiv_L} = [\texttt{cov}]_{\equiv_L} = [\texttt{cove}]_{\equiv_L} = [\texttt{cover}]_{\equiv_L} = \{\texttt{co}, \texttt{cov}, \texttt{cove}, \texttt{cover}\}$, and $[\texttt{r}]_{\equiv_R} = [\texttt{er}]_{\equiv_R} = \{\texttt{r}, \texttt{er}\}$, $[\texttt{o}]_{\equiv_R} = \{\texttt{o}\}$, $[\texttt{over}]_{\equiv_R} = [\texttt{cover}]_{\equiv_R} = \{\texttt{over}, \texttt{cover}\}$.

Definition 3. *For any string x in $Sub(S)$, let $\overrightarrow{x}$ and $\overleftarrow{x}$ denote the unique longest members of $[x]_{\equiv_L}$ and $[x]_{\equiv_R}$, respectively.*

For example, if $S = \{\texttt{discover}, \texttt{cover}, \texttt{November}, \texttt{vertical}\}$, then $\overrightarrow{\varepsilon} = \overleftarrow{\varepsilon} = \varepsilon$, $\overrightarrow{\texttt{o}} = \overrightarrow{\texttt{ov}} = \overrightarrow{\texttt{ove}} = \texttt{ove}$, $\overrightarrow{\texttt{c}} = \texttt{c}$, $\overrightarrow{\texttt{co}} = \overrightarrow{\texttt{cov}} = \overrightarrow{\texttt{cove}} = \overrightarrow{\texttt{cover}} = \texttt{cover}$, and $\overrightarrow{\texttt{r}} = \overrightarrow{\texttt{er}} = \texttt{er}$, $\overleftarrow{\texttt{o}} = \texttt{o}$, $\overleftarrow{\texttt{over}} = \overleftarrow{\texttt{cover}} = \texttt{cover}$.

Definition 4. *For any string x in $Sub(S)$, let $\overleftrightarrow{x}$ be the string $\alpha x \beta$ such that α and β are the strings satisfying $\overleftarrow{x} = x\beta$ and $\overrightarrow{x} = \alpha x$.*

For example, if $S = \{\texttt{discover}, \texttt{cover}, \texttt{November}, \texttt{vertical}\}$, then $\overleftrightarrow{\varepsilon} = \varepsilon$, $\overleftrightarrow{\texttt{o}}$ $= \overleftrightarrow{\texttt{ov}} = \overleftrightarrow{\texttt{ove}} = \texttt{ove}$, $\overleftrightarrow{\texttt{c}} = \texttt{c}$, $\overleftrightarrow{\texttt{r}} = \overleftrightarrow{\texttt{er}} = \texttt{er}$, and $\overleftrightarrow{\texttt{co}} = \overleftrightarrow{\texttt{cov}} = \overleftrightarrow{\texttt{cove}} = \overleftrightarrow{\texttt{over}}$ $= \overleftrightarrow{\texttt{cover}} = \texttt{cover}$.

Intuitively, $\overleftrightarrow{x} = \alpha x \beta$ means that:

- Every time x occurs in S, it is preceded by α and followed by β.
- Strings α and β are as long as possible.

Definition 5. *Strings x and y are said to be equivalent on S if and only if:*

1. $x \notin Sub(S)$ and $y \notin Sub(S)$, or

2. $x, y \in Sub(S)$ and $\overleftrightarrow{x} = \overleftrightarrow{y}$.

This equivalence relation is denoted by $\equiv$. The equivalence class of a string x in $Sub(S)$ with respect to $\equiv$ is denoted by $[x]_\equiv$.

For example, if $S = \{\texttt{discover}, \texttt{cover}, \texttt{November}, \texttt{vertical}\}$, then the strings in $Sub(S)$ are divided into the equivalence classes: $\{\varepsilon\}$, $\{\texttt{o}, \texttt{ov}, \texttt{ove}\}$, $\{\texttt{c}\}$, $\{\texttt{r}, \texttt{er}\}$, and $\{\texttt{co}, \texttt{cov}, \texttt{cove}, \texttt{over}, \texttt{cover}\}$.

A string x in $Sub(S)$ is said to be *prime* if $\overleftrightarrow{x} = x$. Let $Prime(S)$ denote the set of prime substring of S, that is, $Prime(S) = \{\overleftrightarrow{x} \mid x \in Sub(S)\}$. For example, if $S = \{\texttt{discover}, \texttt{cover}, \texttt{November}, \texttt{vertical}\}$, then $Prime(S) = \{\texttt{c}, \texttt{i}, \texttt{er}, \texttt{ve},$ $\texttt{ove}, \texttt{ver}, \texttt{cover}, \texttt{discover}, \texttt{November}, \texttt{vertical}\}$.

We regard each prime string x as the *representative* of the equivalence classes $[x]_\equiv$.

For any x, y in Σ^*, we write $x \preceq y$ if x is a substring of y. For any x in $Prime(S)$, let $Minimal(x)$ denote the set of minimal elements of $[x]_\equiv$, that is, $Minimal(x) = \{y \in [x]_\equiv \mid z \preceq y \text{ and } z \in [x]_\equiv \text{ imply } z = y\}$. Let $Maximin(x)$ denote the maximum length of strings in $Minimal(x)$.

For example, if $S = \{\texttt{discover}, \texttt{cover}, \texttt{November}, \texttt{vertical}\}$, then $Minimal(\texttt{cover}) = \{\texttt{co}, \texttt{over}\}$ and $Maximin(x) = |\texttt{over}| = 4$.

The following lemma states that any equivalence class that contains a substring of S is represented by its representative and its minimal elements.

Lemma 1 ([5]). *For any x in $Prime(S)$, let $y_1, \ldots, y_k$ be the elements of $Minimal(x)$. Then, $[x]_\equiv = Pincer(y_1, x) \cup \cdots \cup Pincer(y_k, x)$, where $Pincer(y_i, x)$ is the set of strings z with $y \preceq z \preceq x$.*

It can also be shown that such representations of all equivalence classes of substrings in S require only linear space [16].

4.2   Measures with Equivalence Classes

In this subsection, we give three measures for quantifying the alienness of equivalence classes. We use a data set obtained from the Web in order to evaluate these measures. The data set consists of postings from the YahooJapanFinance[1]

---

[1] http://quote.yahoo.co.jp/

forum. This forum is surveyed by the forum administrator, and postings are manually deleted if they are judged to be spam. Therefore, we can obtain spam and non-spam document examples by gathering the postings of a given forum over a certain period of time. We regard the postings which have been deleted as spams, and the writings not deleted as nonspams. This data set contains 1087 postings including 226 spam posts and 861 nonspam posts.

We will not regard strings only occurring once as spam, and only consider the equivalence classes whose elements appear at least twice in the documents.

Length: In general, spams are different from natural sentences in that they tend to be lengthy and appear more frequently. Hence, we first consider the length of the representative of an equivalence class as a measure for spam detection:

$$measure_{Length}(x) = |\overleftrightarrow{x}|$$

There seems to exist a power law between the length of the representative of an equivalence class and the number of equivalence classes with the length (see Figure 3-(1)). In this plot, "Spam" denotes that all equivalence classes with that Length measure are substrings of spam documents only. "NonSpam" denotes that all equivalence classes with that Length measure are substrings of nonspam documents only. "Spam and NonSpam" denotes otherwise, and equivalence classes with that Length measure are included in both spam and nonspam documents. In this plot, we can see that spam equivalence classes are distributed on high length parts. We can say that an equivalence class has a high degree of probability for being spam if the length of the representative of the equivalence class is long.

Figure 3-(4) shows the ROC curve for the Length measure. The x-axis is the negative ratio for each equivalence class, that is

$$negative\ ratio = \frac{\#\ of\ detected\ nonspam\ documents}{\#\ of\ detected\ documents},$$

and y-axis is the positive ratio for each equivalence class, that is

$$positive\ ratio = \frac{\#\ of\ detected\ spam\ documents}{\#\ of\ detected\ documents}.$$

The graph is drawn by considering all possible Length measure values as a threshold, and plotting the above values by classifying documents which contain the equivalence class whose length is longer than the threshold value as spam, and nonspam otherwise. As can be seen in Figure 3-(4), the length of the representative of an equivalence class seems to be an effective measure for spam detection.

Size: Next, we consider the size of an equivalence class as a measure for spam detection:

$$measure_{Size}(x) = |[x]_{\equiv}|$$

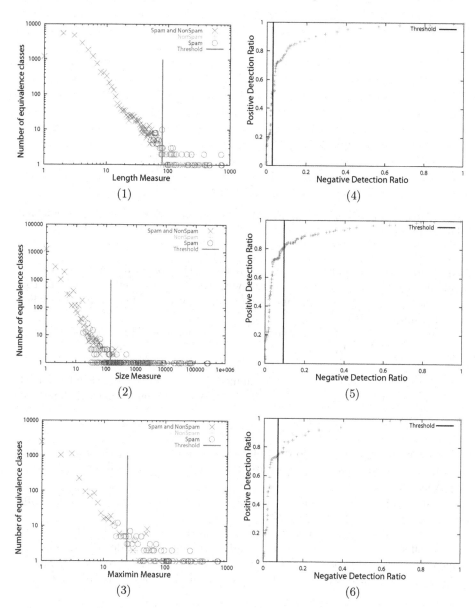

Fig. 3. Measure value distribution of equivalence classes for (1)Length, (2)Size and (3)Maximin. The threshold line is obtained by our method described in Section 4.3. (4)Length, (5)Size and (6)Maximin measure ROC curve for discriminating between spam and nonspam documents.

Although there is a strong relationship between the length of the representative of equivalence classes and the size of equivalence classes (see Figure 4), there are equivalence classes whose representative is long, but whose size is not large.

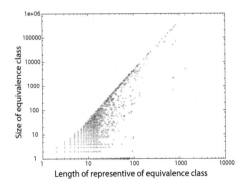

Fig. 4. Relation between the length of the representative of the equivalence class and the size of the equivalence class

There also seems to exist a power law between the size of an equivalence class and the number of equivalence classes with the size (see Figure 3-(2)). In this plot, there are more spam specific points than that of the length measure when the measurement is high.

Figure 3-(5) is the ROC curve for the Size measure. This curve shows that the Size measure is also effective.

Maximin: The Maximin measure is defined as the difference between the length of the representative and the length of the longest minimal element of the equivalence class:

$$measure_{Maximin}(x) = | \overleftrightarrow{x} | - Maximin(x)$$

This measure represents a lower bound on how much a representative string can be shortened (by removing its prefix and/or suffix) and still be in the same equivalence class.

Figure 3-(3) shows the relation between the $measure_{Maximin}(x)$ and the number of the equivalence classes with the measure.

Figure 3-(6) shows the ROC curve for this measure, showing that this measure is also effective for spam detection. In addition, the area under the ROC curve given the negative ratio is less than 0.3 is larger than those of other measures (see Table 1). Hence, we expect that this measure has lower false-positive error.

4.3 Determining the Threshold Value

Our method takes an unlabeled document set as input. We simply use a threshold to determine whether an equivalence class with some measure value is "alien" or not. Documents are then judged as spam if it includes an "alien" substring equivalence class. Below, we describe an unsupervised method for determining the threshold value.

Looking at each of the measure plots, Figure 3-(1), (2), (3), we can see that there are roughly two parts. One is the spam specific part on right area, and

Table 1. The area under the ROC curve for each measure

negative ratio	≤ 0.3	≤ 0.5	≤ 1.0 (all)
Length	0.282	0.485	0.913
Size	0.243	0.469	0.915
Maximin	0.318	0.559	0.919

the rest is on the left area. We propose heuristics to find a point that separates the spam part from the nonspam part, and regard the point as the threshold. More precisely, we model each of the two parts using a linear model, and we look for the point of separation where the two linear models best explains the data points.

Let $S = ((x_1, y_1), \ldots, (x_n, y_n))$ be a sequence of n points, where $x_1 \leq x_2 \leq \cdots \leq x_n$. For $1 \leq k < n$, let S_1^k be the sequence of the first k points in S, and let S_2^k be the remaining sequence in S. We choose the k^*-th point that minimizes the sum of the least square errors in the left and the right sides of points. That is, we choose

$$k^* = \operatorname*{argmin}_{1 \leq k < n}(LSE(S_1^k) + LSE(S_2^k)),$$

where

$$LSE(S') = \min_{a,b} \sum_{i=1,\ldots,n'} (y_i' - ax_i' - b)^2$$

for $S' = ((x_1', y_1'), \ldots, (x_{n'}', y_{n'}'))$. It is well known that

$$LSE(S') = \sum_{i=1,\ldots,n'} (y_i' - \hat{a}x_i' - \hat{b})^2,$$

where

$$\hat{a} = \frac{n' \sum_{i=1}^{n'} x_i' y_i' - \sum_{i=1}^{n'} x_i \sum_{i=1}^{n'} y_i'}{n' \sum_{i=1}^{n'} x_i'^2 - (\sum_{i=1}^{n'} x_i')^2},$$

and

$$\hat{b} = \frac{n \sum_{i=1}^{n'} x_i'^2 \sum_{i=1}^{n'} y_i' - \sum_{i=1}^{n'} x_i' y_i' \sum_{i=1}^{n'} x_i'}{n' \sum_{i=1}^{n'} x_i'^2 - (\sum_{i=1}^{n'} x_i')^2}.$$

5 Computational Experiments

We detect spams in four forums of Yahoo Japan Finance [2], which are collected in the same way as the data used for evaluating the measures in Section 4. If spam is posted in these forums, the spam is manually deleted by the forum administrator. We regard the deleted posts as spam.

---

[2] http://quote.yahoo.co.jp

Table 2. The results of each measure for YahooJapanFinance forum data

forum	# of spam	# of nonspam	measure	Recall	Precision	F-score	nonspam(%)
4314	291	1424	Length	0.45	0.57	0.50	**2.67**
			Size	**0.89**	0.60	0.72	7.51
			Maximin	0.80	**0.80**	**0.80**	6.18
4974	331	1315	Length	0.39	**0.77**	0.52	**4.33**
			Size	**0.63**	0.69	0.66	11.56
			Maximin	0.60	0.75	**0.67**	9.81
6830	317	1613	Length	0.40	**0.72**	0.52	**3.47**
			Size	**0.74**	0.57	0.64	17.73
			Maximin	0.69	0.69	**0.69**	13.95
8473	264	1597	Length	0.57	**0.76**	0.65	**4.32**
			Size	**0.72**	0.63	0.67	8.14
			Maximin	0.67	0.69	**0.68**	9.39

We selected and collected data from four forums: 4314[3], 4974[4], 6830[5], 8473[6]. We detect spams from these four data sets using the three measures proposed in Section 4. The results are shown in Table 2, where the values Recall, Precision, F-score in the table are defined as follows:

$$Recall = \frac{\# \text{ of detected spam documents}}{\# \text{ of spam documents}}$$

$$Precision = \frac{\# \text{ of detected spam documents}}{\# \text{ of detected documents}}$$

$$F\text{-}score = \frac{2 * Recall * Precision}{Recall + Precision}$$

As shown in Table 2, the measure Maximin has the highest F-scores, which vary from 68% to 80%, among three measures for all of the data sets. On the other hand, as for recall values, the measure Size outperforms others, while its F-scores remain close to those of Maximin. We also evaluate the three measures when the inputs are nonspam documents only. The percentages of documents judged as spams in this setting are summarized in the column "nonspam" of Table 2. Note that the value nonspam should be 0% ideally. The measure Length has the lowest nonspam values over all the data sets. In summary, none of the three measures completely outperforms the others.

We examined the false positive strings (nonspams which our method judges to be spams) and the false negative strings (spams which our method judges to be nonspams). In the former case, most of the false positive strings were difficult to distinguish from spams by their contents, even for human. In the latter case, many false negative strings contained abusive language, which are not spams in

---

[3] http://messages.yahoo.co.jp/?action=q&board=4314
[4] http://messages.yahoo.co.jp/?action=q&board=4974
[5] http://messages.yahoo.co.jp/?action=q&board=6830
[6] http://messages.yahoo.co.jp/?action=q&board=8473

general but are deleted by the forum administrator. So, in a practical point of view, our method detects spams well.

6   Conclusions and Discussions

We proposed a new, unsupervised, language independent method for spam detection based on the alienness of strings. We provided three alienness measures, namely Length, Size and Maximin. We observed that spam documents seem to give rise to equivalence classes with measurements larger than nonspam documents, and developed a method for finding a threshold value for discriminating between spam and nonspam. To the best of our knowledge, our method is the only method that is truly unsupervised, and requires no tuning of parameters.

Since our method depends on the redundant information contained in spam documents, most existing benchmark datasets that remove this could not be used. The data used in our experiments was the only data readily available to us that consisted of manually annotated positive and negative examples taken from an unprocessed, "natural" distribution of documents. It should also be noted that Japanese is a very popular language on the Web, and according to [18], it is the most frequently used language for blogs. We are currently collecting more data in order to further evaluate and improve our method.

References

1. Jansen, B.J.: Adversarial information retrieval aspects of sponsored search. In: Proceedings of the First International Workshop on Adversarial Information Retrieval on the Web (AIRWeb) (2006), http://airweb.cse.lehigh.edu/2006/
2. CNETNEWS.COM: Tempted by blogs, spam becomes 'splog' (2005),
   http://news.com.com/2100-1032_3-5903409.html
3. CNETNEWS.COM: Spim, splog on the rise (2006),
   http://news.com.com/2100-7349_3-6091123.html
4. Page, L., Brin, S., Motwani, R., Winograd, T.: The PageRank Citation Ranking: Bringing Order to the Web. Stanford Digital Library Working Paper (1998)
5. Takeda, M., Matsumoto, T., Fukuda, T., Nanri, I.: Discovering characteristic expressions in literary works. Theoretical Computer Science 292(2), 525–546 (2003)
6. Narisawa, K., Yamada, Y., Ikeda, D., Takeda, M.: Detecting blog spams using the vocabulary size of all substrings in their copies. In: Proceedings of the 3rd Annual Workshop on Weblogging Ecosystem(at the 15th World Wide Web Conference) (2006)
7. Benczú, A.A., Csalogány, K., Sarlós, T.: Link-based similarity search to fight web spam. In: Proceedings of the First International Workshop on Adversarial Information Retrieval on the Web (AIRWeb) (2006),
   http://airweb.cse.lehigh.edu/2006/
8. Becchetti, L., Castillo, C., Donato, D., Leonardi, S., Baeza-Yates, R.: Link-based characterization and detection of web spam. In: Proceedings of the First International Workshop on Adversarial Information Retrieval on the Web (AIRWeb) (2006), http://airweb.cse.lehigh.edu/2006/

9. da Costa Carvalho, A.L., Chirita, P.A., de Moura, E.S., Calado, P., Nejdl, W.: Site level noise removal for search engines. In: Proceedings of the 15th World Wide Web Conference (2006)
10. Sahami, M., Dumais, S., Heckerman, D., Horvitz, E.: A Bayesian approach to filtering junk e-mail. In:AAAI Workshop on Learning for Text Categorization (1998), ftp://ftp.research.microsoft.com/pub/ejh/junkfilter.pdf
11. Yoshida, K., Adachi, F., Washio, T., Motoda, H., Homma, T., Nakashima, A., Fujikawa, H., Yamazaki, K.: Density-based spam detector. In: KDD, pp. 486–493 (2004)
12. Zipf, G.K.: Human Behavior and the Principle of Least Effort. Addison-Wesley, Reading (1949)
13. Zipf, G.K.: The Psycho-Biology of Language: An Introduction to Dynamic Philology. Houghton Mifflin (1935)
14. Blumer, A., Blumer, J., Haussler, D., Ehrenfeucht, A., Chen, M.T., Seiferas, J.I.: The smallest automaton recognizing the subwords of a text. Theoretical Computer Science 40, 31–55 (1985)
15. Manber, U., Myers, G.: Suffix arrays: a new method for on-line string searches. SIAM Journal on Computing 22(5), 935–948 (1993)
16. Narisawa, K., Inenaga, S., Bannai, H., Takeda, M.: Efficient computation of substring equivalence classes with suffix arrays. In: Ma, B., Zhang, K. (eds.) CPM 2007. LNCS, vol. 4580, pp. 340–351. Springer, Heidelberg (2007)
17. Kasai, T., Lee, G., Arimura, H., Arikawa, S., Park, K.: Linear-time longest-common-prefix computation in suffix arrays and its applications. In: Amir, A., Landau, G.M. (eds.) CPM 2001. LNCS, vol. 2089, pp. 181–192. Springer, Heidelberg (2001)
18. Technorati: The state of the live web (2007), http://technorati.com/weblog/2007/04/328.html

A Consequence Finding Approach for Full Clausal Abduction

Oliver Ray[1] and Katsumi Inoue[2]

[1] University of Bristol, United Kingdom
oray@cs.bris.ac.uk
[2] National Institute of Informatics, Japan
ki@nii.ac.jp

Abstract. Abductive inference has long been associated with the logic of scientific discovery and automated abduction is now being used in real scientific tasks. But few methods can exploit the full potential of clausal logic and abduce non-ground explanations with indefinite answers. This paper shows how the consequence finding method of Skip Ordered Linear (SOL) resolution can overcome the limitations of existing systems by proposing a method that is sound and complete for finding minimal abductive solutions under a variety of pruning mechanisms. Its utility is shown with an example based on metabolic network modelling.

1 Introduction

The importance of abductive inference in scientific discovery was recognised by C.S. Peirce over a century ago. He saw abduction as a form of reasoning, from known effects to possible causes, which underlies the process of hypotheses formation [6]. Recent advances in Artificial Intelligence have shown that (some aspects of) abductive logic can be automated and exploited in real domains like diagnosis [10, 4] and bioinformatics [21, 17]. The benefit of logic-based methods is their ability to use prior background knowledge and return meaningful testable hypotheses. For example, in [12], abduction was used to infer functional genomic hypotheses that were experimentally tested by a 'Robot Scientist'. The abductive reasoning used in these tasks all follow a simple but very useful logical pattern: given a theory T, a goal G, and a set of possible assumptions A, find a minimal consistent subset of A which can be added to T in order to ensure that (some instance of) G is satisfied.

In general, T, G and the elements of A are arbitrary first order formulae and an abductive solution consists of two parts: a set of formulae Δ, called an *explanation*, stating which assumptions in A should be added to T; and a set of substitutions Θ, called an *answer*, stating which instances of the free variables in G are satisfied. By utilising standard normalisation techniques, it suffices to consider the case when T is clausal theory, G is a conjunction of literals, and A is a set of literals called *abducibles*. But, existing approaches for abduction typically impose additional restrictions that rule out the possibility of unrestricted abductive reasoning in full clausal logic. In particular, most insist

V. Corruble, M. Takeda, and E. Suzuki (Eds.): DS 2007, LNAI 4755, pp. 173–184, 2007.
© Springer-Verlag Berlin Heidelberg 2007

that either the goal or abducibles should be ground, and most only allow Horn clauses to appear in the theory.

The restrictions imposed by existing abductive systems reflect certain issues arising in non-Horn logic. Many use input resolution methods, which are not complete in general and cannot handle indefinite answers [20]. Moreover, non-ground abducibles complicate the relationship between variables in the query and explanation. Finally, the possibility of non-ground explanations and indefinite answers makes it harder to compare alternative solutions and define appropriate notions of minimality. A great deal of recent work in abduction has focussed on the framework of logic programming [11], which uses Negation as Failure (NAF) to avoid the need for non-Horn reasoning. But these approaches inherit a number of more serious logical concerns regarding the semantics of NAF, the difficulty of floundering, and the complexity of reasoning in a non-monotonic formalism where standard pruning strategies are inapplicable.

This paper gives a sound and complete method for finding minimal abductive answers in full clausal (classical) logic that overcomes the syntactic restrictions of existing systems. By treating abduction as a form of conditional query answering, we propose a semantics for abduction and minimality that correctly handles indefinite answers and non-ground explanations. We present a proof procedure, based on the clausal consequence finding approach of Skip Ordered Linear (SOL) resolution [7], which includes a rule for 'skipping' or assuming literals during a proof. To do this, we introduce a method for lifting a previous limitation on the language bias for specifying skipped literals that would otherwise prevent us from using arbitrary sets of abducibles. Previous results ensure our approach is sound and complete under a combination of efficient pruning strategies [8, 9].

The paper is structured as follows. Section 2 gives the relevant notation and background material on abduction, consequence finding and SOL. Section 3 formalises the semantics of minimal abduction in full clausal logic. Section 4 presents our abductive procedure. Section 5 compares our approach with related work. The paper concludes with a summary and directions for future work.

2   Background

2.1   Notation and Terminology

This paper assumes a first-order language $\mathcal{L}$ (without equality) containing the connectives $\wedge$, $\vee$, $\neg$, $\leftarrow$, $\rightarrow$, $\leftrightarrow$, logical constants $\top$, $\bot$, and quantifiers $\forall$, $\exists$. It also assumes standard first-order entailment $\models$ and equivalence $\equiv$ relations whose semantics is purely classical (and not restricted to Herbrand models). The term 'iff' abbreviates 'if and only if' and the term 'wrt' abbreviates 'with respect to'. A literal L is either an atom A or its negation $\neg A$. The complement of L, denoted $\overline{L}$, is defined as $\neg A$ (resp. A) if $L = A$ (resp. $\neg A$). A maximally general literal is one, e.g., $p(X, Y, Z)$, whose arguments are distinct variables. If S is a set of literals then $\bigwedge_{L \in S} L$ (resp. $\bigvee_{L \in S} L$) denotes the conjunction (resp. disjunction) of the literals in S and is defined as $\top$ (resp. $\bot$) when S is empty. A clause C is a disjunction of literals $L_1 \vee \ldots \vee L_m$ that, for convenience, will often be identified with the set $\{L_i | 1 \leq i \leq m\}$ of its disjuncts. As usual, any free variables are implicitly universally quantified at the front of the clause. A clause is Horn iff it has at most one positive literal and is full otherwise. A clause is a tautology iff it contains a literal L and its complement $\overline{L}$. The empty clause is denoted $\square$. A variable binding is an expression of the form X/t where X is a variable and t is a term. In this case, we say that X is bound to t. A substitution is a set of variable bindings for distinct variables. The application of a substitution σ to an expression E is written $E\sigma$ and denotes the expression obtained from E by (simultaneously) replacing each free variable X by the corresponding term t for each binding X/t in σ. For any (set of) expressions E, let $Inst(E)$ denote the set of all instances of (members of) E. A clause D subsumes C, written $D \geq C$, iff D has no more literals than C and there is a substitution θ such that $D\theta \subseteq C$. Moreover D properly subsumes C, written $D > C$, iff $D \geq C$ and $C \not\geq D$. A theory T is a conjunction of clauses $C_1 \wedge \ldots \wedge C_n$ that, for convenience, will often be identified with the set $\{C_i | 1 \leq i \leq n\}$ of its conjuncts. A theory is Horn iff all of its clauses are Horn and is full otherwise. For any theory T, let $\mu(T)$ denote the theory obtained from T by removing all clauses properly subsumed by another clause in T, and let $Th(T)$ denote the set of all clauses logically entailed by T. A goal G is a conjunction of literals. An underscore '_' is sometimes used to denote an anonymous variable.

2.2   Abduction

Abduction is an established AI technique for hypothetical reasoning [11]. In essence, abduction computes the conditions under which a goal G follows from a given theory T. Implicitly, G is understood as an existentially quantified query asking "is some instance of G satisfied in an extension of T"? If so, then the abductive computation should succeed, returning a set of assumptions Δ which must be added to T and a substitution σ stating which instances of the free variables in G are entailed. The assumptions in Δ are usually restricted to the instances of a set A of literals, called abducibles. Intuitively, these are literals whose truth is not specified in the intended domain: e.g., potential faults in a diagnosis task and possible actions in a planning problem. Each explanation Δ is implicitly understood as an existentially quantified conjunction that should be consistent with T and should be minimal in the sense of not containing atoms which could be removed to leave a smaller explanation Δ'. These notions are typically formalised in the literature as shown in Definition 1 below.[1]

Definition 1 (Naive Abduction). *Let T be a theory, G be a goal, and A be a set of literals. A (naive) abductive solution (for G wrt T and A) consists of a set of literals $\Delta \subseteq Inst(A)$ (called an explanation) and a substitution σ for G (called an answer) such that (i) $T \wedge \Delta \models G\sigma$ and (ii) $T \wedge \Delta \not\models \bot$. The explanation Δ is minimal iff there is no other explanation Δ' such that $\Delta' \subset \Delta$.*

---

[1] Analogous characterisations are obtained for logic program formalisms by replacing classical entailment with some appropriate completion or preferred model semantics. For convenience, we assume integrity constraints IC are included in the theory T.

2.3 Consequence Finding

Consequence finding is a general reasoning technique for computing the logical theorems entailed by a set of axioms [7]. Since the deductive closure of a logical theory may be infinite, it is generally infeasible or undesirable to compute all possible consequences. It is often more useful to consider a refinement of this task where it is required to compute the clausal consequences of a given theory that satisfy a given vocabulary and are minimal with respect to subsumption. If the vocabulary is specified by a form of language bias called a production field, the resulting consequences are known as characteristic clauses [7]. A production field P is a pair $\langle L, Cond \rangle$, where L is a set of literals and $Cond$ is a certain condition to be satisfied. When $Cond$ is not specified, P is just written as $\langle L \rangle$. A clause C is said to belong to $P = \langle L, Cond \rangle$ iff (i) every literal in C is an instance of a literal in L and (ii) C satisfies $Cond$. A production field P is stable iff, for any two clauses C and D such that $C \geq D$, the clause D belongs to P only if C belongs to P. If $\mathcal{L}_P$ denotes the set of clauses that belong to P, then the characteristic clauses of a theory T with respect to P are the set of clauses $Carc(T, P) = \mu(Th(T) \cap \mathcal{L}_P)$. The importance of these notions lies in the fact that many reasoning tasks, such as abduction, induction, and theorem proving, can be reduced to the computation of characteristic clauses [7].

2.4 SOL Resolution

Characteristic clauses can be computed a procedure called SOL resolution [7], which can be seen as extending the Model Elimination [14] calculus with a rule for 'skipping' literals. Intuitively, skipped literals represent assumptions that are needed for a proof to succeed. As explained in [7], this feature is needed to ensure the completeness of SOL for consequence finding. SOL deductions are defined using the notion of a structured clause, which is a pair $\langle A, B \rangle$ consisting of two clauses A and B, where the latter may contain so-called framed literals of the form $\boxed{L}$ denoting previously resolved upon literals. SOL deductions are formalised in Definition 2, which is recalled from [8].

Definition 2 (SOL Deduction). *Let T be a theory, S be a clause, and P be a production field. An SOL deduction of S from T and P (of length n) is a sequence of structured clauses $D_0, \dots, D_n$ satisfying rules 1-6 below.*

1. *$D_0 = \langle \Box, C \rangle$ for some clause $C \in T$.*
2. *$D_n = \langle S, \Box \rangle$.*
3. *For each $D_i = \langle A_i, B_i \rangle$ clause $A_i \cup B_i$ is not a tautology.*
4. *For each $D_i = \langle A_i, B_i \rangle$ clause B_i is not subsumed by any B_j with the empty substitution, where $D_j = \langle A_j, B_j \rangle$ is a previous structured clause with $j < i$.*
5. *For each $D_i = \langle A_i, B_i \rangle$ clause A_i belongs to P.*
6. *$D_{i+1} = \langle A_{i+1}, B_{i+1} \rangle$ is obtained from $D_i = \langle A_i, B_i \rangle$ as follows:*
    (a) let L be the left-most literal of B_i. Then A_{i+1} and R_{i+1} are obtained by applying one of the rules:

    i. *Skip: if $A_i \cup \{L\}$ belongs to P, then $A_{i+1} = A_i \cup \{L\}$ and R_{i+1} is the clause obtained by removing L from B_i.*

    ii. *Resolve: if there is a clause E_i from $T \cup \{C\}$ such that $\neg K \in E_i$ and L and K have a most general unifier θ, then $A_{i+1} = A_i\theta$ and R_{i+1} is the clause obtained by concatenating $E_i\theta$ and $B_i\theta$, framing $L\theta$, and removing $\neg K\theta$.*

    iii. *Factoring: A_i or B_i contains an unframed literal K such that L and K have a most general unifier θ, then $A_{i+1} = A_i\theta$ and R_{i+1} is obtained from $B_i\theta$ by deleting $L\theta$.*

    iv. *Reduction: B_i contains a framed literal $\boxed{\neg K}$, and L and K have a most general unifier θ, then $A_{i+1} = A_i\theta$ and R_{i+1} is obtained from $B_i\theta$ by deleting $L\theta$.*

  (b) *B_{i+1} is obtained from R_{i+1} by deleting every framed literal not preceded by an unframed literal in the remainder (truncation).*

It has been shown $Carc(T, P)$ is equal to the set of subsume-minimal clauses derivable by SOL from T and P [7]. Moreover, an efficient implementation of SOL has been developed [16] that uses several pruning mechanisms [9] to further constrain SOL deductions. These include mandatory rules native to SOL, such as merge and regularity for skipped literals, as well as some generic theorem proving methods, such as order preserving reduction, lemma matching, and local failure caching. It has been shown that these pruning strategies do not compromise the completeness of SOL if the production field P is stable [9].

3 Full Clausal Abduction

This section shows full clausal abduction can be efficiently realised by SOL resolution. Section 3.1 presents a semantics for abduction that correctly handles non-ground abducibles and disjunctive answers. It then shows how computing minimal abductive solutions can be reduced to the computation of characteristic clauses. Section 3.2 explains why unstable production fields arise in abductive problems and how this leads to the incompleteness of SOL. It then shows how to overcome this limitation by means of an efficient program transformation.

3.1 Semantics

The naive formulation of abduction in Definition 1 is only satisfactory when G and Δ are ground. Even though the implicit existential quantification on G and Δ correctly specifies if the goal should succeed, it does not adequately constrain the returned solutions. First, just like the answers returned by standard Prolog systems [13], every instance of the computed answer $G\sigma$ should be entailed by T (and Δ). Otherwise it suffices to simply take $\sigma = \emptyset$. Second, as in other non-Horn extensions of Prolog [20], it is necessary to return a *set* of answer substitutions to account for indefinite answers. For example, the goal $q(X)$ should succeed from the theory $q(a) \vee q(b)$ with answer $\{\{X/a\}, \{X/b\}\}$, indicating that X must be bound to either a or b in any model of T.

All of these issues are resolved in Definition 3 below, which gives a general clausal formalisation of abduction.[2] In effect, this formulation treats abduction as a type of conditional query answering [3, 8] with the explanation Δ denoting the assumptions under which the answer σ is valid. Intuitively, in order for $\langle \Delta, \Theta \rangle$ to solve G wrt T, condition (i) states that the conjunction of assumed literals L in Δ must imply the disjunction of answers obtained by applying each substitution σ in Θ to G. All variables are universally quantified at the front of the implication. Analogously, the consistency condition (ii) can be viewed as saying that Δ should not be an explanation for the contradictory goal $G = \bot$.

Definition 3 (General Abduction). *Let T be a theory, G be a goal, and A be a set of (abducible) literals. An abductive solution (for G wrt T and A) is a pair $\langle \Delta, \Theta \rangle$ consisting of a set of literals $\Delta \subseteq Inst(A)$ (called an explanation) and a set of substitutions Θ for G (called an answer) such that*

$$\text{(i) } T \models \forall \left(\bigwedge_{L \in \Delta} L \to \bigvee_{\sigma \in \Theta} G\sigma \right) \text{ and (ii) } T \not\models \forall \left(\bigwedge_{L \in \Delta} L \to \bot \right)$$

Example 1. Let T, G and A be as defined below. Theory T says there is a metabolic pathway from X to Z if there is a reaction from X to Y and a pathway from Y to Z; or if there is a reaction from X to Z. It also says there is a reaction from a to either b or c; and a reaction from either b or c to d; but no reaction from c to b. Goal G asks "from which metabolites U is there a pathway to d"? Abducibles A allow all instances of the predicate *reaction* to be assumed.

$$T = \left\{ \begin{array}{l} \neg reaction(X,Y) \vee \neg pathway(Y,Z) \vee pathway(X,Z) \\ \neg reaction(Y,Z) \vee pathway(Y,Z) \\ reaction(a,b) \vee reaction(a,c) \\ reaction(b,d) \vee reaction(c,d) \\ \neg reaction(c,b) \end{array} \right\}$$

$$G = \{ pathway(U,d) \}$$

$$A = \{ reaction(V,W) \}$$

From the 4th clause in T, there is a pathway from b or c to d: i.e., $\langle \Delta = \emptyset, \Theta = \{\{U/b\}, \{U/c\}\} \rangle$ is a solution. Moreover, there is a pathway from b to d if we assume a reaction from b to c: i.e., $\langle \Delta = \{reaction(b,c)\}, \Theta = \{\{U/b\}\} \rangle$ is also a solution. But, $\langle \Delta = \{reaction(c,b)\}, \Theta = \{\{U/c\}\} \rangle$ is not valid as it contradicts the 5th clause in T. However, there is a pathway from a to d given reactions from both b and c to d: i.e., $\langle \Delta = \{reaction(b,c), reaction(b,d)\}, \Theta = \{\{U/a\}\} \rangle$ is a solution. Similarly, we have $\langle \Delta = \{reaction(a,b), reaction(a,c)\}, \Theta = \{\{U/a\}\} \rangle$. In general, there is a pathway from any X to d if there is a reaction from X to d: i.e., we have $\langle \Delta = \{reaction(X,d)\}, \Theta = \{\{U/X\}\} \rangle$. Additionally, there is a pathway X or b to d if there is a reaction from X to c: i.e., we have $\langle \Delta = \{reaction(X,c)\}, \Theta = \{\{U/X\}, \{U/b\}\} \rangle$. In this case, Definition 3 is satisfied since (i) $T \models \forall X \left(reaction(X,c) \to pathway(X,d) \vee pathway(b,d) \right)$, and (ii) $T \not\models \forall X \left(reaction(X,c) \to \bot \right)$ as it is consistent to let $X = a$, for example.

---

[2] We assume that all instances of the abducible literals in A can be added to Δ and we assume Θ only binds variables in G. If G is ground, we can always set $\Theta = \{\emptyset\}$.

While the view of abduction presented above imposes the correct semantics on Δ and Θ, this pathway example shows that, in general, a large number of solutions are possible. It is therefore desirable to introduce a minimality criterion for eliminating redundant hypotheses. A simple subsumption test, formalised in Definition 5, can be used based on the notion of a solution clause, given in Definition 4.[3] This clause contains the complement of each abducible in Δ and an answer literal [5], with predicate ϕ, for each substitution in Θ. An abductive solution $\langle \Delta, \Theta \rangle$ can then be defined as minimal iff its solution clause is not strictly subsumed by the solution clause of another solution $\langle \Delta', \Theta' \rangle$.

Definition 4 (Solution Clause). *Given a theory T, goal G, and abducibles A, let $\Delta \subseteq Inst(A)$ be a set of literals, and let Θ be a set of substitutions for G. Let $X_1, \ldots, X_n$ be the variables in G, and let ϕ be a predicate not appearing in G, T or A. A solution clause for $\langle \Delta, \Theta \rangle$, denoted $Soln(\Delta, \Theta)$, is a clause of the form:*

$$Soln(\Delta, \Theta) = \bigvee_{L \in \Delta} \overline{L} \vee \bigvee_{\sigma \in \Theta} \phi(X_1, \ldots, X_n)\sigma.$$

Definition 5 (Minimal Solution). *Let T be a theory, G be goal, and A be a set of literals. A minimal solution for G wrt T and A is an abductive solution $\langle \Delta, \Theta \rangle$ (for G wrt T and A) for which there is no other abductive answer $\langle \Delta', \Theta' \rangle$ (for G wrt T and A) such that $Soln(\Delta', \Theta') > Soln(\Delta, \Theta)$.*

Example 2. All of the solutions mentioned in Example 1 are minimal except $\langle \Delta = \{reaction(a, b), reaction(a, c)\}, \Theta = \{\{U/a\}\} \rangle$ — as its solution clause $Soln(\Delta, \Theta) = \neg reaction(a, b) \vee \neg reaction(a, c) \vee \phi(a)$ is subsumed by the solution clause $Soln(\Delta', \Theta') = \neg reaction(X, b) \vee \neg reaction(X, c) \vee \phi(X)$ of the minimal solution $\langle \Delta' = \{reaction(X_1, b), reaction(X_1, c)\}, \Theta' = \{\{U/X_1\}\} \rangle$. There are infinitely many other minimal solutions of the form $\langle \Delta_n = \{reaction(X_1, X_2), \ldots, reaction(X_{n-1}, X_n), reaction(X_n, b), reaction(X_n, c)\}, \Theta_n = \{\{U/X_1\}\} \rangle$.

Proposition 1 shows that the solution clause $Soln(\Delta, \Theta)$ of any abductive solution $\langle \Delta, \Theta \rangle$ can be deduced from the theory T augmented with a so-called answer clause. As formalised in Definition 6, the answer clause $Ansr(G)$ of a goal G is composed of the complement of each literal in G and an answer literal, with predicate ϕ and arguments corresponding to the variables in G. Thus, in Example 1, $Ansr(G)$ is the clause $\neg pathway(U, d) \vee \phi(U)$.

Definition 6 (Answer Clause). *Let G be a goal with the variables $X_1, \ldots, X_n$. An answer clause for G, denoted $Ansr(G)$, is a clause of the form:*

$$Ansr(G) = \bigvee_{L \in G} \overline{L} \vee \phi(X_1, \ldots, X_n).$$

Proposition 1. *Let T be a theory, G be a goal, and A be a set of literals. Let $\Delta \subseteq Inst(A)$ be a set of literals, and Θ be a set of substitutions for G. Then $\langle \Delta, \Theta \rangle$ is an abductive solution for G wrt T and A iff*

(i) $T \wedge Ansr(G) \models Soln(\Delta, \Theta)$ and (ii) $T \wedge Ansr(G) \not\models Soln(\Delta, \emptyset)$

---

[3] We assume the variables $X_1, \ldots, X_n$ are always written in some standard order and we assume that the predicate ϕ does not appear in T, G or A.

3.2   Proof Procedure

Proposition 1 shows that the computation of abductive solutions can be reduced
to a consequence finding problem by adding an answer clause to the theory
and searching for any solution clauses that are entailed (i). This can be done
using a production field containing the answer literal and the complement of
every abducible. In this case, subsume minimal consequences, i.e., characteristic
clauses, correspond to minimal solutions and thus can be computed by SOL
resolution. Moreover, Proposition 1 also shows that inconsistent explanations
can be avoided by simply rejecting solution clauses with no answer literals (ii).

Example 3. Let T, G and A be as defined in Example 1 and apply the method
suggested above. First form the answer clause $Ansr(G) = \neg pathway(U, d) \vee \phi(U)$
and the production field $P = \langle \{\neg reaction(\_, \_), \phi(\_)\} \rangle$. Then search for all SOL
deductions from $T \wedge Ansr(G)$ and P. One such deduction is shown below. For
clarity, each step is annotated with its index and predicates are abbreviated.
The deduction starts with $Ansr(G)$. Referring to Definition 2, clauses 2,3,4,5
& 10 are obtained by resolution; 7,9 & 12 are obtained by truncation; 8 is
obtained by reduction; while 6 & 13 are obtained by skipping. The derived clause
$\neg r(c, d) \vee \neg r(b, d) \vee \phi(a)$ gives the abductive solution $\Delta = \{r(c, d), r(b, d)\}$ and
$\Theta = \{\{U/a\}\}$. Similar deductions yield the other minimal solutions. Note that
the inconsistent explanation $\Delta = \{reaction(c, b)\}$ is easily recognised by the lack
of answer literals in its solution clause.

$^1\langle \Box, \underline{\neg p(U, d)} \vee \phi(U) \rangle$

$^2\langle \Box, \underline{\neg r(U, Y)} \vee \neg p(Y, d) \vee \boxed{\neg p(U, d)} \vee \phi(U) \rangle$

$^3\langle \Box, \underline{r(a, c)} \vee \boxed{\neg r(a, b)} \vee \neg p(b, d) \vee \boxed{\neg p(a, d)} \vee \phi(a) \rangle$

$^4\langle \Box, \underline{\neg p(c, Z)} \vee p(a, Z) \vee \boxed{r(a, c)} \vee \boxed{\neg r(a, b)} \vee \neg p(b, d) \vee \boxed{\neg p(a, d)} \vee \phi(a) \rangle$

$^5\langle \Box, \underline{\neg r(c, Z)} \vee \boxed{\neg p(c, Z)} \vee p(a, Z) \vee \boxed{r(a, c)} \vee \boxed{\neg r(a, b)} \vee \neg p(b, d) \vee \boxed{\neg p(a, d)} \vee \phi(a) \rangle$

$^6\langle \neg r(c, Z), \boxed{\neg p(c, Z)} \vee p(a, Z) \vee \boxed{r(a, c)} \vee \boxed{\neg r(a, b)} \vee \neg p(b, d) \vee \boxed{\neg p(a, d)} \vee \phi(a) \rangle$

$^7\langle \neg r(c, Z), \underline{p(a, Z)} \vee \boxed{r(a, c)} \vee \boxed{\neg r(a, b)} \vee \neg p(b, d) \vee \boxed{\neg p(a, d)} \vee \phi(a) \rangle$

$^8\langle \neg r(c, d), \boxed{r(a, c)} \vee \boxed{\neg r(a, b)} \vee \neg p(b, d) \vee \boxed{\neg p(a, d)} \vee \phi(a) \rangle$

$^9\langle \neg r(c, d), \underline{\neg p(b, d)} \vee \boxed{\neg p(a, d)} \vee \phi(a) \rangle$

$^{10}\langle \neg r(c, d), \underline{\neg r(b, d)} \vee \boxed{\neg p(b, d)} \vee \boxed{\neg p(a, d)} \vee \phi(a) \rangle$

$^{11}\langle \neg r(c, d) \vee \neg r(b, d), \boxed{\neg p(b, d)} \vee \boxed{\neg p(a, d)} \vee \phi(a) \rangle$

$^{12}\langle \neg r(c, d) \vee \neg r(b, d), \underline{\phi(a)} \rangle$

$^{13}\langle \neg r(c, d) \vee \neg r(b, d) \vee \phi(a), \Box \rangle$

Example 3 shows SOL can be used for abductive query answering. But, a potential flaw is that the completeness of SOL is only true for stable production fields. This holds only if all abducibles are maximally general literals, which, in practice, will not be the case. For instance, suppose that we only wish to assume the existence of reactions that produce d. The most efficient way to do this is by using the abducibles $A' = \{\, reaction(V, d) \,\}$. But the new production field $P' = \langle \{\neg reaction(\_, d), \phi(\_)\} \rangle$ is not stable as $\neg reaction(X, Y)$ does not belong to P' even though it subsumes a clause $\neg reaction(X, d)$ that does.

However, using an unstable production field results in a loss of completeness. In particular the explanation $\Delta = \{reaction(c, d), reaction(b, d)\}$ is no longer computed even though it falls within the given language bias. Under this new production field P', the SOL deduction shown in Example 3 is invalid, as it becomes impossible to apply the skip operation after step 5. While it is possible, in this particular case, to construct a deduction by postponing the skip until after the reduction binding Z to d between steps 7 and 8, such re-orderings are not considered as they would dramatically increase the search space.

Fortunately, completeness is restored by the procedure in Definition 9 which uses a transformation formalised in Definition 7. For each literal L in A, a bridge clause is added to T which contains L and the negation of an atom $p_L(X_1, \ldots, X_n)$, which becomes a new abducible in place of L. The predicate p_L is a new predicate symbol which, just like an answer literal, represents any bindings to the variables X_i in L.[4] This transformation ensures all abducibles are maximally general and that abductive solutions to the transformed and original problems are isomorphic by simply propagating the bindings from p_L back to L.

Definition 7 (Bridge Theory). *Let A be a set of literals. A bridge theory for A, denoted $Brdg(A)$ is a theory of the form*

$$Brdg(A) = \bigwedge_{L \in A} \neg p_L(X_1, \ldots, X_n) \vee L$$

where $X_1, \ldots, X_n$ are the variables in L, and where p_L is a new predicate symbol.

Definition 8 (SOL Procedure). *Let T be a theory, P be a production field, and k be a positive integer. $SOL(T, P, k)$ denotes the set of clauses S for which there exists an SOL deduction (from T and P) of length $n \leq k$.*

Definition 9 (Abduce Procedure). *Given a theory T, goal G, literals A, and integer k, let $Abduce(T, G, A, k)$ denote the set S computed as follows:*

1. *Let B be the theory $T \cup Ansr(G) \cup Brdg(A)$*
2. *Let N be the answer literal $\phi(X_1, \ldots, X_n)$ appearing in $Ansr(G)$*
3. *Let M be the set of bridge literals $\neg p_L(X_1, \ldots, X_m)$ appearing in $Brdg(A)$*
4. *Let P be the production field $\langle \{N\} \cup M \rangle$*
5. *Let Q be the set of clauses obtained from $SOL(B, P, k)$ by replacing each bridge literal of the form $p_L(t_1, \ldots, t_m)$ with the literal $L\{X_1/t_1, \ldots, X_m/t_m\}$*
6. *Let S be the set of solutions $\langle \Delta, \Theta \rangle$ such that $Ansr(\Delta, \Theta) \in \mu Q$ and $\Theta \neq \emptyset$*

---

[4] A new (predicate) proposition p_L is used for each (non-) ground abducible L.

The soundness and completeness result in Proposition 2 follows immediately from analogous results on the soundness and completeness of SOL for minimal conditional answers under stable production fields [8]. Note that minimality is ensured by applying the μ operator in step 6 *after* the computed bindings are transferred back to the original abducibles in step 5. Note also how consistency is ensured by removing in step 6 any explanations with an empty answer. Finally note that the transformation needed to ensure a stable production field is very efficient, resulting in the addition of only one theory clause for each abducible and incurring just one additional resolution step per abduced literal.

Proposition 2. *Let T be a theory, G be a goal, and A be a set of literals. $\langle \Delta, \Theta \rangle$ is a minimal solution of G wrt T and A iff there exists an integer k such that $\langle \Delta, \Theta \rangle \in Abduce(T, G, A, n)$ for all $n \geq k$.*

This formulation reflects the fact that, in practice, some bound k must be imposed on the depth of SOL derivations. A minimal solution is a solution that is computed at some depth k and not later subsumed by another solution of higher depth. While the existence of k is guaranteed, it follows from previous work [8] that its value may be undecidable. This is due to the undecidability of (consistency and) minimality checking. In practice, the procedure can only return those solutions that are minimal with respect to the depth bound k. But this is true of any abductive procedure and is often good enough in practice.

Returning to the previous example, the abducibles $A' = \{ reaction(V, d) \}$ result in the bridge theory $\{\neg p_r(V) \vee reaction(V, d)\}$ and production field $\langle \{\neg p_r(\_), \phi(\_)\} \rangle$. The literal $\neg reaction(c, Z)$, selected at step 5 of the earlier derivation now resolves with the bridge clause to leave the goal $\neg p_r(Z)$ which can be skipped. The SOL derivation, which proceeds just as before, results in the clause $\neg p_r(c) \vee \neg p_r(b) \vee \phi(a)$, which is subsequently replaced by the clause $\neg reaction(c, d) \vee \neg reaction(b, d) \vee \phi(a)$. Thus, our transformation overcomes the restriction of SOL to stable production fields and ensures the completeness of our abductive procedure for computing minimal solutions.

4   Related Work

Several other procedures have been proposed for abductive reasoning. Poole [18] describes a method for compiling full clausal abductive problems into Prolog. This approach handles indefinite answers, but only returns ground explanations. Kakas et al. [11] review several procedures for abductive logic programs. Some of these include constraint solvers, which allow the computation of non-ground explanations, but none of them compute indefinite answers. Mayer and Pirri [15] propose a general first-order abductive proof procedure, based on the tableaux and sequent calculi, which uses dynamic Skolemisation and anti-Skolemisation to avoid a conversion to clausal form. Their approach is complete for finding entailment-minimal explanations, but does not return answer substitutions and does not exploit the many resolution pruning strategies used by SOL.

Abductive reasoning is closely related to conditional query answering, which could, in principle, be used to implement abduction [2]. Baumgartner et al. [1] give a calculus for conditional answer computation, but this is incomplete for finding minimal solutions [8]. Demolombe [3] gives a calculus for computing minimal conditional answers, but this does not support function symbols or indefinite answers. Iwanuma and Inoue [8] show SOL resolution is complete for minimal answer computation, but only under the assumption of a stable production field and without addressing the issue of consistency.

In effect, the approach introduced in this paper overcomes the restriction of SOL resolution to stable production fields and can potentially be built directly into SOL so that other applications can benefit from this generalisation. We have also developed a slightly extended transformation that allows abducibles to be annotated with goals in order to impose typing and validity constraints on the terms in abduced literals, e.g., $A = \{reaction(U, V) : node(U), node(V)\}$. These goal literals are simply added to the bridge clause along with the abducible and thereby afford a finer degree of control over the abductive bias.

5   Conclusions

This paper presented a proof procedure and semantics for full clausal abduction that caters for indefinite answers and non-ground abducibles. Viewing abduction as a form of conditional query answering, it showed how the consequence finding approach of SOL resolution can be used to overcome the syntactic restrictions imposed by other systems. In so doing, it revealed the significance of a stability restriction underlying all previous work on SOL, and gave an efficient program transformation to overcome this assumption. The approach was illustrated on a small example motivated by metabolic pathway analysis which also showed the potential utility of logical abduction in Discovery Science. In this example, non-Horn clauses were used to represent incomplete knowledge and non-ground solutions were used to suggest possible refinements of the initial knowledge by inferring the presence of missing nodes and arcs. While our approach is not yet mature enough to tackle real scientific discovery tasks, we intend to validate our methodology more fully on a more realistic model of biochemical networks. We are also developing an extension of our method to allow the induction of general laws from examples and a background theory. This extended approach, called Hybrid Abductive Inductive Learning [19], uses abductive explanations returned by (a simplified version of) our procedure to seed the formation of a ground unit theory that is subsequently generalised by an inductive search procedure. In this way we eventually hope to utilise our abductive method in the process scientific knowledge discovery.

Acknowledgements

This work is supported by the Research Councils UK (RCUK) and the Japan Society for the Promotion of Science (JSPS).

184   O. Ray and K. Inoue

References

[1] Baumgartner, P., Furbach, U., Stolzenburg, F.: Computing Answers with Model Elimination. Artif. Intel. 90(1-2), 135–176 (1997)

[2] Burhans, D., Shapiro, S.: Abduction and Question Answering. In: Proc. the IJCAI'01 Workshop on Abductive Reasoning, pp. 905–908 (2001)

[3] Demolombe, R.: A Strategy for the Computation of Conditional Answers. In: Proc. 10th Europ. Conf. on Artif. Intel. pp. 134–138 (1992)

[4] Gartner, J., Swift, T., Tien, A., Damásio, C., Pereira, L.: Psychiatric Diagnosis from the Viewpoint of Computational Logic. In: Palamidessi, C., Moniz Pereira, L., Lloyd, J.W., Dahl, V., Furbach, U., Kerber, M., Lau, K.-K., Sagiv, Y., Stuckey, P.J. (eds.) CL 2000. LNCS (LNAI), vol. 1861, pp. 1362–1376. Springer, Heidelberg (2000)

[5] Green, C.: Theorem-Proving by Resolution as a Basis for Question-Answering Systems. Mach. Intel. 4, 183–205 (1969)

[6] Hartshorne, C., Weiss, P., Burks, A.W. (eds.): Collected papers of Charles Sanders Peirce. Harvard University Press (1931–1958)

[7] Inoue, K.: Linear resolution for Consequence Finding. Artif. Intel. 56(2-3), 301–353 (1992)

[8] Iwanuma, K., Inoue, K.: Minimal Answer Computation and SOL. In: Flesca, S., Greco, S., Leone, N., Ianni, G. (eds.) JELIA 2002. LNCS (LNAI), vol. 2424, pp. 245–257. Springer, Heidelberg (2002)

[9] Iwanuma, K., Inoue, K., Satoh, K.: Completeness of Pruning Methods for Consequence Finding Procedure SOL. In: Proc. 3rd Int. Workshop on First-Order Theorem Proving, pp. 89–100 (2000)

[10] Josephson, J., Josephson, S. (eds.): Abductive Inference: Computation, Philosophy, Technology. Cambridge University Press, Cambridge (1994)

[11] Kakas, A., Kowalski, R., Toni, F.: Abductive Logic Programming. Journal of Logic and Computation 2(6), 719–770 (1992)

[12] King, R., Whelan, K., Jones, F., Reiser, P., Bryant, C., Muggleton, S., Kell, D., Oliver, S.: Functional Genomic Hypothesis Generation and Experimentation by a Robot Scientist. Nature 427, 247–252 (2004)

[13] Lloyd, J.: Foundations of Logic Programming. Springer, Heidelberg (1987)

[14] Loveland, D.: Automated Theorem Proving: A Logical Basis, North Holland (1978)

[15] Mayer, M., Pirri, F.: First-Order Abduction via Tableau and Sequent Calculi. Bulletin of the IPGL 1(1), 99–117 (1993)

[16] Nabeshima, H., Iwanuma, K., Inoue, K.: SOLAR: A Consequence Finding System for Advanced Reasoning. In: Mayer, M.C., Pirri, F. (eds.) TABLEAUX 2003. LNCS, vol. 2796, pp. 257–263. Springer, Heidelberg (2003)

[17] Papatheodorou, I., Kakas, A., Sergot, M.: Inference of Gene Relations from Microarray Data by Abduction. In: Baral, C., Greco, G., Leone, N., Terracina, G. (eds.) LPNMR 2005. LNCS (LNAI), vol. 3662, pp. 389–393. Springer, Heidelberg (2005)

[18] Poole, D.: Compiling a Default Reasoning System into Prolog. New Generation Computing 9(1), 3–38 (1991)

[19] Ray, O., Inoue, K.: Mode-Directed Inverse Entailment for Full Clausal Theories. In: Blockeel, H., Shavlik, J., Tadepalli, P. (eds.) ILP 2007. LNCS, Springer, Heidelberg (2007)

[20] Stickel, M.: A Prolog Technology Theorem Prover: A New Exposition and Implementation in Prolog. Theoretical Computer Science 104(1), 109–128 (1992)

[21] Zupan, B., Demsar, J., Bratko, I., Juvan, P., Halter, J., Kuspa, A., Shaulsky, G.: GenePath: a System for Automated Construction of Genetic Networks from Mutant Data. Bioinformatics 19(3), 383–389 (2003)

Literature-Based Discovery by an Enhanced Information Retrieval Model

Kazuhiro Seki[1] and Javed Mostafa[2]

[1] Kobe University, Hyogo 657-8501, Japan
seki@cs.kobe-u.ac.jp
[2] Indiana University, Bloomington, IN 47405, USA

Abstract. The massive, ever-growing literature in life science makes it increasingly difficult for individuals to grasp all the information relevant to their interests. Since even experts' knowledge is likely to be incomplete, important findings or associations among key concepts may remain unnoticed in the flood of information. This paper brings and extends a formal model from information retrieval in order to discover those implicit, hidden knowledge. Focusing on the biomedical domain, specifically, gene-disease associations, this paper demonstrates that our proposed model can identify not-yet-reported genetic associations and that the model can be enhanced by existing domain ontology.

Keywords: Hypothesis discovery, Text data mining, Inference network, Implicit association, Gene Ontology.

1 Introduction

With the advance of computer technologies, the amount of scientific knowledge is rapidly growing beyond the pace we could digest. For example, Medline[1]— the most comprehensive bibliographic database in life science—currently indexes over 17 million articles and the number keeps increasing by 1,500–3,000 per day. Given the substantial volume of the publications, it is virtually impossible to deal with the information without the aid of intelligent information processing techniques, such as information retrieval (IR), information extraction (IE), and text data mining (TDM).

In contrast to IR and IE, which find information explicitly stated in documents, TDM aims to discover heretofore unknown knowledge through an automatic analysis on textual data [1]. A pioneering work in TDM, also known as literature-based discovery, was conducted by Swanson in the 1980's. He argued that there were two premises logically connected but the connection had been unnoticed due to overwhelming publications and/or over-specialization. To demonstrate the validity of the basic idea, he manually analyzed numbers of articles and identified logical connections implying a hypothesis that fish oil was effective for clinical treatment of Raynaud's disease [2]. The hypothesis was later supported by experimental evidence.

---

[1] http://www.ncbi.nlm.nih.gov/sites/entrez?db=pubmed

V. Corruble, M. Takeda, and E. Suzuki (Eds.): DS 2007, LNAI 4755, pp. 185–196, 2007.
© Springer-Verlag Berlin Heidelberg 2007

This study is motivated by Swanson's work and attempts to advance the research in literature-based discovery. Specifically, we target implicit associations between genes and hereditary diseases as a test bed. Gene-disease associations are the links between genetic variants and diseases to which the genetic variants influence the susceptibility. For example, BRCA1 is a human gene encoding a protein that suppresses tumor formation. A mutation of this gene increases a risk of breast cancer. Identification of these genetic associations has tremendous importance for prevention, prediction, and treatment of diseases. To this end, we develop a discovery framework by extending the models and techniques developed for IR. Furthermore, we propose the use of domain ontologies for more robust predictions. To demonstrate the effectiveness of the proposed framework, we conduct various evaluative experiments on realistic benchmark data.

2   Related Work

Over two decades, Swanson has argued the potential use of a literature to discover new knowledge that has *implicitly* existed for years but has not been noticed by anybody. His discovery framework is based on a syllogism; i.e., two premises, "A causes B" and "B causes C," suggest a potential association, "A causes C," where A and C do not have a known, explicit relation. Such an association can be seen as a hypothesis testable for verification to produce new knowledge, such as the above-mentioned association between Raynaud's disease and fish oil. For this particular example, Swanson manually inspected two sets of articles concerning Raynaud's disease and fish oil and identified premises that "Raynaud's disease is characterized by high platelet affregability, high blood viscosity, and vasoconstriction" and that "dietary fish oil reduces blood lipids, platelet affregability, blood viscosity, and vascular reactivity," which together suggest a potential benefit of fish oil for Raynaud's patients.

Based on the groundwork, Swanson himself and other researchers developed computer programs to aid hypothesis discovery. The following briefly introduces some of the representative studies.

Weeber et al. [3] implemented a system, called DAD-system, taking advantage of a natural language processing tool. The key feature of their system is that the Unified Medical Language System (UMLS) Metathesaurus[2] was incorporated for knowledge representation and pruning. While the previous work focused on words or phrases appearing in Medline records for reasoning, DAD-system maps them to a set of concepts defined in the UMLS Metathesaurus using MetaMap [4]. An advantage of using MetaMap is that it can automatically collapse different wordforms (e.g., inflections) and synonyms to a single concept. In addition, using *semantic types* (e.g., "Body location or region") under which each Metathesaurus concept is categorized, irrelevant concepts can be excluded from further exploration if particular semantic types of interest are given. This

---

[2] UMLS is an NLM's project to develop and distribute multi-purpose, electronic knowledge sources and its associated lexical programs.

filtering step can drastically reduce the number of potential associations, enabling more focused knowledge discovery.

Srinivasan [5] developed another system, called Manjal, for literature-based discovery. A key difference of Manjal from the previous work is that it solely relies on MeSH terms assigned to Medline records, disregarding all textual information, so as to study the utility of MeSH terms for hypothesis discovery. Manjal conducts a Medline search for a given concept and extracts MeSH terms from the retrieved articles. Then, according to a predefined mapping, the MeSH terms are grouped into their corresponding UMLS semantic types. Similar to DAD-system, the subsequent processes can be applied only to the concepts under particular semantic types of interest, so as to narrow down the potential associations. Manjal uses the semantic types also for grouping resultant concepts to help its user browse system output. With Manjal, Srinivasan demonstrated that most of the hypotheses Swanson had proposed were successfully replicated.

Despite the prolonged efforts partly mentioned above, however, the research in literature-based discovery can be seen to be at an early stage of development in terms of the models, approaches, and evaluation methodologies. Most of the previous work was largely heuristic without a formal model and their evaluation was limited only on a small number of Swanson's hypotheses. In contrast, this study adapts a formal IR model to literature-based discovery and conducts quantitative experiments based on real-world data.

3   Our Proposed Approach

Focusing on gene-disease associations, we extend a formal IR model, specifically, the inference network [6] for this related but different problem targeting unknown associations. This section details the proposed model and how to estimate the probabilities involved in the model.

In this study, we assume a disease name and known causative genes, if any, as system input. In addition, a target region in the human genome may be specified to limit search space. Given such input, we attempt to predict an unknown causative gene and produce a ranked list of candidate genes.

3.1   An Inference Network for Gene-Disease Associations

In the original IR model, a user query and documents are represented as nodes in a network and are connected via intermediate nodes representing keywords that compose the query and documents. To adapt the model to represent gene-disease associations, we treat disease as query and genes as documents and use two types of intermediate nodes: gene functions and phenotypes which characterize genes and disease, respectively (Fig. 1). An advantage of using this particular IR model is that it is essentially capable of incorporating multiple layers of intermediate nodes. Other popular IR models, such as the vector space models, are not easily applicable as documents and queries are represented by a single layer of the same vocabularies.

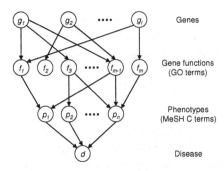

Fig. 1. Inference network for representing gene-disease associations

The network consists of four types of nodes: genes (g), gene functions (f) represented by Gene Ontology (GO) terms,[3] phenotypes (p) represented by MeSH C terms,[4] and disease (d). Each gene node g represents a gene and corresponds to the event that the gene is found in the search for the causative genes underlying d. Each gene function node f represents a function of gene products. There are directed arcs from genes to functions, representing that instantiating a gene increases the belief in its functions. Likewise, each phenotype node p represents a phenotype of d and corresponds to the event that the phenotype is observed. The belief in p is dependent on the belief in f's since phenotypes are (partly) determined by gene functions. Finally, observing certain phenotypes increases the belief in d. As described in the followings, the associations between genes and gene functions ($g \to f$) are obtained from an existing database, Entrez Gene,[5] whereas both the associations between gene functions and phenotypes ($f \to p$) and the associations between phenotypes and disease ($p \to d$) are derived from the biomedical literature.

Given the inference network model, disease-causing genes can be predicted based on the probability defined below.

$$P(d|G) = \sum_i \sum_j P(d|\boldsymbol{p_i}) \times P(\boldsymbol{p_i}|\boldsymbol{f_j}) \times P(\boldsymbol{f_j}|G) \qquad (1)$$

Eq. (1) quantifies how much a given set of genes, $G \subseteq \{g_1, g_2, \cdots, g_l\}$, increases the belief in the development of disease d. In the equation, $\boldsymbol{p_i}$ (or $\boldsymbol{f_j}$) is defined as a vector of random variables with i-th (or j-th) element being positive (1) and all others negative (0). By applying Bayes' theorem and some independence assumptions discussed later, we derive

$$P(d|G) \propto \sum_i \sum_j \left(\frac{P(p_i|d)}{P(\bar{p}_i|d)} \times \frac{P(f_j|p_i)P(\bar{f}_j|\bar{p}_i)}{P(\bar{f}_j|p_i)P(f_j|\bar{p}_i)} \times F(p_i) \times F(f_j) \times P(f_j|G) \right) \qquad (2)$$

[3] http://www.geneontology.org
[4] http://www.nlm.nih.gov/mesh
[5] http://www.ncbi.nlm.nih.gov/entrez/query.fcgi?DB=gene

where

$$F(p_i) = \prod_{h=1}^{m} \frac{P(\bar{f}_h|p_i)}{P(\bar{f}_h|\bar{p}_i)}, \ \ F(f_j) = \prod_{k=1}^{n} \frac{P(\bar{f}_j)P(f_j|\bar{p}_k)}{P(f_j)P(\bar{f}_j|\bar{p}_k)} \tag{3}$$

The first factor of the right-hand side of Eq. (2) represents the interaction between disease d and phenotype p_i, and the second factor represents the interaction between p_i and gene function f_j, which is equivalent to the odds ratio of $P(f_j|p_i)$ and $P(f_j|\bar{p}_i)$. The third and fourth factors are functions of p_i and f_j, respectively, representing their main effects. The last factor takes either 0 or 1, indicating whether f_j is a function of any gene in G under consideration.

The inference network described above assumes independence among phenotypes, among gene functions, and among genes. We assert that, however, the effects of such associations are minimal in the proposed model. Although there may be strong associations among phenotypes (e.g., phenotype p_x is often observed with phenotype p_y), the model does not intend to capture those associations. That is, phenotypes are attributes of the disease in question and we only need to know those that are frequently observed with disease d so as to characterize d. The same applies to gene functions; they are only attributes of the genes to be examined and are simply used as features to represent the genes under consideration.

3.2   Probability Estimation

Conditional Probability $P(p|d)$. This probability can be seen as a degree of belief that phenotype p is observed when disease d has developed. To estimate the probability, we take advantage of the literature data. Briefly, given a disease name d, a Medline search is conducted to retrieve articles relevant to d and, within the retrieved articles, we identify phenotypes (MeSH C terms) strongly associated with the disease based on chi-square statistics. Given disease d and phenotype p, the chi-square statistic is computed as

$$\chi^2(d,p) = \frac{N(n_{11} \cdot n_{22} - n_{21} \cdot n_{12})^2}{(n_{11}+n_{21})(n_{12}+n_{22})(n_{11}+n_{12})(n_{21}+n_{22})} \tag{4}$$

where N is the total number of articles in Medline, n_{11} is the number of articles assigned p and included in the retrieved set (denoted as R), n_{22} is the number of articles not assigned p and not included in R, n_{21} is the number of articles not assigned p and included in R, and n_{12} is the number of articles assigned p and not in R. The resulting chi-square statistics are normalized by the maximum to treat them as probabilities $P(p|d)$.

Incidentally, for the reason described later, the Medline search is limited to the articles published up to 6/30/2003.

Conditional Probability $P(f|p)$. This probability indicates the degree of belief that gene function f underlies phenotype p. For probability estimation, we adopt the framework similar to the one proposed by Perez-Iratxeta et al. [7].

Unlike them, however, this study focuses on the use of textual data and domain ontologies and investigate their effects for literature-based discovery.

We estimate these probabilities by exploiting the Medline records that were used as the source of the Genetic Association Database (GAD) [8] entries. GAD is a manually-curated archive of human genetic studies, containing pairs of gene and disease that are known to have causative relations. Since each of those Medline records reports experimental evidence indicating causative genetic associations, it is likely to contain descriptions regarding causative associations between phenotypes and gene functions. We can obtain a set of phenotypes (MeSH C terms) associated with a given disease by the same procedure for estimating $P(p|d)$ and can obtain a set of gene functions associated with the gene paired with the disease by consulting the Entrez Gene database. Given the sets of phenotypes and gene functions, our task is to identify which phenotypes and which gene functions have true causative associations and to what degree.

We estimate these associations using three different schemes, i.e., *SchemeK*, *SchemeT*, and *SchemeK+T*. *SchemeK* simply assumes a link between every pair of the phenotypes and gene functions with equal strength, whereas *SchemeT* seeks for evidence in the textual portion of the Medline record to better estimate the strength of associations. Lastly, *SchemeK+T* combines the two schemes by linearly interpolating association scores, $S(f,p)$, described shortly.

SchemeT essentially searches for co-occurrences of gene functions (GO terms) and phenotypes (MeSH terms) in a sliding window, assuming that associated concepts tend to co-occur more often in proximity than unassociated ones. However, a problem is that gene functions and phenotypes are descriptive by nature and may not be expressed in concise GO or MeSH term. To deal with it, we apply the idea of query expansion, a technique used in IR to enrich a query by adding related terms. If GO and MeSH terms are expanded, there is more chance that they could co-occur in text. For this purpose, we use the definitions (or scope notes) of GO and MeSH terms and identify representative terms by inverse document frequencies (IDF), which have been used in IR to quantify term specificity in a document collection. We treat term definitions as documents and define IDF for term t as $\log(N/Freq(t))$, where N denotes the total number of MeSH C (or GO) terms and $Freq(\cdot)$ denotes the number of MeSH C (or GO) terms whose definitions contain term t. Only the terms with high IDF are used as *proxy terms* to represent the original concept, i.e., gene function or phenotype.

Each co-occurrence of the two sets of proxy terms (one representing a gene function and the other representing a phenotype) can be seen as evidence that supports the association between the gene function and phenotype, increasing the strength of their association. We define the increased strength by the product of the term weights, w, for the two co-occurring proxy terms. Then, the strength of the association between gene function f and phenotype p within article a, denoted as $S(f,p,a)$, can be defined as the sum of the increases for all co-occurrences of the proxy terms in a. That is,

$$S(f,p,a) = \sum_{(t_f,t_p,a)} \frac{w(t_f) \cdot w(t_p)}{|Proxy(f)| \cdot |Proxy(p)|} \tag{5}$$

where t_f and t_p denote any terms in the proxy term sets for f and p, respectively, and (t_f, t_p, a) denotes a set of all co-occurrences of t_f and t_p within a. The product of the term weights is normalized by the proxy size, $|Proxy(\cdot)|$, to eliminate the effect of different proxy size. (Note that a larger proxy size generally produces a greater numerator.) As term weight w, we used the TF·IDF weighting scheme. For term t_p for instance, we define $\mathrm{TF}(t_p)$ as $1 + \log Freq(t_p, Def(p))$, where $Def(p)$ denote p's definition and $Freq(t_p, Def(p))$ denotes the number of occurrences of t_p in $Def(p)$.

The association scores, $S(f, p, a)$, are computed for each GAD entry by either *SchemeK* or *SchemeT* and are accumulated over all entries to estimate the associations between f's and p's, denoted as $S(f, p)$. Based on the associations, we define probability $P(f|p)$ as $S(f, p) / \sum_p S(f, p)$.

A possible shortcoming of the approach described above is that the obtained associations $S(f, p)$ are symmetric despite the fact that the network in Fig. 1 is directional. However, since it is well-known that an organism's genotype (in part) determines its phenotype, we assume the estimated associations between gene functions and phenotypes to be directed from the former to the latter.

Enhancing $P(f|p)$ by Domain Ontology. The proposed framework may not be able to establish true associations between gene functions and phenotypes for various reasons, e.g., the amount of training data may be insufficient. Those true associations may be uncovered using the structure of MeSH and/or GO. MeSH and GO have a hierarchical structure[6] and those located nearby in the hierarchy are semantically close to each other. Taking advantage of these properties, we enhance the estimated probabilities $P(f|p)$ as follows.

Let A denote the matrix whose element a_{ij} is probability estimate $P(f_j|p_i)$ and A' denote the enhanced or updated matrix. Then, A' is formalized as $A' = W_p A W_f$, where W_p denotes an $n \times n$ matrix with element $w_p(i, j)$ indicating a proportion of a probability to be transmitted from phenotypes p_j to p_i. Similarly, W_f is an $m \times m$ matrix with $w_f(i, j)$ indicating a proportion transmitted from gene functions f_i to f_j. This study experimentally uses only direct child-to-parent and parent-to-child relations and defines the weight function $w_p(i, j)$ as

$$w_p(i, j) = \begin{cases} 1 & \text{if } i = j \\ \dfrac{1}{\#\text{ of children of } p_j} & \text{if } p_i \text{ is a child of } p_j \\ \dfrac{1}{\#\text{ of parents of } p_j} & \text{if } p_i \text{ is a parent of } p_j \\ 0 & \text{otherwise} \end{cases} \tag{6}$$

Eq. (6) means that the amount of probability is equally split among its children (or parents). Similarly, $w_f(i, j)$ is defined by replacing i and j in the right-hand side of Eq. (6). Note that this enhancement can be iteratively applied to take advantage of more distant relationships than children/parents.

---

[6] To be precise, GO's structure is directed acyclic graph, allowing multiple parents.

Table 1. Number of gene-disease associations in the benchmark data

	Cancer	Cardio-vascular	Immune	Metabolic	Psych	Unknown	Total
Training	258	305	376	313	172	864	2,288
Test	45	36	61	23	12	80	257
Total	303	341	437	336	184	944	2,545

4   Empirical Evaluation

4.1   Benchmark Data

To evaluate the validity of the proposed approach, we implemented a prototype system and conducted various experiments on the benchmark data sets created from GAD. The following details the creation of the benchmark data.

1. Associate a gene-disease pair from each GAD entry with the publication date of the article from which the entry was created. The date can be seen as the time when the causative relation became public knowledge.
2. Group gene-disease pairs based on disease names. (As GAD deals with complex diseases, a single disease may be paired with multiple genes.)
3. For each pair of a disease and its causative genes,
   (a) Identify the gene whose relation to the disease was most recently reported based on the publication date. If the date is on or after 7/1/2003, the gene will be used as the target (i.e., new knowledge), and the disease and the rest of the causative genes will be used as system input (i.e., old knowledge). In other words, the target-input pair will be an instance composing test data. If the date is before 7/1/2003, the pair of the disease and the gene is added to training data.
   (b) Remove the most recently reported gene from the set of causative genes and return to step (3a).

The separation by publication dates ensures that a training phase does not use new knowledge in order to simulate gene-disease association discovery. The particular date was arbitrarily chosen by considering the size of the resulting data sets. Table 1 shows the number of gene-disease associations in the resulting data sets under six disease classes defined in GAD. In the following experiments, the cancer class is used for system development and parameter tuning.

4.2   Experimental Setup

Given input (disease name d, known causative genes C, and target region r), the system computes the probability $P(d|G)$ as in Eq. (3) for each candidate gene g located in r, where G is C plus g. For instance, d, C, and r might be hepatocellular carcinoma, {APC,IL1}, and 8q24, respectively. The candidate genes are then output in a decreasing order of their probabilities.

Table 2. System performance in AUC for each disease class. The figures in the parentheses indicate percent increase/decrease relative to *SchemeK*.

	Cardio-vascular	Immune	Metabolic	Psych	Unknown	Overall
K	0.707	0.612	0.681	0.628	0.684	0.661
T	0.731	0.611	0.614	0.667	0.761	0.686
	(3.4%)	(-0.2%)	(-9.9%)	(6.2%)	(11.2%)	(3.8%)
$K+T$	0.697	0.656	0.682	0.702	0.743	0.699
	(-1.4%)	(7.2%)	(0.1%)	(11.8%)	(8.5%)	(5.9%)

As evaluation metrics, we use *area under the ROC curve* (AUC) for its attractive property as compared to the *F*-score measure (see Fawcett [9] for details). ROC curves are two dimensional measure for system performance with x axis being true positive proportion (TPP) and y axis being false positive proportion (FPP). TPP is defined as TP/(TP+FN), and FPP as FP/(FP+TN), where TP, FP, FN, and FP denote the number of true positives, false positives, false negatives, and false positives, respectively. AUC takes a value between 0 and 1 with 1 being the best. Intuitively AUC indicates the probability that a gene randomly picked from positive set is scored more highly than one from negative set.

Probabilities $P(f|p)$ were pre-computed using the training data. Then the test data in the cancer class were used to determine several parameters for each scheme, including the number of Medline articles as the source of phenotypes (n_m), threshold for chi-square statistics (t_c), threshold for IDF to determine proxy terms (t_t), and window size for co-occurrences (w_s). For example, for *SchemeT*, they were set as n_m=400, t_c=2.706, t_t=4.0, and w_s=100 (words) by testing numbers of their combinations.

4.3   Results and Discussions

Overall Performance. With the optimal parameter settings identified with the cancer class, the system was applied to all the other disease classes. Table 2 summarizes the system performance in AUC.

All the schemes achieved significantly higher AUC than 0.5 (which corresponds to a random guess), indicating the validity of the general framework using the inference network for discovering implicit associations. For individual disease classes, it is observed that *SchemeT* yielded the best AUC for the *Cardiovascular* and *Unknown* classes and *SchemeK+T* for the others. Overall, *SchemeK+T* works the best, followed by *SchemeT*. The difference between *SchemeK+T* and *SchemeK* is significant ($p < 0.01$), which proves the benefit of textual information.

Enhancing $P(f|p)$ by Domain Ontology. Section 3.2 discussed that ontology could be exploited to enhance probability estimates $P(f|p)$. In brief, considering parent-to-child (P-to-C) and child-to-parent (C-to-P) relations between two concepts defined in MeSH and GO, one could deduce associations between

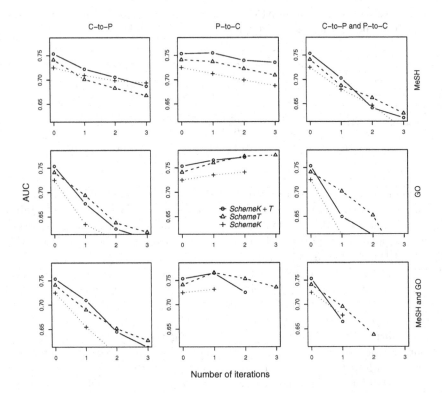

Fig. 2. Transitions of AUC for different source and direction of relations

the concepts that were *not* observed in the training phase. Again, using the cancer class data, we investigated an effective use of domain ontologies.

Eq. (6) was defined to use both MeSH and GO and both P-to-C and C-to-P relations. However, it is expected that each knowledge source and each direction of the relations would have different effects on the outcome. To determine the best strategy, we compare the combination of the following alternative settings: only MeSH, only GO, or both for the source of the semantic relations, and only P-to-C, only C-to-P, or both for the direction of the relations. Because these two properties are independent, there are $3 \times 3 = 9$ different combinations to be examined. Fig. 2 shows plots for these combinations, where x and y axes represent the number of iterations and AUC, respectively. Note that, due to the limitation of computer memory used for this experiment, we could iterate the computation only once or twice for some cases.

Contrary to our expectation, the use of the ontologies rather deteriorated AUC for many cases. Especially, when C-to-P relations were considered (the left and right columns), AUC dropped as the number of iterations increased regardless of the scheme used. On the other hand, when GO and only P-to-C relations were used (the center and bottom middle), AUC mildly improved at

Table 3. System performance in AUC after enhancing probability estimates with only parent-to-child relations in GO hierarchy. The figures in the parentheses indicate percent increase/decrease relative to the corresponding cells in Table 2.

	# of iterations	Cardio-vascular	Immune	Metabolic	Psych	Unknown	Overall
K	1	0.707 (0.0%)	0.601 (-1.8%)	0.702 (3.0%)	0.672 (7.0%)	0.706 (3.2%)	0.673 (1.8%)
T	1	0.727 (-0.5%)	0.609 (-0.4%)	0.618 (0.6%)	0.750 (12.4%)	0.774 (1.7%)	0.695 (1.4%)
	2	0.726 (-0.6%)	0.613 (0.4%)	0.633 (3.1%)	0.768 (15.0%)	0.782 (2.7%)	0.703 (2.5%)
K+T	1	0.703 (0.9%)	0.631 (-3.8%)	0.698 (2.3%)	0.765 (8.9%)	0.763 (2.7%)	0.708 (1.3%)

least at the first iteration. These results suggest that the associations between gene functions and phenotypes could be safely enhanced only downwards in the hierarchies. Among the two plots, using only GO hierarchy (the center) shows constant improvement of AUC with the number of iterations, whereas the other (the bottom middle) gradually declines from the second iteration. The best AUC (=0.776) was achieved with *SchemeT* after three iterations using only GO and P-to-C relations.

Based on these observations, the same strategy (i.e., GO with P-to-C) was applied to all the other disease classes; The results are summarized in Table 3.

As shown, the system performance more or less improved except for the *Cardiovascular* and *Immune* classes. Overall, AUC marginally increased irrespective of the schemes. After applying two times of iterations to *SchemeT*, it further improved to 0.703. (It could not be applied to the other two due to the memory limitation.) These experiments verify that the strategy of using P-to-C relations in the GO hierarchy is generally effective in other types of diseases and that system performance slightly but steadily increases with the number of iterations. The improvement of *SchemeT* is statistically significant at the 5% level.

In the experiments above, considering the MeSH hierarchy was found harmful in enhancing $P(f|p)$. It may have been caused by a possible difference in the nature of the MeSH and GO hierarchies. To investigate, we compared their organizational structures (e.g., the number of children per node) but were not able to find notable difference in this regard. Another cause of the problem may be possible spurious phenotypes associated with a query disease. Remember that while GO terms are obtained from Entrez Gene given a candidate gene (i.e., a simple database lookup), MeSH terms are harvested from Medline search results with a disease name being a query, assuming that MeSH terms annotated with the retrieved articles are representative phenotypes of the disease. Thus, some of those MeSH terms may not be associated with the disease at all. Enhancing associations based on those spurious phenotypes, if any, would degrade system prediction. More work needs to be done to determine the benefit of MeSH.

5   Concluding Remarks

This paper explored a novel discovery framework targeting implicit gene-disease associations and proposed an extension of IR models/techniques in conjunction with domain-specific resources, such as the literature, gene database, and ontology. To examine the validity of the framework, we created realistic benchmark data, where old and new knowledge were carefully separated to simulate knowledge discovery. The key findings identified by empirical observations include that a) the consideration of textual information improved system prediction by 5.9% in AUC over simply relying on co-annotations of keywords, and b) semantic relations defined in domain ontologies could be leveraged to enhance probability estimates, where MeSH were found rather harmful in the current scheme.

For future work, we plan to investigate more sophisticated schemes, e.g., the semantic distance [10], in propagating the probabilities $P(f|p)$. In addition, we would like to compare the proposed framework with the previous work and with other IR models so as to study the characteristics/advantages of our model.

Acknowledgments. This work was partially supported by the Artificial Intelligence Research Promotion Foundation grant #18AI-255, the Nakajima Foundation, the Japanese Ministry of Education, Culture, Sports, Science and Technology, and the NSF grant #0549313.

References

1. Hearst, M.A.: Untangling text data mining. In: Proceedings of the 37th Annual Meeting of the Association for Computational Linguistics, pp. 3–10 (1999)
2. Swanson, D.R.: Fish oil, Raynaud's syndrome, and undiscovered public knowledge. Perspectives in Biology and Medicine 30(1), 7–18 (1986)
3. Weeber, M., Klein, H., de Jong-van Berg den, L.T.W., Vos, R.: Using concepts in literature-based discovery: simulating Swanson's Raynaud-fish oil and migraine-magnesium discoveries. Journal of the American Society for Information Science and Technology 52(7), 548–557 (2001)
4. Aronson, A.R.: Effective mapping of biomedical text to the UMLS metathesaurus: The MetaMap program. In: Proceedings of American Medical Informatics 2001 Annual Symposium, pp. 17–21 (2001)
5. Srinivasan, P.: Text mining: generating hypotheses from Medline. Journal of the American Society for Information Science and Technology 55(5), 396–413 (2004)
6. Turtle, H., Croft, W.B.: Evaluation of an inference network-based retrieval model. ACM Transactions on Information Systems 9(3), 187–222 (1991)
7. Perez-Iratxeta, C., Wjst, M., Bork, P., Andrade, M.: G2D: a tool for mining genes associated with disease. BMC Genetics 6(1), 45 (2005)
8. Becker, K.G., Barnes, K.C., Bright, T.J., Wang, S.A.: The genetic association database. Nature Genetics 36, 431–432 (2004)
9. Fawcett, T.: ROC graphs: Notes and practical considerations for researchers. Technical Report HPL-2003-4, HP Laboratories (2004)
10. Resnik, P.: Semantic similarity in a taxonomy: An information-based measure and its application to problems of ambiguity in natural language. Journal of Artificial Intelligence Research 11, 95–130 (1999)

Discovering Mentorship Information from Author Collaboration Networks

V. Suresh[1,*], Narayanan Raghupathy[2], B. Shekar[3], and C.E. Veni Madhavan[1]

[1] Dept. of Computer Science and Automation, Indian Institute of Science,
Bangalore - 560 012, India
{vsuresh,cevm}@csa.iisc.ernet.in
[2] Dept. of Biological Sciences, Carnegie Mellon University, Pittsburgh, PA 15213
rnarayan@cmu.edu
[3] Indian Institute of Management Bangalore, Bannerghatta Road, Bangalore-560076,
India
shek@iimb.ernet.in

Abstract. Researchers are assessed from a *researcher-centric* perspective — by quantifying a researcher's contribution to the field. Citation and publication counts are some typical examples. We propose a *student-centric* measure to assess researchers on their mentoring abilities. Our approach quantifies benefits bestowed by researchers upon their students by characterizing the publication dynamics of research advisor-student interactions in author collaboration networks. We show that our measures could help aspiring students identify research advisors with proven mentoring skills. Our measures also help in stratification of researchers with similar ranks based on typical indices like publication and citation counts while being independent of their direct influences.

1 Introduction

In scientific research, rankings of journals are decided through impact factors, and researchers are rated based on their publication and citation counts. Most of these approaches are *researcher-centric*. They do not capture the context in which a ranking attribute was acquired. For example, they do not reveal if highly cited authors actually contributed to the growth and success of their junior collaborators. We present a student-centric approach to quantifying benefits derived by students from their associations with their advisors. We show that mentorship credentials, an important socio-academic aspect of research, could be inferred and quantified by a meaningful data mining of publication databases.

Research on citation networks and ranking of journals and conferences dates back to the influential works authored by Price [1] and Garfield [2]. Availability of large bibliographic social-network data has resulted in a renewed interest to understand the underpinnings of research collaborations and citations [3,4,5,6,7]. Impact factor [2] is used to rank the importance of journals. This is based on the number of publications appearing in a journal and the number of citations they attract. Bollen [8,9] devised a measure called Y-factor that combines Google's

---

* Contact author.

V. Corruble, M. Takeda, and E. Suzuki (Eds.): DS 2007, LNAI 4755, pp. 197–208, 2007.

PageRank algorithm [10] and impact factor [2]. Publications and citations are traditionally associated with a scientist's research output. Recently, Hirsch [11], proposed h-index to quantify a scientist's research output that is not affected by the publication or citation counts of the researcher. Mohan's [12] heuristic assigned ranks to researchers in Computer Science based on the *nurturing* provided to their students. Till date, this is the only student-centric approach.

The road map for the rest of the paper is as follows. In the following section, we provide motivation for our work. Section 3 describes the scoring mechanism that quantifies mentoring skills of researchers. Section 4 contains the results obtained from applying the proposed mechanism on real-world scientometric databases. This is followed by a comparative analysis with relevant work in the literature. Finally we present our conclusions in Section 6.

2   Assessing Research Mentorship

Research advisor selection is a complex procedure involving several factors including: time constraints, large number of universities and a larger pool of researchers, financial constraints and penalty of making wrong choices. While choosing their prospective research advisors, applicants usually consult information available in home pages of researchers along with graduate school rankings. In addition to this, knowledge about mentoring abilities of researchers would also be of immense help in making such choices.

Good Mentoring: The Premise. Academic research outputs are usually presented in the form of peer-reviewed publications. Thus a student's career in research depends on the publications with his advisor. Apart from helping students get publications, the advisor should ensure that the quality of publications gives a status to the students thus helping them acquire future research collaborators. In addition, a student may gain immensely by way of research introductions, the process by which research advisors leverage their relationships in the academic community to enable their students acquire new co-authors. These are the general expectations of a student from his research mentor. Thus there is a natural student-centric formulation to assessing the credentials of a researcher as a mentor. In the next section we show that these expectations are quantifiable thereby arriving at a scoring mechanism to assess mentorship.

Other Conventions for Mentoring. Defining and quantifying the qualities of good mentoring is a difficult task [13]. Although there is no single convention for capturing all aspects of student-advisor interactions, the available publication data makes it possible to perform a quantitative analysis for measuring the impact of mentorship of advisors on students. There could be many non-publication oriented conventions and practices that may influence mentorship. However, these otherwise valuable information are not available to a prospective student who has limited time and resources. Our approach assumes publications as an important base for assessing advisor-student interaction because of its amenability to computational analysis. We study the prospects of a student to get established in the scientific community by acquiring publications on the basis of his foundational interactions with the advisor.

3   Scoring Mechanism

We quantify the attributes of good mentorship into two broad categories: *Direct Benefits* and *Indirect Benefits*. In real world, these benefits usually take the form of internships, postdoctoral fellowships and faculty positions. It is reasonable to assume that these benefits could form a basis for future success in a student's research career. Our mechanism computes scores for direct and indirect benefits that are bestowed by researchers upon their students. A researcher's potential for mentoring students is quantified based on the researchers interactions with his current as well as past students.

3.1   Direct Benefits

A student's direct benefit is computed with respect to each new joint-publication with the research advisor. These publications are outcomes of the student's past association with the research advisor. Strength of the past association is captured by the number of prior publications the student has had with the advisor. Consider a publication

$$pub : \{a_1, a_2, \ldots, a_n, t, id\}$$

where a_i's are the authors of the paper, t is the time of publication and id denotes information such as title of the paper and category of publication (conference/journal). Let a_1 be the student and a_2 be the advisor. a_2's coauthoring a new publication with a_1 could be deemed beneficial to a_1. Suppose a_2 has p prior publications with a_1, we assign a_2 the score

$$\mathcal{DB} = p + 1,$$

where $\mathcal{DB}$ denotes the direct benefit derived by a_1 from a_2, and it takes the new joint-publication into account. We note that this score does not depend on the publication count of the advisor. The advisor might have a high publication count. However, only a minuscule part of it would overlap with the initial publications of a single student. As far as the student is concerned joint-publication count is all that matters. This aspect is inbuilt in our simple update rule.

3.2   Indirect Benefits

Research collaboration with the advisor helps a student secure new positions and collaborators. This becomes visible when the student stops publishing with the advisor and starts publishing with a new set of co-authors. An advisor's mentoring effect in this direction may be classified as: association benefits and introduction benefits. In the following sections, we use the terms "advisors", "co-authors" and "past co-authors" interchangeably. It is often unknown from publication databases whether a co-author of a given publication is a research advisor or not. We overcome this by assigning roles to authors of a publication. Those with publication counts above a certain threshold are considered as advisors and others as students. This is in accordance with the intuitions about the publication counts of research advisors. This is applied in computing $\mathcal{DB}$ scores as well. The importance of this is elaborated in section 3.3.

Association Benefits. Students publish with researchers other than their research advisors during internships and postdoctoral fellowships. The process of acquiring new co-authors is a result of the quality of the student's past work. A new collaboration arising from joint publications with the advisor can be viewed as the student deriving *association benefit* from the advisor. This benefit is intuitively captured by the publication count of the new co-author. We assign each past co-author a score computed as a function $f(nP)$, where n is the number of joint-publications this past coauthor has had with the student and P is the publication count of the new co-author. The intuition for this is as follows. The new co-author's academic reputation is reflected in the publication count. This co-author's association with the student is founded on the joint-publication count of the advisor with the student. One might consider other parameters such as citation count or h-index as better ways to capture the new co-author's reputation. As for this experiment, we consider only publication count as it is much simpler to obtain and handle. We explain association benefits for a publication

$$pub : \{a_1, a_2, \ldots, a_n, t, id\}.$$

Suppose a_1 is the student and a_2 is a new co-author of a_1 with P publications. Let $\{b_1, b_2, \ldots, b_m\}$ be a_1's past co-authors, each of them with $\{n_1, n_2, \ldots, n_m\}$ past joint-publications with a_1 respectively. Then each of the past co-author b_i is assigned a score of $f(n_i P)$ corresponding to student a_1 deriving an association benefit by acquiring a new co-author a_2. This is computed for each (student, new co-author) pair. We denote scores arising out of an association benefit as $\mathcal{AB}$. Note that the scores assigned to the past-coauthors are independent of their publication counts.

Choice for $f(nP)$. From our trials we found that *square root* is a good candidate for $f(nP)$ in computing $\mathcal{AB}$ scores as it helps to bring down the intensity of disparities induced by the linear product nP. This is illustrated with the following example. Let a student acquire a new co-author who has 300 publications. Suppose this is founded on a joint-publication count of 5 with his advisor. One would like to compare this with benefits derived by another student acquiring two new co-authors each having 50 and 60 publications, founded on a joint-publication count of 3 with his advisor. Linear products are not appropriate for such intuitive comparisons. Square root serves as a suitable sublinear function for this purpose. For the present experiment $f(nP) = \sqrt{nP}$. Other sub-linear functions in addition to square root may be suited for this task — we found $\log_2(nP)$ being one such. $\log_{10}(nP)$ on the other hand, compresses scores into a narrow band thus making it unsuitable for meaningful comparisons.

Introduction Benefits. In the above publication, let a_2, the new co-author of a_1, also be a past co-author of b_1. One could assume b_1 *introduced* a_2 to a_1. The strength of this introduction is quantified by the total joint-publications, q, shared by b_1 and a_2. and the publication count P of a_2. b_1 is assigned a score $\sqrt{qP}$ for helping a_1 get a new co-author by way of introduction. We denote this as $\mathcal{IB}$. If the author list of the present publication has authors in common with b_i's past collaborators then this score is computed for each b_i. $\mathcal{AB}$ and $\mathcal{IB}$ scores are computed only with respect to students acquiring new co-authors. The co-authors are considered *new* only for a period of one year. Then onwards they become past co-authors and stop generating $\mathcal{AB}$ and $\mathcal{IB}$ scores. This convention is adopted due to lack of fine-grained time-stamps in publication databases.

3.3   Data Mining Parameters

$\mathcal{DB}$, $\mathcal{AB}$ and $\mathcal{IB}$ scores for advisors are computed with the help of a curated public domain databases like DBLP (http://dblp.uni-trier.de). It is straightforward to compute the benefit scores and find good mentors if we know the exact roles of authors in a publication. However, publication databases lack such social context. This makes mentorship assessment a difficult problem. Therefore, we make reasonable assumptions as an aide to compute scores for research-advisors. These assumptions can be easily relaxed to test their effect on $\mathcal{DB}$, $\mathcal{AB}$ and $\mathcal{IB}$ scores. These assumptions are also made with the purpose of removing certain inequities that might result in lopsided scores. We have presented these assumptions in an order that progressively refines the social context. Coarse graining of the context may be obtained by truncating this list. The conducted experiment applies our heuristic to assess mentorships in Computer Science on DBLP. However, the assumptions we make are not specific to DBLP.

Assessment Period for Advisors. We compute $\mathcal{DB}$, $\mathcal{AB}$ and $\mathcal{IB}$ scores for advisors over a specific time period. In our experiment we assess mentorship over the 10 year period $1996 - 2005$. This period is chosen primarily to study the mentorship of researchers who are still active. By choosing this period, we hope to capture a summary of mentorship of researchers in the recent past.

Roles. These databases are role neutral in that the roles played by the co-authors (student/research-advisor/co-worker) in the real-world are not documented. To overcome this limitation, we assign roles to the authors by using a simple thumb-rule: authors who have publication count above a threshold number at the start of the assessment period are deemed "advisors" and those with counts below this threshold are deemed "students". For our experiments this is taken as 10. This might result in publications that involve more than one "advisor" and more than one "student". However this does not limit our approach. Direct, Association and Introduction benefits are still computable, but now should be computed for all advisor-student pairs in a publication. It is common knowledge that, in the social context, students benefit from co-authors other than their advisors. Our simple role assignment heuristic gives credit to this social phenomenon. From now on, the terms student and advisor refer to the roles assigned to the authors.

Early-Phase. Students contribute to scores of their advisors only until they are in the *early-phase* of their research careers. The early-phase of a researcher needs to be defined in terms of time and publications so that benefits derived from the advisor are within reasonable bounds. This is to capture the notion that the student is independent and requires little mentoring after a certain stage. Since information regarding the early-phase is not available from the publication databases, we assign the role: "student" for a maximum period of 10 years. This heuristic estimate is arrived at in the following way: 5 years for graduation, a post doctoral position for 2 years and a margin of 2 − 3 years.

Skewness in scores could be generated by students who get inordinately large number of publications within the early phase. Hence we refine the early phase by imposing a threshold on the number of publications. We assume 10 as a reasonable number of publications achievable in the early phase (amounting to one publication per year in this phase). A students exits early phase once he attains 10 publications even if he is yet to complete 10 years.

Not all students who enter the assessment window (1996 − 2005 for the present experiment) are without prior publications. Students who enter the window with prior publications should not be given the same 10 year period as they are already in their early-phase. We apply a simple rule: every prior publication reduces the early-phase by a year. We use this linear scheme as an approximation since we do not have models for the non-uniform spread of publications with respect to time. This normalization is liberal in the following sense. A student's present publications usually have a logical flow from the publications of the immediate past. In general, a student who has prior publications is more likely to acquire larger number of publications than students who enter the window without any publications.

Wean Period. While computing association and introduction benefits, we assume a *wean period* to erode the strength of a student's past relationship with the advisor. If a student has not published for a certain number of years with the advisor, the effect of acquiring a new co-author is not attributable to the advisor. Otherwise, long dormant relationships could derive skewed benefits despite lack of part in the student's acquisition of new collaborators. For example, a student's Master's advisor could get inordinate amount of $\mathcal{AB}$ and $\mathcal{IB}$ scores from the student's post-doctoral work.

Capturing weaning periods through functions that diminish with time pose difficulties. Publications are results of social interactions, new collaborators become new co-authors after a gestation period. For example, a dormancy with respect to doctoral advisors appears when a doctoral student graduates and joins a research lab or a gets a post doctoral position. Thus some associations are inevitably dormant for short intervals. We simplify the situation by applying a fixed threshold of 5 years as wean-period. This margin is 50% of the year component of the early-phase we chose for the present experiment (10 years).

4   Experiments and Results

We quantified mentorship of researchers in Computer Science by applying our heuristic on DBLP. The efficacy of our approach was also tested on physics publication data from arXiv [3]. Similar characteristics were observed in the mentorship dynamics of physicists (data not shown) despite the presence of more multi-author publications as against computer science. This suggests that the heuristics may be applicable to diverse fields. However, interpretation of the results may require specific post processing to account for the area specific characteristics (such as tendency towards multiauthorships). Experiments were conducted on DBLP for the period 1996 − 2005. A publication threshold of 10 determined the role of an author (advisor or student). The first 10 publications of students were tracked for a maximum period of 10 years with a wean period of 5 years. $\mathcal{DB}$, $\mathcal{AB}$ and $\mathcal{IB}$ scores were computed for 7752 advisors during this time period.

4.1   Analysis

Results reveal a good correlation between $\mathcal{DB}$ and $\mathcal{AB}$ scores of researchers. This implies students who get a good number of publications with their advisors during their graduate studies acquire new co-authors after graduation. This may seem common sense. However, while choosing an advisor one would typically underestimate the number of researchers possessing such characteristics. As we describe later, this number is good enough to give a wider scope to genuine applicants. We observe that, high $\mathcal{DB}$ scores do not necessarily imply high $\mathcal{IB}$ scores. This means that not all students derive the benefit of getting introduced to the co-authors of their advisors. Hence advisors possessing high $\mathcal{IB}$ could be valued by prospective students. These may be inferred from Figure 1.

Figure 2 shows that $\mathcal{DB}$, $\mathcal{AB}$ and $\mathcal{IB}$ scores do not necessarily correlate with publication counts. This suggests that publication counts of research-advisors need not translate into opportunities for their students. These figures also stratify

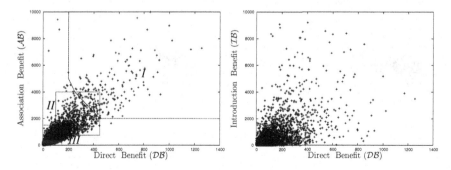

Fig. 1. Left: $\mathcal{DB}$ and $\mathcal{AB}$ scores correlate; deviations though present are not the norm. Right: $\mathcal{IB}$ scores seem to be individual characteristic of researchers and do not correlate with $\mathcal{DB}$ scores.

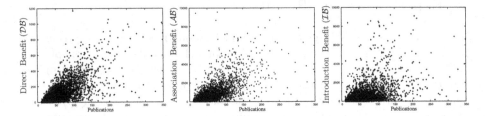

Fig. 2. Publications Vs: $\mathcal{DB}$ scores (Left), $\mathcal{AB}$ scores (Middle) and $\mathcal{IB}$ scores (Right). High publication count need not necessarily translate into $\mathcal{DB}$, $\mathcal{AB}$ or $\mathcal{IB}$ benefits for students.

authors with same or comparable number of publications, on three counts: $\mathcal{DB}$, $\mathcal{AB}$ and $\mathcal{IB}$ scores. Thus a socially neutral information like publication count can be resolved into components using our heuristics and be given a social context.

$\mathcal{AB}$ and $\mathcal{IB}$ scores are not mutually exclusive. This is again due to the lack of social information in publication databases. If a new co-author of the student happens to be a past co-author of the advisor, we are not sure if this new co-author was introduced by the advisor or was the result of the student's individual effort. For this reason, such publications fetch advisors both $\mathcal{AB}$ and $\mathcal{IB}$ scores. Thus some $\mathcal{AB}$ scores could be the result of introductions. A better way to study the Association scores of researchers would be by computing $\mathcal{AB} - \mathcal{IB}$. This removes the effect of Introductions from the $\mathcal{AB}$ scores of the researchers. We observe that for some researchers, $\mathcal{IB}$ scores dominate their $\mathcal{AB}$ scores making $\mathcal{AB} - \mathcal{IB}$ values negative. This indicates presence of cliques. This trend is on the increase in terms of the number of researchers when compared to similar values computed for $1986 - 1995$. These are shown in Figure 3. We also note that multiauthorships is on the ascent. Increase in the *author-cliques* could be attributed to the current trend towards multiauthorship.

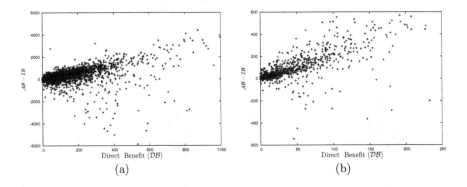

(a)   (b)

Fig. 3. Comparison of $\mathcal{DB}$ with $\mathcal{AB} - \mathcal{IB}$ scores for period (a) 1996-2005 (7752 researchers) (b) 1986-1995 (1191 researchers)

4.2 An Advisory on Choosing Research Mentors

Let us take a closer look at Figure 1. The graph has been demarcated into three regions. The region labeled **I** comprises advisors who offer high DB and AB benefits to their students. This is sparsely populated — has around 150 researchers ($\sim 2\%$ of all advisors) — and hence competitive. The region close to the origin (labeled **III**) comprises advisors who offer low DB and AB benefits to their students. This is densely populated with roughly 80% of the researchers clustered here. On the other hand, the region in the middle (**II**) comprises researchers who offer reasonable enough benefits to students towards furthering their research careers. This region comprises around 15% of all advisors. Thus our heuristic gives scope for students to make informed choices, lack of which could lead to choosing advisors from inappropriate regions.

Advisors in region **III** could be those without tenure or in early periods of their tenure. Some highly successful mentors who have stopped taking students in the present time window could also fall in this region. In any case, their mentorships are not available to prospective students. Hence their getting clustered in this region is not unreasonable. High DB scores and low AB scores imply that their students do not enjoy association benefits by working with them. This does not mean that their students do not have successful research careers. If they do, then it is despite the lack of association benefits from their mentors. Alternately, it could be that such students have crossed their *early-phase* before beginning to contribute to their advisors' AB scores. This crossover could be rectified by increasing the number of publications related to the *early-phase* for such students. In any case, it shows lack of association benefits before these students acquire a reasonable number of publications. The analysis thus far is summarized in Table 1.

Introduction of old research contacts who have been dormant in the recent past with advisors could result in high IB scores. We checked for the presence of this effect by computing scores after ignoring publications earlier to 1986. This is to observe the effect of old publications (pre-1986) on the IB scores

Table 1. Categorizing mentorship through DB, AB, IB score ranges

	DB	AB	IB	$\|AB - IB\|$	Advisory
1	low	low	low	low	wary
2	low	low	high	low	wary
3	low	high	low	high	caution
4	low	high	high	low high	caution/clique-like caution
5	high	low	low	low	wary
6	high	low	high	low	caution/wary/clique-like
7	high	high	low	high	optimistic
8	high	high	high	low high	clique-like/optimistic optimistic

of researchers. We found that the $\mathcal{IB}$ scores of authors are largely unaffected by this. This means researchers are more likely to collaborate with their recent co-authors and benefits derived by students are largely dependent on the advisors' more recent collaborations.

It might seem that interactions present in inter-disciplinary research pose a problem to our scoring mechanism. Consider the case of an advisor who is an expert in Mathematics, mentoring a research student working in Computer Science. The advisor's research contributions may not be visible in a publication database like DBLP that compiles bibliographic records of Computer Science publications. Our mining parameters would even ascribe the role of *student* to such researchers. This scenario has more to do with the limitations of database coverage rather than our scoring mechanism or mining parameters. Such discrepancies may be rectified fully by an integrated study of publication databases cutting across research areas involved in these interactions.

5   Comparisons to Related Work

Researchers are ranked based on parameters such as h-index, citation and publication counts. These researcher-centric measures quantify researchers' impact on the field. These measures were not intended to be advisory for prospective students. These are scalar ranking indices. Ours is a multi-dimensional characterization of researchers' social ability to mentor research students thus giving them a head start towards successful research careers. We observe that these characteristics are essentially independent of h-index, publication and citation counts of researchers. We also show that researchers with similar researcher-centric measures could be stratified using our approach, thereby adding social context to such rankings. Table 2 is a summary of these aspects of our approach. Here we identify "top 10 mentors" from region I of Figure 1 based on their

Table 2. Top 10 mentors from Region I of Figure 1 — listing based on publication count. For each mentor in the list: P refers to the total publication count, h is the h-index, n is the number of students interacted within the period 1996-2005 and P_s is the number of publications coauthored with the students during this period.

P	h	Mentors	n	P_s	$\mathcal{DB}$	$\mathcal{AB}$	$\mathcal{IB}$
349	33	Kang G. Shin	95	187	885	7206	1031
346	50	A. Sangiovanni-Vincentelli	171	156	1017	13775	15787
326	17	Edwin R. Hancock	37	189	931	3833	1086
318	32	Elisa Bertino	96	157	590	6444	7022
305	46	Thomas S. Huang	129	207	1021	7973	2633
295	18	Chin-Chen Chang	99	211	832	4296	749
277	57	Anil K. Jain	109	169	874	5186	744
260	8	Donald F. Towsley	102	154	812	10433	13234
258	17	Sajal K. Das	94	166	701	5565	2481
256	29	HongJiang Zhang	158	204	961	11416	10036

publication counts. Despite having comparable publication counts, these highly successful researchers differ in the number of students they have interacted with and the number of publications that resulted from these interactions. It can be seen that the h-index of these mentors are unrelated to the number of students and the joint publications counts. Parameters such as these, indicative of the social context, are not included in researcher-centric parameters like publication count and h-index. The $\mathcal{DB}$, $\mathcal{AB}$ and $\mathcal{IB}$ scores provided by our approach add relevant social context to these conventional indices and help in identifying good mentors.

Another approach to quantify researchers is in terms of the study of author collaboration networks [3,5]. One may view our approach as counting two paths ($\mathcal{AB}$ scores) and triangles ($\mathcal{IB}$ scores) on such hyper-graphs. In a broader sense, our approach analyzes the interactions of high degree nodes with low degree nodes thus unraveling hidden or implicit social contexts. Our approach could supplement existing network measures to understand the topology of author collaboration networks and other social networks.

Mohan [12] proposed *Nurturer Heuristic* to rank researchers for their ability to nurture their young associates. This is the only student-centric measure available in the literature. Our approach computes scores that cover a wide range of student-advisor interaction dynamics in the form of $\mathcal{DB}$, $\mathcal{AB}$ and $\mathcal{IB}$. Nurturer heuristic lacks the equivalent of direct and introduction benefits. *Tribute* of nurturer heuristic, a distant equivalent of our $\mathcal{AB}$ score, is scaled by $1/n$ for an $n-$author publication. This unwittingly discriminates against multiauthorships. Further, it assumes that students with single author publications do not require as much mentoring as other students. Unlike Nurturer Heuristic, our approach mines the social context in terms of parameters such as *roles, early-phase* and *wean period*. Nurturer heuristic, like other assessment mechanisms, falls in the class of ranking metric while ours computes a much richer dynamics of the student-advisor interaction on author collaboration networks. We show how mentoring skills of researchers could be assessed meaningfully without resorting to explicit rankings as this could inadvertently obscure important details. As against Nurturer Heuristic, our scoring scheme is non-iterative, therefore, it is adequate to have information corresponding to the present students and the advisors concerned for computing scores. Further, a non-iterative scheme like ours is resistant to perturbations in publication databases. The proposed heuristic is suited for distributed computation of scores and could enable real-time systems to compute scores for different mining parameters.

6   Conclusions

We studied the dynamics of collaborations between students and advisors on author collaboration networks and presented a new heuristic measure to quantify mentoring abilities of advisors. This is a student-centric index for assessing a researcher as against indices like publication and citation counts that do not take into account a researcher's mentoring skills. We computed our measures on author collaborations (availabe from publication databases), studied their characteristics and discussed their usefulness in assessing mentorship.

Additional empirical data on mentors as outlined by Lee *et. al* [13] will be useful to extend and validate our approach. Apart from author collaboration networks, the present approach offers scope to evolve mechanisms to assess benefits in other social networks like friendships, weblinks etc., that could be fashioned into recommendor systems.

Acknowledgment

We thank Nan Song (CMU) for helping with the manuscript preparation and Mark Newman (U.Mich) for providing arXiv data.

References

1. Derek, J.S.P.: Networks of scientific papers. Science 149, 510–515 (1965)
2. Garfield, E.: Citation analysis as a tool in journal evaluation. Science 178(4060), 471–479 (1972)
3. Newman, M.E.: The structure of scientific collaboration networks. Proc.Natl.Acad.Sci. USA 98(2), 404–409 (2001)
4. Nascimento, M.A., Sander, J., Pound, J.: Analysis of SIGMOD's co-authorship graph. SIGMOD Record 32(3), 8–10 (2003)
5. Newman, M.E.: Coauthorship networks and patterns of scientific collaboration. Proc.Natl.Acad.Sci. USA 101(Suppl.1), 5200–5205 (2004)
6. Börner, K., Maru, J.T., Goldstone, R.L.: The simultaneous evolution of author and paper networks. Proc.Natl.Acad.Sci. USA 101(Suppl.1), 5266–5273 (2004)
7. Liu, X., Bollen, J., Nelson, M.L., Sompel, H.V.: Co-authorship networks in the digital library research community. Inf. Process. Manage. 41(6), 1462–1480 (2005)
8. Bollen, J., Rodriguez, M.A., Sompel, H.V.: Journal status. Scientometrics 69 (2006)
9. Ball, P.: Index aims for fair ranking of scientists. Nature 436(7053), 900 (2005)
10. Page, L., Brin, S., Motwani, R., Winograd, T.: The pagerank citation ranking: Bringing order to the web. Technical report, Stanford Digital Library Technologies Project (1998)
11. Hirsch, J.E.: An index to quantify an individual's scientific research output. Proc.Natl.Acad.Sci. USA 102(46), 16569–16572 (2005)
12. Mohan, B.K.: Searching association networks for nurturers. Computer 38(10), 54–60 (2005)
13. Lee, A., Dennis, C., Campbell, P.: Nature's guide for mentors. Nature 447(7146), 791–797 (2007)

Active Contours as Knowledge Discovery Methods

Arkadiusz Tomczyk[1], Piotr S. Szczepaniak[1,2], and Michal Pryczek[1]

[1] Institute of Computer Science, Technical University of Lodz,
Wolczanska 215, 90-924 Lodz, Poland
tomczyk@ics.p.lodz.pl
[2] Systems Research Institute, Polish Academy of Sciences,
Newelska6, 01-447 Warsaw, Poland

Abstract. In the paper we show that active contour methods can be interpreted as knowledge discovery methods. Application area is not restricted only to image segmentation, but it covers also classification of any other objects, even objects of higher granulation. Additional power of the presented method is that expert knowledge of almost any type can be used to classifier construction, which is not always possible in case of classic techniques. Moreover, the method introduced by the authors, earlier used only for supervised classification, is here applied in an unsupervised case (clustering) and examined on examples.

1 Introduction

The classification techniques are widely used e.g. in diagnostic or decision systems. So far many different methods of classification have been developed. All of them solve the problem of the optimal classifier construction. In this paper a new approach, *potential active hypercontour*, to the problem of supervised and unsupervised classification is proposed. The idea of this approach comes from active contour methods, which are used for image segmentation. The main advantage of this technique is that it is quite intuitive and allows to use any expert knowledge during optimal classifier construction (which is not always possible in case of classic methods). That additional knowledge can significantly improve classification results, which is illustrated by the examples.

The paper is organized as follows: in section 2 the relationship between active contour methods and classifier construction is presented, in section 3 *potential active hypercontour* method is described, the next three sections presents the examples of the proposed approach and finally the paper concludes with the summary of the proposed method.

V. Corruble, M. Takeda, and E. Suzuki (Eds.): DS 2007, LNAI 4755, pp. 209–218, 2007.
© Springer-Verlag Berlin Heidelberg 2007

2   Relationship Between Active Contours and Classifier Construction

2.1   Active Contours

Active contour (AC) methods were firstly introduced and applied for image segmentation in [9]. Their aim is to find the optimal contour $c \in C$ describing object in the image by optimization of energy function $E : C \to \mathbb{R}$ where C denotes a space of all considered contours. Energy evaluates quality of contour which is changing its shape. The specific feature of the method is the use of higher-level knowledge for detection of objects composed of lower-level image elements (pixels). Since the first description of the *snakes* method ([9]) a variety of different techniques has been proposed: *active shape models* ([10]), *geodesic active contours* ([11]), *Brownian strings* ([12]), etc.

2.2   Classifier Construction

Classifiers can be considered as functions $k : \mathcal{X} \to \mathbb{L}(\mathsf{L})$ where $\mathcal{X}$ denotes the feature space (each object has uniquely assigned element of feature space), $\mathbb{L}(\mathsf{L})$ denotes set of labels and L denotes the number of labels (in the whole paper $\mathbb{L}(\mathsf{L}) = \{1, \ldots, \mathsf{N}\}$ for $\mathsf{N} \in \mathbb{N}$ will be used). It is obvious, that many correctly constructed classifiers can be found. The problem is, however, to find the optimal one from the set of all possible classifiers $\mathcal{K}$ that map $\mathcal{X}$ into $\mathbb{L}(\mathsf{L})$. The optimality criterion used here should express the available expert knowledge (e.g. a training set with correctly labeled objects) and can be formulated as a performance index $Q : \mathcal{K} \to \mathbb{R}$ evaluating the quality of each classifier. Consequently, the problem of optimal classifier construction is optimization problem (e.g. neural network training).

Unsupervised Classification. When the available knowledge used for construction of the classifier does not contain any information about the expected object labels and is based only on inner similarities or dissimilarities of considered objects, one speaks about unsupervised classification (clustering). The goal here is to divide the given set of objects into groups where homogeneity criteria inside the groups and heterogeneity criteria between them are fulfilled. Additionally, the number of classes L can be unknown. There exist many approaches to clustering, such as: *single-link, complete-link* and *average-link clustering, Hard C-Means* (HCM), *Fuzzy C-Means* (FCM), *Possibilistic C-Means* (PCM) clustering ([6,8]), *Kohonen neural network* ([1,3,4]), *knowledge-based clustering* ([6]), etc.

Supervised Classification. If information about the expected assignments of labels to the objects is available and consequently the number of classes L is precisely known, the supervised classification (or simply classification) is considered. There are also many techniques of supervised classification, for example: *k nearest neighbors* (k-NN) ([3,5]), *Multi-layer Perceptron* (MLP) ([1,3,5]), *probabilistic classification* ([2,5]), *Support Vector Machines* (SVM) ([4]), etc.

2.3   Relationship

In [16], relationship between active contour methods and classification techniques has been discussed. It was shown that contours can be considered as contextual classifiers of pixels and contour evolution can be understood as optimization method of classifier parameters where energy function E is a performance index Q. The result of this notice is introduction of active hypercontour (AH) concept in [17] which is a generalization of AC (hypercontours are the boundaries between regions in feature space that contain objects with different labels). That relationship allows also to exchange experiences (methods, algorithms) between those, so far separately developed, domains. Practical realization of active hypercontours are *potential active hypercontours* described in this work. The strongest advantage of AC, which is its ability to use external expert knowledge of any type during classifier's construction, can also be of use here.

3   Potential Active Hypercontours

Potential active contours (PAC) and their generalization *potential active hypercontours* (PAH) were firstly proposed in [18]. They base on the well known *potential function* method of classification ([5]) where the label assigned to the object depends on the distribution of other known and already classified objects. Their influence on the classification results depends on the distances between them and the object that is classified In the classic formulation of this approach, the known objects are fixed (e.g. the training set can be of use here). In the PAH algorithm, their position and parameters are subject of optimization. Formally, the can be described by the classifier:

$$k(\mathbf{x}) = \arg\max_{l \in \mathbb{L}(L)} \sum_{i=1}^{N} P_{\Psi_i, \mu_i}(\rho(\mathbf{x}_i, \mathbf{x}))\, \delta(l_i, l) \qquad (1)$$

where $\mathbf{x}_i \in \mathcal{X}$ and $l_i \in \mathbb{L}(L)$ for $i \in \mathbb{L}(N)$ denote features of known objects and corresponding to them labels, respectively, $P : \mathbb{R} \to \mathbb{R}$ denotes a strictly decreasing potential function, $\rho : \mathcal{X} \times \mathcal{X} \to \mathbb{R}$ is a metric in the space $\mathcal{X}$, $\delta : \mathbb{N} \times \mathbb{N} \to \{0,1\}$ denotes Kronecker delta and finally Ψ_i and μ_i for $i \in \mathbb{L}(N)$ are parameters of potential functions. Here, the Euclidean metric and inverse potential function were used, where the latter was defined as follows:

$$P_{\Psi,\mu}(d) = \frac{\Psi}{1 + \mu d^2} \qquad (2)$$

Optimization of the distribution of labeled objects and the parameters of potential functions has been performed using *simulated annealing* (SA) ([14]) with exponential cooling scheme (parameters were randomly modified to stay in the neighborhood of their previous values).

So far in [18] the supervised classification was considered where the external knowledge was gathered in training set of N^{tr} objects $\mathbf{x}_i^{tr} \in \mathcal{X}$ with labels $l_i^{tr} \in$

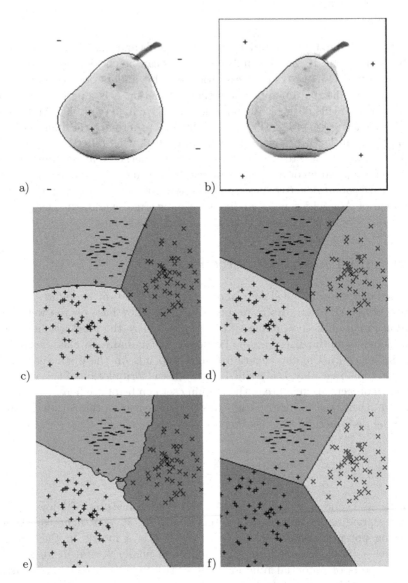

Fig. 1. Sample results of the supervised and unsupervised classification used for image segmentation and for classification of data: (a) - supervised segmentation using PAH, (b) - unsupervised segmentation using PAH, (c) - PAH algorithm used for supervised segmentation (E^s, $F^s = 0.986$), (d) - PAH algorithm used for unsupervised classification (E^u, $F^u = 0.889$), (e) - result of k-NN algorithm ($k = 7$, $F^s = 0.986$), (f) - result of HCM algorithm ($F^u = 0.889$)

$\mathbb{L}(L)$ for $i \in \mathbb{L}(N^{tr})$ and where the optimization objective function (the energy of the hypercontour and consequently the energy of the classifier) was was defined in the following way:

$$E^s(k) = \frac{1}{N^{tr}} \sum_{i=1}^{N^{tr}} \left(1 - \delta(l_i^{tr}, k(\mathbf{x}_i^{tr}))\right) \tag{3}$$

Unsupervised classification can be performed by the use of *potential active hypercontour* controlled by the energy function of the following form:

$$E^u(k) = \frac{1}{N^{tr2}} \sum_{i=1}^{N^{tr}} \sum_{j=1}^{N^{tr}} \rho(\mathbf{x}_i^{tr}, \mathbf{x}_j^{tr}) \, \delta(k(\mathbf{x}_i^{tr}), k(\mathbf{x}_j^{tr})) + \\ \frac{1}{1 + \frac{1}{N^{tr2}} \sum_{i=1}^{N^{tr}} \sum_{j=1}^{N^{tr}} \rho(\mathbf{x}_i^{tr}, \mathbf{x}_j^{tr}) \, (1 - \delta(k(\mathbf{x}_i^{tr}), k(\mathbf{x}_j^{tr})))} \tag{4}$$

It assures that the dissimilarity (in this case distance) between objects in the same group is minimal and between objects from different groups is maximal. Both formula (3) and (4) are only the examples, which allow to present the idea of the approach.

To verify the results of both supervised and unsupervised classification some methods of the quality evaluation of the constructed classifiers must be used. In this work for supervised classification, the number of correct classifications in the testing set, which contains N^{te} objects $\mathbf{x}_i^{te} \in \mathcal{X}$ with labels $l_i^{te} \in \mathbb{L}(L)$ for $i \in \mathbb{L}(N^{te})$, was used:

$$F^s(k) = \frac{1}{N^{te}} \sum_{i=1}^{N^{te}} \delta(l_i^{te}, k(\mathbf{x}_i^{te})) \tag{5}$$

In case of the unsupervised classification, the quality measure described in [15] was applied. It compares the achieved clusters with the clusters determined by known labeling of objects in the testing set (in this case training and testing sets are identical and consequently $N^{te} = N^{tr}$):

$$F^u(k) = \frac{2}{N^{te}} \sum_{l^{tr}=1}^{L^{tr}} \sum_{l^{te}=1}^{L^{te}} n_{l^{tr}}^{l^{te}} \log_{L^{tr}L^{te}} \left(\frac{n_{l^{tr}}^{l^{te}} N^{te}}{n_{l^{tr}} n^{l^{te}}} \right) \tag{6}$$

where:

- $n_{l^{tr}}^{l^{te}}$ denotes the number of objects from group l^{tr} with label l^{te}

$$n_{l^{tr}}^{l^{te}} = \sum_{i=1}^{N^{te}} \delta(l^{tr}, k(\mathbf{x}_i^{te})) \delta(l^{te}, l_i^{te}) \tag{7}$$

- $n_{l^{tr}}$ denotes the number of objects in group l^{tr}

$$n_{l^{tr}} = \sum_{i=1}^{N^{te}} \delta(l^{tr}, k(\mathbf{x}_i^{te})) \tag{8}$$

- $n^{l^{te}}$ denotes the number of objects with label l^{te}

$$n^{l^{te}} = \sum_{i=1}^{N^{te}} \delta(l^{te}, l_i^{te}) \tag{9}$$

In this paper, the number of groups L^{tr} and the number of expected classes L^{te} are equal L but PAH algorithm can be of use in more general case, as well.

In our previous works ([18]) the extension of the PAH method, namely *adaptive potential active contours* (APAH) were proposed. Its main novel feature is that classification abilities of the hypercontour can be modified during the optimization process (the number of objects N is changing). This concept can also be applied for unsupervised classification but is not discussed here.

4   Examples

4.1   Segmentation

Active contour methods are the special case of active hypercontours where the classified objects are pixels. The only features of the pixel that can be considered are its color components and coordinates. In the presented example only the intensity of the pixel was considered (it was the gray-scale image), consequently $\mathcal{X} = \mathbb{R}$. Of course $L = 2$ because usually image is divided into two regions: object and background:

In the Fig. 1a and Fig. 1b two sample images with localized contours are presented. Those contours are not a direct visualization of hypercontour boundaries in feature space. They only reflect the division of that space which is possible because the space of image coordinates has larger discriminative power than space $\mathcal{X}$ (pixel coordinates identify them uniquely). In both cases the training set contained all the image pixels. In Fig. 1a the result of supervised classification is presented (E^s) where the information about the expected labels of pixels, depending on their color intensity, was utilized. And in the the Fig. 1b the unsupervised segmentation was used (E^u) where only the mutual similarity of pixels' color was considered. It is worth mentioning that during unsupervised segmentation, depending on initial distribution of points defining contour, the contour can describe either object (as in the first example) or the background (the second one).

The presented examples and energy functions are of course very simple and in case of the image segmentation they cannot be used because of many practical reasons (e.g. calculation of E^u even for the image of average size would be very time-consuming and in presented paper it had to be optimized). They are good enough, to illustrate the presented concept. In general, however, in practical applications energy usually possesses information both about the expected shape of the contour (internal energy) and about the desired position of the contour in the image (external energy) ([9,10,11,12]). It allows to take into account the relations between color and coordinates of pixels (context). Those aspects were described in [16] and are not discussed in this paper.

4.2   Classification

The second example presented also in Fig. 1 considers randomly generated data where $\mathcal{X} = \mathbb{R}^2$ (it can be visualized) and $L = 3$. In Fig. 1c and Fig. 1e the results of supervised classification (E^s) are presented. They were obtained with PAH and k-NN classifier ($k = 7$), respectively (in the figures the data from training set are shown but the test set was also prepared to compare results by means of quality measure F^s). Similarly, in Fig. 1d and Fig. 1f the results of clustering are presented. The first one was obtained with PAH algorithm while in the second the HCM algorithm was used. In all the experiments the set of points controlling the hypercontour contained 3 objects, one for each class (it can be changed for example by the mentioned above APAH algorithm). In both cases here, though hypercontours are different, the results of classification measured by F^s and F^u are identical, which proves that PAH can be used for supervised and unsupervised data classification task. Here again it can be noticed that in the case of grouping the the initial distribution of points generating hypercntour causes that assignment of the labels to the groups can differ.

The presented PAH approach has other advantages, which are not analyzed here but are still under investigation. One of the is fact it can be used for any metric space $\mathcal{X}$. The other is that potential hypercontours can have different topology which can be of use in case of more complicated data (for example the other similarity measure than metric used for hypercontour description can be applied in the clustering algorithm). Finally, the number of expected groups need not be known as it is in the examples (the PAH algorithm could allow to adapt it during optimization).

4.3   Expert Knowledge

In some cases the knowledge gathered in the training set may be insufficient to obtain satisfactory results. In such situations the other additional information must be used. Classic algorithms of classification like k-NN or HMC have problems with utilization of other types of knowledge. The presented PAH algorithm does not possess such limitations. It can relatively easily incorporate any information into energy function.

In Fig. 2a and Fig. 2c the results of k-NN ($k = 7$) and PAH algorithms are presented for other (but also randomly generated) set of training data. The problem is that data for one of the classes are not representative enough. There are only a few points covering important, for that class, part of the feature space. Consequently the classifiers considered so far have problems with proper classification of the testing set where underrepresented objects can occur. This problem can be solved thanks to additional information about the importance of certain training points. That knowledge can be used in the energy function of the following form:

$$E^{sk}(k) = \tfrac{1}{N^{tr}} \textstyle\sum_{i=1}^{N^{tr}} u_i^{tr}(1 - \delta(l_i^{tr}, k(\mathbf{x}_i^{tr}))) \tag{10}$$

where $u_i^{tr} \in [0, 1]$ denotes the weight of the object $\mathbf{x}_i^{tr}$ for $i \in \mathbb{L}(\mathbf{N}^{tr})$. The results obtained by means of that energy are presented in Fig. 2e.

Similar approach can be used for unsupervised classification. The method described below bases on the approach proposed in [7]. The problem that can appear is that the distribution of training points in feature space not always reflects the desired grouping of those points (it can be caused by the inability

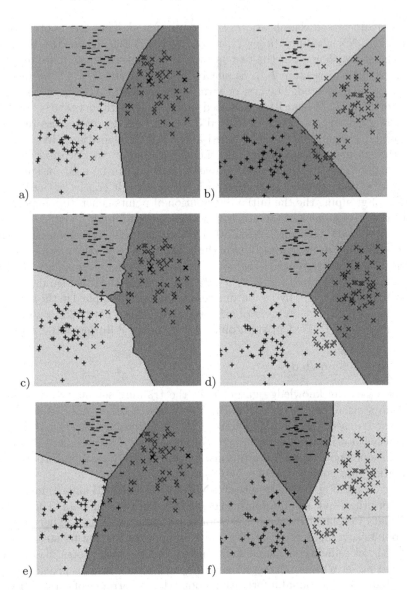

Fig. 2. The influence of additional expert knowledge on the classification results: (a) - PAH algorithm (E^s, $F^s = 0.888$), (b) - PAH algorithm (E^u, $F^u = 0.710$), (c) - k-NN method ($k = 7$, $F^s = 0.894$), (d) - HCM method ($F^u = 0.667$), (e) - PAH with additional knowledge (E^{sk}, $F^s = 0.958$), (f) - PAH with additional knowledge (E^{uk}, $F^u = 0.908$)

to extract better features). Such a problem is presented in Fig. 2b and Fig. 2d together with the solutions generated by HCM and PAH (E^u) methods. To solve it again the additional knowledge must be used:

$$E^{uk}(k) = \frac{1}{N^{tr2}} \sum_{i=1}^{N^{tr}} \sum_{j=1}^{N^{tr}} u_{ij}^{tr} \rho(\mathbf{x}_i^{tr}, \mathbf{x}_j^{tr}) \, \delta(k(\mathbf{x}_i^{tr}), k(\mathbf{x}_j^{tr})) +$$

$$\frac{1}{1 + \frac{1}{N^{tr2}} \sum_{i=1}^{N^{tr}} \sum_{j=1}^{N^{tr}} u_{ij}^{tr} \rho(\mathbf{x}_i^{tr}, \mathbf{x}_j^{tr}) \, (1 - \delta(k(\mathbf{x}_i^{tr}), k(\mathbf{x}_j^{tr})))} \tag{11}$$

where $u_{ij}^{tr} \in [0,1]$ denotes the similarity of objects $\mathbf{x}_i^{tr}$ and $\mathbf{x}_j^{tr}$ for $i \in \mathbb{L}(\mathbf{N}^{tr})$ and $j \in \mathbb{L}(\mathbf{N}^{tr})$. The sample result of PAH algorithm after such modification of energy function together with corresponding quality measures is presented in Fig. 2f.

The presented above usage of additional weights to improve classification results is not entirely a new concept because this form of expert knowledge can be used also in traditional techniques. It presents, however, how other types of knowledge could be utilized by simple redefinition of energy function (E^s or E^u). Other, novel approaches can be considered here: neural networks trained on the examples to properly evaluate classifiers and fuzzy expert systems able to use expert knowledge expressed in linguistic form. These concepts are still being investigated.

5 Summary

Potential active contours and their extension called hypercontours can be used for both tasks: classification and grouping. To the best knowledge of authors, they are novel, powerful methods, which can operate on objects of diverse granulation. Consequently, image segmentation performed on pixels and grouping performed on vectors become similar tasks. The additional power of the method is that expert knowledge can be incorporated and used for modification of hypercontour during its automatic modification, which can significantly improve supervised and unsupervised classification result there where classic techniques fail.

Acknowledgment

This work has been partly supported by the Ministry of Science and Higher Education, Republic of Poland, under project no. N 519 007 32/0978, decision no. 0978/T02/2007/32.

References

1. Bishop, C.: Neural Networks for Pattern Recognition. Clarendon Press, Oxford (1993)
2. Kwiatkowski, W.: Methods of Automatic Pattern Recognition. WAT, Warsaw (2001)(in Polish)
3. Looney, C.: Pattern Recognition Using Neural Networks. In: Theory and Algorithms for Engineers and Scientists, Oxford University Press, New York (1997)
4. Szczepaniak, P.S.: Soft computing, fast transforms and classifiers, Exit, Warsaw (2004)

5. Tadeusiewicz R., Flasinski M.: Pattern Recognition, PWN, Warsaw, (1991) (in Polish)
6. Pedrycz, W.: Knowledge-Based Clustering. Wiley-Interscience, New Jersey (2005)ISBN 0-471-46966-1
7. Pedrycz, W., Loia, V., Senatore, S.: P-FCM: A proximity-based fuzzy clustering. Fuzzy Sets and Systems 128, 21–41 (2004)
8. Rutkowski, L.: Methods and Techniques of Artificial Intelligence, PWN, Warsaw (2005) (in Polish)
9. Kass, M., Witkin, W., Terzopoulos, D.: Snakes: Active Contour Models. International Journal of Computer Vision, 321–331 (1988)
10. Cootes, T., Taylor, C., Cooper, D., Graham, J.: Active Shape Model - Their Training and Application. CVGIP Image Understanding 61(1), 38–59 (1994)
11. Caselles, V., Kimmel, R., Sapiro, G.: Geodesic Active Contours. International Journal of Computer Vision 22(1), 61–79 (1997)
12. Grzeszczuk, R., Levin, D., Strings, B.: Segmenting Images with Stochastically Deformable Models. IEEE Transactions on Pattern Analysis and Machine Intelligence 19(10), 1100–1013 (1997)
13. Xu, C., Yezzi, A., Prince, J.: On the Relationship between Parametric and Geometric Active Contours. In: Proc. of 34th Asilomar Conference on Signals, Systems and Computers, pp. 483–489 (2000)
14. Kirkpatrick, S., Gerlatt Jr., C.D., Vecchi, M.P.: Optimization by Simulated Annealing. Science 220, 671–680 (1983)
15. Strehl, A.: Relationship-based Clustering and Cluster Ensembles for High dimensional Data Mining PhD thesis, The University of Texas at Austin (2002)
16. Tomczyk, A., Szczepaniak, P.S.: On the Relationship between Active Contours and Contextual Classification. Computer Recognition Systems. In: Proceedings of the 4th International Conference on Computer Recognition Systems, CORES'05, Poland, pp. 303–311. Springer, Heidelberg (2005)
17. Tomczyk, A.: Active Hypercontours and Contextual Classification. In: Intelligent Systems Design and Applications Proceedings of 5th International Conference on Intelligent Systems Design and Applications, ISDA'05, pp. 256–261. IEEE Computer Society Press, Los Alamitos (2005)
18. Tomczyk, A., Szczepaniak, P.S.: Adaptive Potential Active Hypercontours. In: Rutkowski, L., Tadeusiewicz, R., Zadeh, L.A., Zurada, J.M. (eds.) ICAISC 2006. LNCS (LNAI), vol. 4029, pp. 692–701. Springer, Heidelberg (2006)

An Efficient Polynomial Delay Algorithm for Pseudo Frequent Itemset Mining

Takeaki Uno[1] and Hiroki Arimura[2]

[1] National Institute of Informatics,
2-1-2, Hitotsubashi, Chiyoda-ku, Tokyo 101-8430, Japan
uno@nii.jp
[2] Graduate School of Information Science and Technology, Hokkaido University,
Kita 14 Nishi 9, Sapporo 060-0814, Japan
arim@ist.hokudai.ac.jp

Abstract. Mining frequently appearing patterns in a database is a basic problem in informatics, especially in data mining. Particularly, when the input database is a collection of subsets of an itemset, the problem is called the frequent itemset mining problem, and has been extensively studied. In the real-world use, one of difficulties of frequent itemset mining is that real-world data is often incorrect, or missing some parts. It causes that some records which should include a pattern do not have it. To deal with real-world problems, one can use an ambiguous inclusion relation and find patterns which are mostly included in many records. However, computational difficulty have prevented such problems from being actively used in practice. In this paper, we use an alternative inclusion relation in which we consider an itemset P to be included in an itemset T if at most k items of P are not included in T, i.e., $|P \setminus T| \leq k$. We address the problem of enumerating frequent itemsets under this inclusion relation and propose an efficient polynomial delay polynomial space algorithm. Moreover, To enable us to skip many small non-valuable frequent itemsets, we propose an algorithm for directly enumerating frequent itemsets of a certain size.

1 Introduction

The frequent pattern mining problem is to find patterns frequently appearing in a given database. It is one of the central tasks in data mining, and has been a focus of recent informatics studies. Particularly, when the database is a collection of transactions and the patterns to be found are also subsets of itemsets, the problem is called the frequent itemset mining problem[1,4,10,11,12]D Precisely, we define the *frequency* of an itemset by the number of transactions including the pattern, and say an itemset is a *frequent itemset* if its frequency is no less than the given threshold value σ, called *minimum support*.

Frequent pattern mining is often used especially for data analysis. For data so huge that humans can not get any intuition from an overview of it, the frequent pattern mining is a useful way to capture the features of the data's features, both

V. Corruble, M. Takeda, and E. Suzuki (Eds.): DS 2007, LNAI 4755, pp. 219–230, 2007.

in a global sense and in a local sense. However, in the real world use, we often encounter difficulties in trying to use the frequent pattern mining on real-world data. One difficulty is that data are often incorrect or missing parts. Such errors mean that some records that should include a pattern P do not include P, thus P may be overlooked because its frequency appears to be too low. A way to deal with this difficulty is to consider an ambiguous inclusion relation whereby we consider that a transaction T includes a pattern P if most items of P are included in T.

There are several studies on the frequent pattern mining with ambiguous inclusions. In some contexts, these patterns are called fault-tolerant frequent itemsets[5,6,7,8,14]. In some of these studies, ambiguous inclusion is defined such that an itemset P is included in a transaction T if the fraction of items of P included in T is no less than a given threshold θ, i.e., $|P \cap T|/|P| \geq \theta$[14]. Given this definition, the family of frequent itemsets is not always anti-monotone; thus the usual apriori based algorithms are not output sensitive in the sense of time complexity.

If an item of the itemset is not included in a transaction, then it can be considered to be a fault. Some studies, such as Boulicaut et al., Liu et al., and Seppanen et al.[5,6,7,8], treat mining pairs of an itemset and a transaction set such that there are few faults between their elements. When the size of the transaction set is large, we can regard the itemset as a frequent pattern with ambiguous inclusions. Many mining algorithms have been devised for solving both problems, but enumeration difficulties prevent them from having the completeness that ensures that they output exactly all frequent patterns.

On the other hand, in sequence pattern mining and text mining, ambiguous matching is used to define the occurrence of a pattern, i.e., if a pattern is homogeneous to a substring of the input string, then we regard that the pattern appears at the position. Such patterns are called degenerate patterns in some contexts, especially in genome sciences, and several algorithms have been proposed[9,13].

There are possibly several models for such ambiguous inclusions. In this paper, we define our ambiguous relation with a constant k that a pattern P is included in a transaction (or pattern) T if at most k items are not included in T. In this paper, we address the problem of enumerating all frequent itemsets under this inclusion relation, for given a transaction database, minimum support σ, and k. When σ is large, such as 90% of the number of transactions, the problem can be considered to be one to find combinations of items $i_1, \ldots, i_h$ such that at least $h-k$ items are included in 90% of transactions, thus such combinations characterize the database. These combinations can also be used as rules separating the database from other database, thus has applications to learning theory and practice.

For the frequent itemset mining with our ambiguous relation, we propose a polynomial delay polynomial space algorithm. To best of our knowledge, this is the first result of even output polynomial time algorithm for this problem. Although the algorithm is polynomial time, we still encounter a problem in the real-world applications, that is, quite many uninteresting small patterns are frequent in our ambiguous inclusion relation. We can avoid this problem by directly enumerating all frequent patterns of given size l, and we propose an efficient algorithm for this task.

The organization of the paper is as follows. We introduce several notations and notions concerning to our ambiguous inclusion in Section 2, and propose a basic algorithm for frequent itemset mining under the condition of he ambiguous inclusion in Section 3. The algorithm is improved in Section 4. Section 5 describes an algorithm for directly enumerating those of size l, and Section 6 is for the conclusion.

2 Preliminary

Let an *itemset* I be a set of items $1, \ldots n$. A *transaction database* $\mathcal{D}$ is a collection of transactions where a *transaction* is a subset of I^1. We denote the number of transactions in $\mathcal{D}$ by $|\mathcal{D}|$, and the size of $\mathcal{D}$ by $||\mathcal{D}||$. Here the size of $\mathcal{D}$ is the sum of $|\mathcal{D}|$ and the sizes of the transactions in $\mathcal{D}$, i.e., $||\mathcal{D}|| = |\mathcal{D}| + \sum_{T \in \mathcal{D}} |T|$. Note that $||\mathcal{D}||$ is not defined in the usual sense. The aim of this definition is to consider the computation time for empty transactions, which is $O(1)$. Hereafter, we fix the database $\mathcal{D}$ for the input of the algorithm.

A *pattern* P is a subset of itemset I. The largest item in P is called the *tail* of P and is denoted by $tail(P)$. A transaction of $\mathcal{D}$ including P is called an *occurrence* of P. We denote the set of occurrences of P by $Occ(P)$. The *frequency* $frq(P)$ of a pattern P is defined by the number of transactions including P, i.e., $|Occ(P)|$. Given a transaction database $\mathcal{D}$ and constant number σ, a pattern with frequency no less than σ is called a *frequent itemset*. The frequency is often called *support*, and σ is called the *minimum support*. The problem of finding all frequent itemsets for given a transaction database and minimum support is called the *frequent itemset enumeration problem*[2].

For a constant k and two patterns $P, T \subseteq I$, we write $P \subseteq_k T$ if $|P \setminus T| \leq k$ holds. We call the binary relation $\subseteq_k$ the *k-pseudo inclusion relation*. For a pattern P, a transaction T is a *k-pseudo occurrence* of P if $P \subseteq_k T$. We denote the set of k-pseudo occurrences of P by $Occ_{\leq k}(P)$. Particularly, the set of transactions satisfying $|P \setminus T| = k$ is denoted by $Occ_{=k}(P)$. We have $Occ(P), Occ_{=k}(P) \subseteq Occ_{\leq k}(P)$. See the example in Fig. 1. We define the *k-pseudo frequency* of P by $|Occ_{\leq k}(P)|$, and denote it by $frq_k(P)$. A *k-pseudo frequent itemset* is a pattern P such that its k-pseudo frequency is no less than σ. Here, we define our problem as follows.

Pseudo Frequent Itemset Enumeration Problem
Input: transaction database $\mathcal{D}$, minimum support σ, constant k
Output: all k-pseudo frequent itemsets in $\mathcal{D}$

---

[1] In the literatures, a transaction is often defined by a pair of an item subset and its ID. However, we will here omit the ID since ID has no meaning in the arguments in this paper.

[2] This problem is also called frequent itemset/pattern mining/discovery. Usually, the terms mining and discovery do not require the output to be complete, thus here we use the term enumeration which is used in the problem of outputting all the solutions completely.

A: 1,2,4,5,6	$Occ_{\leq 0}(\{2,7\})$ $= \{E,F\}$
B: 2,3,4	$Occ_{\leq 1}(\{1,2,4\})$ $= \{A,B,C\}$
C: 1,2,7	$Occ_{-1}(\{1,3,7\})$ $= \{C,E\}$
D: 1,5	
E: 2,3,7	$Occ_{\leq 2}(\{1,2,4,7\}) = \{A,B,C,E,F\}$
F: 2,7	$Occ_{-2}(\{1,2,4,7\}) = \{B,C,E,F\}$
G: 4	$Occ_{\leq 2}(\{1,2,4,7\} \cup \{3\})$
H: 6	$= Occ_{\leq 1}(\{1,2,4,7\}) \cup (Occ_{-2}(\{1,2,4,7\}) \cap Occ(\{3\}))$
	$= \{A,E\}$

Fig. 1. Examples of pseudo occurrences and update by addition of an item

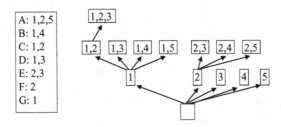

A: 1,2,5
B: 1,4
C: 1,2
D: 1,3
E: 2,3
F: 2
G: 1

Fig. 2. An example of backtrack algorithm execution for minimum support $\sigma = 4$

If an algorithm terminates in polynomial time for both the input size and the output size, the algorithm is called *output polynomial*. Output polynomiality is a popular measure of the theoretical efficiency of the algorithm. If the computation time between any two consecutive output solutions is bounded by a polynomial of the input size, the algorithm is called *polynomial delay*. If an algorithm is polynomial delay, the computation time is linear in the number of outputs, and hence better in practice. If the memory usage of the algorithm is bounded by a polynomial of the input size, the algorithm is called *polynomial space*. Here our goal is to develop an efficient polynomial delay polynomial space algorithm for solving the pseudo frequent itemset enumeration problem.

3   Basic Algorithm

The frequent itemset enumeration problem is, from the viewpoint of complexity theory, an easy problem. The reason is that the frequency has a monotone property, thus obviously any frequent itemset can be obtained by iteratively adding items to the emptyset by passing through only frequent itemsets. Although a naive implementation may produce duplicate solutions, we can avoid duplications by using tail extension. For a frequent itemset P, a pattern obtained from P by adding an item larger than the tail of P is called a *tail extension* of P. By generating frequent patterns only via tail extensions, each pattern is generated only from the pattern obtained by removing its tail, thus we can enumerate all frequent itemsets without duplicates. A backtrack algorithm generates tail extensions in a depth-first manner, and thus is a polynomial time delay polynomial space algorithm. Precisely, the computation time for each frequent itemset is linear in the size of the database, i.e., $O(||\mathcal{D}||)$. The space complexity is also optimal, that is, $O(||\mathcal{D}||)$.

Regarding the practical use of frequent itemset enumeration, the number of frequent itemsets is usually not so large compared to the input size, but the input size is usually large. Thus, linear time in the input size for each solutionis too long. To reduce the practical computation time, several techniques have been proposed. One of the most efficient techniques is called database reduction.

Consider the following operation: Remove all items included in less than σ transactions, and unify the same transactions into one transaction. Then, the database shrinks, and its size becomes small. We can further reduce the size by using trie or prefix tree. This operation is called *database reduction*. Database reduction performs well in practice, especially when σ is large, Moreover, if we apply database reduction to *conditional databases* to recursively reduce their sizes, we can further reduce the computation time. Here a conditional database is the database restricted to items larger than the tail of the current operating pattern P and transactions including P, which is the input of an iteration with respect to P. This technique is called *iterated database reduction*.

These techniques can be applied to pseudo frequent itemsets in a similar way. We begin with the following proposition to see the monotonicity.

Proposition 1. *For any patterns P and P' satisfying $P \subseteq P'$, $Occ_{\leq k}(P') \subseteq Occ_{\leq k}(P)$ holds.*

The statement holds since any transaction $T \in Occ_{\leq k}(P')$ does not include at most k items in P. From this proposition, we can see that the family of k-pseudo frequent itemsets satisfies anti-monotonicity. Hereafter, we assume that the minimum support is from 1 to $|\mathcal{D}|$ thus the emptyset is always k-pseudo frequent. For a k-pseudo frequent itemset P, let the children set of P, denoted by CHD(P), be the set of items i such that $i > tail(P)$ and $P \cup \{i\}$ is k-pseudo frequent. The monotone property leads to the following backtrack algorithm. By calling **Backtrack**($\emptyset$), we can enumerate all k-pseudo frequent itemsets.

backtrack(P)
1. Output P
2. Compute CHD(P)
3. **for each** $i \in$ CHD(P) **call backtrack** $(P \cup \{i\})$

It is easy to see the correctness of this algorithm. Figure 2 shows an execution of the backtrack algorithm. Each iteration inputs a k-pseudo frequent itemset P, outputs P, computes the k-pseudo frequency for all tail extensions of P to obtain CHD(P), and generates recursive calls for each item in CHD(P). Thus, any iteration outputs a k-pseudo frequent itemset, and the computation time for each k-pseudo frequent itemset is bounded by the maximum computation time of an iteration. Computing k-pseudo frequency of each tail extension takes $O(||\mathcal{D}||)$ time thus the computation time of an iteration is $O(n||\mathcal{D}||)$. This can be shortened as follows.

Suppose that we have $Occ_{=0}(P), ..., Occ_{=k}(P)$ for the current processing pattern P. Here we consider the computation of $Occ_{=0}(P \cup \{i\}), ..., Occ_{=k}(P \cup \{i\})$ for all $i > tail(P)$. First, we prove the following proposition.

Proposition 2. *For a transaction T included in $Occ_{=h}(P)$ for some $h, 0 \leq h \leq k$, $T \in Occ_{=h}(P \cup \{i\})$ holds if T includes i. Otherwise, $T \in Occ_{=h+1}(P \cup \{i\})$.*

Proof. Since $T \in Occ_{=h}(P)$, T does not include exactly h items of $(P \cup \{i\})$ if T includes i, and exactly h items otherwise. Then the statement follows.   □

Now we have the following lemma.

Lemma 1. *The following two equations hold,*
(a) $Occ_{=0}(P \cup \{i\}) = Occ_{=0}(P) \cap Occ(\{i\})$
(b) $Occ_{=h}(P \cup \{i\}) = (Occ_{=h}(P) \cap Occ(\{i\})) \cup (Occ_{=h-1}(P) \setminus Occ(\{i\}))$ for any $h \geq 1$.

Proof. Any transaction $T \in Occ_{=h}(P \cup \{i\}), 0 \leq h \leq k$, includes at least $h - 1$ and at most h items of P. This implies that T is included in $Occ_{=h}(P)$ or $Occ_{=h-1}(P)$ only when $h > 0$. On the other hand, from Proposition 2, we have

$$Occ_{=h}(P \cup \{i\}) \cap Occ_{=h}(P) = Occ_{=h}(P) \cap Occ(\{i\}), \text{and}$$

$$Occ_{=h}(P \cup \{i\}) \cap Occ_{=h-1}(P) = Occ_{=h-1}(P) \setminus Occ(\{i\}), \text{ for } h > 0.$$

Thus, the statement of the lemma holds.   □

The next proposition is a consequence of the lemma.

Proposition 3. $Occ_{\leq k}(P \cup \{i\}) = Occ_{\leq k-1}(P) \cup (Occ_{=k}(P) \cap Occ(\{i\}))$.

From Lemma 1, we can see that all we have to do is take intersection of occurrences for all i. For this task, the technique so called *occurrence deliver* described in [10,12,11] is efficient.

Let us consider the task of computing $Occ_{=k}(P \cup \{i\})$ for all $i > tail(P)$. First, we prepare an empty bucket for each item i. Next, for each transaction T in $Occ_{=k}(P)$, we do "insert T into the bucket of i for each item $i \in T, i > tail(P)$". After performing this operation for all transactions in $Occ_{=k}(P)$, the content of the bucket of i is equal to $Occ_{=k}(P \cup \{i\})$. The pseudo code of occurrence deliver is described as follows. The code inputs a set of transactions S and pattern P, then sets $bucket[i]$ to $S \cap Occ(\{i\})$ for all $i > tail(P)$. We suppose that the bucket of any item i is initialized, and thus is empty at the beginning.

Occurrence deliver(S, P)
1. **for each** $T \in S$ **do**
2.   **for each** $i \in T, i > tail(P)$ **do**
3.   insert T into $bucket[i]$
4.   **end for**
5. **end for**

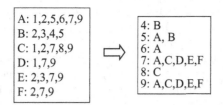

Fig. 3. Example execution of occurrence deliver

Fig. 3 shows an example of the execution of occurrence deliver. Let $\mathcal{S}_{>h} = \{T \cap \{h + 1, \ldots, |I|\} \mid T \in \mathcal{S}\})$. Hereafter, we assume that each transaction T is stored in memory so that the items in T are sorted in increasing order of items. Bucket sort or radix sort to all transactions at once can be done in $O(||\mathcal{D}|| + |I|)$ time. The following proposition is proved in [10,11,12].

Lemma 2. *Algorithm* **Occurrence deliver** *takes* $O(||\mathcal{S}_{>tail(P)}||)$ *time and computes* $\mathcal{S} \cap Occ(\{i\})$ *for all* $i > tail(P)$.

Lemma 2 leads in turn to the following proposition.

Proposition 4. *For pattern* P, *we can compute the* k-*pseudo frequency of all* $P \cup \{i\}, i > tail(P)$ *having non-zero* k-*pseudo frequency in* $O(||Occ_{=k}(P)_{>tail(P)}||)$ *time.*

From Proposition 4, we can see that computation of CHD(P) can be done in $O(||Occ_{=k}(P)_{>tail(P)}||) = O(||\mathcal{D}||)$ time. Next let us consider the cost of computing $Occ_{=0}(P \cup \{i\}), \ldots, Occ_{=k}(P \cup \{i\})$ for each $i \in$ CHD(P). From Lemma 1, we can see that it can be computed by taking the intersection of $Occ_{\leq k}(P)$ and $Occ(\{i\})$ in $O(|Occ_{\leq k}(P)| + |Occ(\{i\})|)$ time. The following proposition is stated for the memory use[10,12,11].

Proposition 5. *For any set* $\mathcal{S} \subseteq \mathcal{D}$ *of transactions and item* i, *the size of the bucket of* i *does not exceed* $|Occ(\{i\})|$ *after applying occurrence deliver.*

We can see from Proposition 5 that the memory used by an iteration is bounded by $O(||\mathcal{D}||)$. The depth of the recursion of **Backtrack** is at most n, and the accumulated memory usage is $O(n||\mathcal{D}||)$.

Theorem 1. *For given a database* $\mathcal{D}$, *minimum support* σ *and constant* k, *algorithm* **Backtrack** *enumerates* k-*pseudo frequent itemsets in* $O(N \cdot ||\mathcal{D}||)$ *time with using* $O(n||\mathcal{D}||)$ *memory, where* N *is the number of* k-*pseudo frequent itemsets.*

Corollary 1. *Algorithm* **Backtrack** *is a polynomial delay polynomial space algorithm for enumerating all* k-*pseudo frequent itemsets.*

4   Reducing Computational Cost

In this section, we improve the efficiency of the algorithm proposed in the previous section by reducing both time and space complexities. Our basic idea is to re-use one bucket in all iterations. This results in a reduction of memory usage.

Here we denote a transaction T in $Occ_{=h}(P \cup \{i\})$ by a pair (T, h). Instead of having all $Occ_{=0}(P \cup \{i\}), ..., Occ_{=k}(P \cup \{i\})$, we maintain $Occ'_{\leq k}(P) = \{(T, h) \mid T \in Occ_{=h}(P \cup \{i\})\}$ keeping that all elements (T, h) in $\bar{O}cc'_{\leq k}(P)$ are sorted in increasing order of h. Then, by applying occurrence deliver to $Occ'_{\leq k}(P)$, we can obtain $Occ'_{\leq k}(P \cup \{i\})$ while keeping the order in $O(\|Occ_{=k}(P)_{>tail(P)}\|)$ time. By looking at the bottom of each bucket, we can easily take $Occ_{=k}(P)$ in $O(|Occ_{=k}(P)|)$ time. This simplifies the operation to maintain the $Occ_{=h}$ for all h.

A technique called *rightmost sweep* is useful for the re-use of buckets[10]. The following propositions and lemmas regard the availability of buckets.

Proposition 6. *For an iteration inputting pattern P, no bucket of $i \leq tail(P)$ is accessed from the beginning of the iteration to the termination of the iteration, including the execution of the recursive calls.*

An iteration adds items i larger than $tail(P)$, and $tail(P \cup \{i\}) > tail(P)$ always holds. Occurrence deliver accesses only the buckets of i satisfying $i > tail(P)$; thus the statement holds. Proposition 6 indicates that when we generate a recursive call with respect to $P \cup \{i\}$, the bucket of any $j < i$ is preserved until the end of the recursive call. Thus, we consider the following algorithm PFIM (Pseudo Frequent Itemset minor) that generates recursive calls in decreasing order of indices.

PFIM$(P, Occ'_{\leq k}(P))$
1. Output P
2. Apply occurrence deliver to $Occ'_{\leq k}(P)$
3. **if** $|Occ_{\leq k-1}(P)| \geq \sigma$ **then** $L := \{tail(P) + 1, ..., n\}$
4. **else** $L := \{i \mid |Occ_{=k}(P \cup \{i\})| > 0\}$
   remove $i \notin \mathrm{CHD}(P)$ from L and initialize the bucket of i
5. **end if**
6. sort items in L in the decreasing order
7. **while** $L \neq \emptyset$ **do**
8.   extract the head i of L
9.   **call PFIM** $(P \cup \{i\}, Occ'_{\leq k}(P \cup \{i\}))$
10.   initialize the bucket of i
11. **end while**

This algorithm re-uses buckets; thus the buckets to be used seem to be not initialized at the beginning of an iteration. However, if we can prove that those buckets are actually initialized at the beginning, we can be assured of the correctness of the algorithm.

Lemma 3. *If all buckets of $i > tail(P)$ are initialized at the beginning of an iteration of **PFIM** inputting pattern P, then the buckets of $i > tail(P)$ are also initialized at the termination of the iteration.*

Proof. We prove the statement by the induction, starting from the leaves of the computation tree of the algorithm. For any iteration, we define its height by 0 if it generates no recursive call, and the maximum height plus one otherwise. The height is the distance to the farthest leaf among its descendants in the computation tree.

First, we consider an iteration that generates no recursive call. In such an iteration, all buckets inserted some elements in step 2 are initialized in step 4, thereby L has no element. Thus, the statement holds.

Next, we suppose that for any iteration of height at most h satisfies the statement, and we consider an iteration I of height $h + 1$. Let P be the input pattern of I and suppose that at the beginning of I, the bucket of any $i > tail(P)$ is initialized. Of the buckets holding some elements in step2, the buckets of $i \notin \text{CHD}(P)$ are initialized in step 4. Several recursive calls are generated in the loop from step 7 to step 12. Suppose that i is the head of L. When we generate the recursive call with $P \cup \{i\}$, the bucket of any $j > tail(P \cup \{i_1\})$ is initialized since L is sorted in decreasing order. From the assumption of the induction, the bucket of any $j > tail(P \cup \{i\})$ is initialized after the termination of the recursive call. Then, the bucket of i is initialized. Since i is extracted from L, for the new head i' of L, the bucket of any $j > i'$ is again initialized. In this way, recursive calls are generated with satisfying the assumption of the statement. Thus, after generating recursive calls for all items in L, the bucket of any $j > tail(P)$ is initialized. □

Form the lemma, we obtain the following theorem.

Theorem 2. *Algorithm* **PFIM** *uses* $O(||\mathcal{D}||)$ *memory and enumerates all k-pseudo frequent itemsets in $\mathcal{D}$ in* $O(\sum_{P \in \mathcal{F}} ||Occ_{\leq k}(P)_{>tail(P)}|| + \log n) = O(|\mathcal{F}| \times ||\mathcal{D}||)$ *time, where $\mathcal{F}$ is the family of k-pseudo frequent itemsets.*

Proof. The correctness of the algorithm is obvious from the correctness of Algorithm **Backtrack** and Lemma 3. The statement for the memory usage is clear from the re-use of buckets.

Next, we discuss the computation time. Step 2 is done in $O(||Occ_{\leq k}(P)_{>tail(P)}||)$ time, and step 6 is done in $O(|\text{CHD}(P)| \log n)$ time. Other steps can be done in $O(|\text{CHD}(P)|)$ time. Thus, by taking the sum over all k-pseudo frequent itemsets, the total computation time is bounded by $O(\sum_{P \in \mathcal{F}}(||Occ_{\leq k}(P)_{>tail(P)}|| + \log n)) = O(|\mathcal{F}| \times ||\mathcal{D}||)$. □

The structure of the algorithms is almost equal to that of LCM[10,11,12] for the frequent itemset enumeration. Our algorithm can be used together with practical efficient techniques such as database reduction, thus our algorithm should perform well in practice.

5 Efficient Computation in Practice

In this paper, we use the k-pseudo inclusion relation as a model of ambiguous inclusion. Although this is a natural modeling, it has a weak point in practice; that is, many small patterns are k-pseudo frequent. For example, any pattern whose size is no greater than k is a k-pseudo frequent itemset, and an addition of any item to a $(k-1)$-pseudo frequent itemset also yields a k-pseudo frequent itemset. In the real-world problems, we may not have much interest in these small patterns.

To cope with this difficulty, we often enumerate only the maximal patterns in the sense of set inclusion. However, possibly so many small itemsets have k-pseudo frequencies close to the minimum support, many of these small patterns become maximal. Moreover, we lose non-maximal but large k-pseudo frequent itemsets. Thus, we here address the method for enumerating k-pseudo frequent itemsets of given size l directly. For a pattern P and its item i, let $Occ^*_{=k}(P, i)$ be the set of k-pseudo occurrences T of P such that T does not include i, i.e., $Occ^*_{=k}(P, i) = \{T \mid T \in Occ_{=k}(P), i \notin T\}$.

Lemma 4. *For any pattern P, there exists a sequence of its items $(i_1, i_2, \ldots, i_{|P|})$ such that for any y, $|Occ_{\leq k-1}(\{i_1, \ldots, i_y\})| \geq |Occ_{\leq k}(P)| \frac{|P|-y}{|P|}$ holds.*

Proof. Let $(i_1, i_2, \ldots, i_{|P|})$ be the items of P sorted in increasing order of $|Occ^*_{=k}(P, i_j)|$, i.e., for any $1 \leq y < |P|$, $|Occ^*_{=k}(P, i_y)| \leq |Occ^*_{=k}(P, i_{y+1})|$ holds. Consider the $(k-1)$-pseudo frequency of pattern $\{1, \ldots, y\}$ for some $1 \leq y < |P|$. For any $j > y$, $\{1, \ldots, y\}$ is included in any transaction of $Occ^*_{=k}(P, i_j)$ in the sense of $(k-1)$-pseudo inclusion. Observe that the average of $|Occ^*_{=k}(P, i_j)|, 1 \leq j \leq |P|$ is at most $|Occ_{=k}(P)| \frac{k}{|P|}$, and one transaction is included in $Occ^*_{=k}(P, i_j)$ at most k j's. Thus, we see that the cardinality of $\bigcup_{j=y+1}^{|P|} Occ^*_{=k}(P, i_j)$ is at least $(|P| - y) \times |Occ_{=k}(P)| \frac{k}{|P|} / k = |Occ_{=k}(P)| \frac{|P|-y}{|P|}$. Since $|Occ_{\leq k-1}(\{i_1, \ldots, i_y\})| = |Occ_{\leq k-1}(P)| + |\bigcup_{j=y+1}^{|P|} Occ^*_{=k}(P, i_j)|$, the sequence $(i_1, \ldots, i_{|P|})$ satisfies the statement. □

For given a constant l and a pattern P such that $|P| < l$, we call the condition $|Occ_{\leq k-1}(P)| \geq \sigma \frac{l-|P|}{l}$ the *partial frequency condition*, and we denote by $\mathcal{K}$ the set of all k-pseudo frequent itemsets of size less than l satisfying the partial frequency condition. From the lemma, we can see that any k-pseudo frequent itemset of size l can be generated by adding items by passing through only patterns in $\mathcal{K}$, thus we can use the condition for pruning the iterations. The size of $\mathcal{K}$ is expected to be smaller than that of k-pseudo frequent itemsets of sizes of at most l, thus the computation time will be short.

For such a generation, we can not use the usual tail extension, since for some $P \in \mathcal{K}$, $P \setminus \{tail(P)\}$ may not be in $\mathcal{K}$. On the other hand, if we add items smaller than the tail, we may produce a pattern $P = \{i_1, \ldots, i_h\} \in \mathcal{K}$ twice from $P \setminus \{i_j\}$, and $P \setminus \{i_g\}$ for some $j \neq g$, Thus, we consider the following generation rule to avoid duplicates.

Generation Rule: Generate each pattern $P \in \mathcal{K}$ only from the pattern $P \setminus \{i\}, i \in P$ maximizing $|Occ_{k-1}(P \setminus \{i\})|$ among all patterns obtained by removing an item from P. Ties are broken by lexicographical order.

Lemma 5. *Adding items under the generation rule, any $P \in \mathcal{K}$ is generated exactly once.*

An enumeration algorithm using such a generation rule is called *reverse search*[3]. The algorithm is as follows.

ReverseSearch (P)
1. **if** $|P| = l$ **then output** P ; **return**
2. **for** each $i \notin P$ **do**
3.   **if** $|Occ_{\leq k}(P \cup \{i\})| \geq \sigma$ **then** // k-pseudo frequency check
4.   **if** $|Occ_{\leq k-1}(P \cup \{i\})| \geq \frac{\sigma}{l}(l - |P|)$ **then** // partial frequency check
5.   **if** P and $P \cup \{i\}$ satisfy the generation rule **then**
        call ReverseSearch $(P \cup \{i\})$
6. **end for**

Lemma 6. *The computation time of an iteration of the algorithm* **ReverseSearch** *is* $O(|P| \times ||D||)$.

Proof. The key to the computation time is steps 3, 4 and 5. For steps 3 and 4, we explained that they can be done in $O(||D||)$ time, thus we have to consider only step 5. It checks the generation rule, by computing $|Occ_{\leq k-1}(P \cup \{i\} \setminus \{j\})|$ for all $j \in P$. This takes $O(||D|| \times |P|)$ in a straightforward way, we thereby explain how to decrease it.

Observe that $|Occ_{\leq k-1}(P \cup \{i\} \setminus \{j\})| = |Occ_{\leq k-1}(P \cup \{i\})| + |Occ_{= k}(P \cup \{i\}) \setminus Occ(\{j\})|$. Since $Occ_{\leq k-1}(P \cup \{i\})$ can be obtained in $O(||D||)$ time, all we have to do is to compute $|Occ_{= k}(P \cup \{i\}) \setminus Occ(\{j\})|$ quickly. For the task, we maintain the set $Occ_{= k}(P) \cap Occ(\{j\})$ for $j \in P$ in memory, and update them in each iteration. This takes $O(||D||)$ time by occurrence deliver. Using these, we can compute $|Occ_{= k}(P \cup \{i\}) \setminus Occ(\{j\})|$ for all $j \in P$ in $O(||Occ_{= k}(P \cup \{i\})|| \times |P|)$ time. Since the sum of $||Occ_{= k}(P \cup \{i\})||$ over all $i \notin P$ never exceed $||D||$, the time to compute $|Occ_{= k}(P \cup \{i\}) \setminus Occ(\{j\})|$ for all pairs of i and j is $O(|P| \times ||D||)$. □

6 Conclusion and Future Work

In this paper, we introduced an ambiguous inclusion relation to the frequent itemset mining as a meaning of dealing with errors and ambiguities. We chose a model for ambiguous inclusion by relaxing the inclusion relation so that several items can be excluded, and formulated the pseudo frequent itemset enumeration problem by the inclusion relation. To solve the problem, we proposed an efficient polynomial delay polynomial space algorithm. The algorithm inherits the structure from the existing efficient frequent itemset mining algorithms, thus we expect that it will have high performance in practical use. To skip many small and non-valuable frequent itemsets, we propose an algorithm for directly enumerating frequent itemsets of a certain size. As future works, to evaluate the efficiency in the real-world problems implementation of the algorithm and computational experiments are crucial. Another interesting research topic is extensions of the technique in this paper to other frequent pattern mining problems.

Acknowledgments

Part of this research was supported by Grant-in-Aid for Scientific Research of Japan "Developing efficient and accurate algorithms for large-scale data processing in genome science", and and joint-research funds of the National Institute of Informatics.

References

1. Agrawal, R., Mannila, H., Srikant, R., Toivonen, H., Verkamo, A.I.: Fast Discovery of Association Rules. Advances in Knowledge Discovery and Data Mining , 307–328 (1996)
2. Asai, T., Abe, K., Kawasoe, S., Arimura, H., Sakamoto, H., Arikawa, S.: Efficient Substructure Discovery from Large Semi-structured Data. In: SDM 2002 (2002)
3. Avis, D., Fukuda, K.: Reverse Search for Enumeration. Discrete App. Math. 65, 21–46 (1996)
4. Bayardo Jr., R.J.: Efficiently Mining Long Patterns from Databases. In: Proc. SIGMOD'98, pp. 85–93 (1998)
5. Besson, J., Robardet, C., Boulicaut, J.F.: Mining Formal Concepts with a Bounded Number of Exceptions from Transactional Data. In: Goethals, B., Siebes, A. (eds.) KDID 2004. LNCS, vol. 3377, pp. 33–45. Springer, Heidelberg (2005)
6. Liu, J., Paulsen, S., Wang, W., Nobel, A., Prins, J.: Mining Approximate Frequent Itemsets from Noisy Data. In: 5th IEEE International Conference on Data Mining (ICDM'05), pp. 721–724 (2005)
7. Seppanen, J.K., Mannila, H.: Dense Itemsets. In: SIGKDD 2004 (2004)
8. Shen-Shung, W., Suh-Yin, L.: Mining Fault-Tolerant Frequent Patterns in Large Databases. In: ICS2002 (2002)
9. Takeda, M., Inenaga, S., Bannai, H., Shinohara, A., Arikawa, S.: Discovering Most Classificatory Patterns for Very Expressive Pattern Classes. In: Grieser, G., Tanaka, Y., Yamamoto, A. (eds.) DS 2003. LNCS (LNAI), vol. 2843, pp. 486–493. Springer, Heidelberg (2003)
10. Uno, T., Asai, T., Uchida, Y., Arimura, H.: LCM: An Efficient Algorithm for Enumerating Frequent Closed Item Sets. In: Proc. IEEE ICDM 2003 Workshop FIMI 2003 (2003)
11. Uno, T., Asai, T., Uchida, Y., Arimura, H.: An Efficient Algorithm for Enumerating Closed Patterns in Transaction Databases. In: Suzuki, E., Arikawa, S. (eds.) DS 2004. LNCS (LNAI), vol. 3245, pp. 16–31. Springer, Heidelberg (2004)
12. Uno, T., Kiyomi, M., Arimura, H.: LCM ver. 2: Efficient Mining Algorithms for Frequent/Closed/Maximal Itemsets. In: Proc. IEEE ICDM'04 Workshop FIMI'04 (2004)
13. Wang, J.T.L., Chirn, G.W., Marr, T.G., Shapiro, B., Shasha, D., Zhang, K.: Combinatorial pattern discovery for scientific data: some preliminary results. In: Proceedings of the 1994 ACM SIGMOD international conference on Management of data, pp. 115–125 (1994)
14. Yang, C., Fayyad, U., Bradley, P.S.: Efficient Discovery of Error-Tolerant Frequent Itemsets in High Dimensions. In: SIGKDD 2001 (2001)

Discovering Implicit Feedbacks from Search Engine Log Files

Ashok Veilumuthu and Parthasarathy Ramachandran

Indian Institute of Science, Bangalore 560 012 India
{ashok,parthar}@mgmt.iisc.ernet.in

Abstract. A number of explicit and implicit feedback mechanisms have been proposed to improve the quality of the search engine results. The current approaches to information retrieval depends heavily on the web linkage structure which is a form of relevance judgment by the page authors. However, to overcome spamming attempts and the huge volumes of data, it is important to also incorporate the user feedback on the page relevance of a document. Since users hardly give explicit/direct feedback on search quality, it becomes necessary to consider implicit feedback that can be collected from search engine logs. In this article we evaluate two implicit feedback measures, namely click sequence and time spent in reading a document. We develop a mathematical programming model to collate the feedback collected from different sessions into a partial rank ordering of documents. The two implicit feedback measures, namely the click sequence and time spent in reading a document are compared for their feedback information content using Kendall's τ measure. Experimental results based on actual log data from **AlltheWeb.com** demonstrate that these two relevance judgment measures are not in perfect aggrement and hence incremental information can be derived from them.

Keywords: implicit feedback, search engines, relevance judgment.

1 Introduction

The World Wide Web (WWW) is a very important tool to locate information. In the early 90's the number of pages in the web was of the order of thousands and web directories was quite sufficient to locate the needed information. But the growth of WWW has complicated the information retrieval process. The April 2007 web server survey estimated a total of 113,658,468 active sites in the web [1]. Given this volume of data, browsing through all the sites even with the help of directories becomes impossible. Search engines came into existence in the mid 90's to overcome this problem. Search engines are information retrieval systems which help the user to find the needed information by posting queries.

Initially, search engines were using traditional information retrieval (IR) techniques, in which the keyword similarity between the query and the documents was used to identify the required documents [2]. These techniques suffered from

V. Corruble, M. Takeda, and E. Suzuki (Eds.): DS 2007, LNAI 4755, pp. 231–242, 2007.
© Springer-Verlag Berlin Heidelberg 2007

problems such as, lack of coherence, lack of self description and manipulations done by site owners to mislead the search engines (spamming) [3]. To overcome these problems, web linkage structures were used in Page Rank [4] and HITS algorithms [5] in addition to the traditional IR techniques. Though the web linkage structure captures the importance of the pages to a larger extent, it still does not satisfy user requirements in many situations . It still demands the users to reformulate their queries until they identify the information.

Hence, it would be preferable, if the search engine learns to incorporate the thinking process with which a user reformulates his queries and the intelligence with which he/she selects certain pages to visit from the big list of results. User feedback (explicit and implicit) can be used to decipher this user intelligence. Since most of the users cannot be convinced to give direct feedback (explicit) on search results it becomes imperative to use implicit feedback. Whenever a user searches the web using a search engine the user will leave a trace of his activities in the log files of the search engine server. This log can be used as a source of information for relevance feedback (i.e., the relative importance of the pages with respect to the query posed). But the most challenging part of using the log file lies is in interpreting the log files. In this article we evaluate the information content of the different fields in the search engine log file. Specifically we evaluate the information content of click sequence and the time spent by the user in reading through a document for its utility as implicit relevance feedbacks.

The rest of this paper is structured as follows: It starts with a literature review that briefly discusses about implicit and explicit feedback mechanisms explored by other researchers. Section 3 formally introduces the problem and it also describes how the click sequence and time spent in reading a document can be converted in to preference orders. Section 4 develops a mathematical programming model that aggregates the user preferences across multiple sessions into a combined ranking. Finally in Section 5 we discuss the experimental results that demonstrates the differential information content in the.click sequence and time sequence ranks.

2   Literature Review

A search engine is an automated system which will search for the documents matching a query on behalf of the user. Initially information retrieval (IR) techniques like term matching was used for the document retrieval [6][7][2]. Later the linkage structure of the web was used in addition to the IR techniques [4][5]. Active spamming by site owners reduces the quality of the results presented by the search engine. The web linkages structure only captures the relevance judgment of authors of web-pages and not that of the readers/users. The users know the best about their information needs and hence are the right persons to give judgment on it. Hence the need is to identify and incorporate user experience/feedback in the search engine.

The idea of relevance feedback has been used to incorporate the user experience in improving the query and the results for the future sessions. It is the information given by the user during and/or after his search and it could be explicit or implicit. In explicit feedback, the users could be asked to rank the documents. Such a task imposes an increased burden and increased cognitive load on the users and many users may not be willing to share this information [8][9]. The alternate solution is to acquire implicit ratings by watching user behavior. Implicit ratings include measures of interest such as whether the user read a document and, if so, the order in which users read the documents and how much time the user spent reading them. White et.al [9] and Claypool et.al [10] had done experiments to compare the explicit and implicit relevance feedback systems. They concluded that implicit feedbacks are viable to substitute the explicit counterparts. Since they are hard to come by.

The implicit feedback mechanism can be used for query expansion during short-term modeling of users' immediate information need and user profiling during the long-term modeling of users' persistent interests and preferences [11]. The chain of queries posted by the user and the subsequent documents visited by them can be stored in the log files of the search engines and this information can be used to develop a recommender system which will help reformulating the queries [12] [13] [14]. Radlinski et.al [15] used the set of queries used by the users within a session to help refining the queries for the future users. But in this paper only the users' relevance judgment of documents presented by search engines gathered through implicit feedback is studied.

Goecks et.al [8] used the hyperlinks clicked, mouse and scroll activity of the user as an implicit feedback to learn his/her profile. Kim et.al [16] used the reading time and the printing behavior of the user as a proxy to the users' document preference. Kelly et.al [17] discussed the possibility of using the display time of a document in the browser as an implicit feedback to measure the document preference. In all these works, the implicit feedbacks were recorded at the browser end by suitably designing the browser and the feedbacks were used to model the user profile.

Joachims [18] used the set of pages selected by the users to develop a partial ranking and SVM has been used to convert these preference orders into weights which forms the ranking function. It was assumed that the pages requested by the users are relatively more important than the other pages in the results. A variation of this technique has been proposed by Tan et.al [19]. It includes the co-training of SVM weights by considering different features and augmenting them into single preference weight. But information in form of preference relations within the set of pages selected by the user has been ignored in these two models.

Even though many researchers have confirmed the fact that the time is an effective implicit feedback measure, they all recorded it only at the browser end [16][17]. Only few attempts had been made to use the time entry of the log file, as a feedback [20]. Ramachandran studies the use of time measures as relevance feedback and identifies the issues to be addressed in using them [21]. This paper attempts to extract the time spent by the users on the selected documents from the log file. It also contrasts this feedback information with the order in which the user visited the pages to discover user relevance judgments.

3   Mining User Preferences from Log Files

Usually, the user of a search engine will pose a query based on his/her information needs. Once that need is satisfied he/she will stop searching further. Unfortunately not all the users are experts to be able to identify the needed information in their first attempt. The order in which the user visits the URLs can be taken as a measure to judge the relative preference of the user. The time spent in those selected pages is also a measure of importance of the page for the query posed. These two information (i.e., sequence of pages visited and time spent on each page) can be stored in the search engine log files. The click sequence is an indication of the relevance judgment given by the user based on the short abstract accompanying the links. The time spent on the pages selected gives the relevance judgment of the user after visiting the page. So, intuitively the rank orders based on these two pieces of information individually need not be same. This is because a person who felt a URL to be interesting by reading the abstract, may feel the page to be irrelevant after spending some time on it. Also these two ranking schemes can be different from the rankings published by the search engines [18] [19] [20]. The objective of this article is to generate two ranking schemes based on these two feedback measured and to contrast the two using the standard Kendels τ measure.

Let Q be the the set of queries, D be the set of documents visited through those queries, S be set of sessions and $S_q \subseteq S$ be the set of sessions in which the same query $q \in Q$ has been posted. Then the preference order or rank order r_s given by the user in the session $s \in S$ in response to query q can be represented as a binary relation $\preceq r_s$ over $D \times D$ that will establish the properties of strict weak ordering (i.e., anti-symmetric and transitive). For any two documents $d_i, d_k \in D$, if $d_i \preceq r_s d_k$, then it means that the user prefers d_i than d_k in session s. These preferences across various sessions need to be aggregated to get a single ranking for a particular query.

3.1   Preference Orders Based on Click Sequence

The click sequence is the order in which the user has visited the documents presented by the search engine. In this paper it is assumed that the user takes a partially informed decision based on the short abstract accompanying the link in the results page. So, if an user visits a *link*, it means that the user has selected that *link* only after reading the short abstract of all the *links* that was presented before selecting and visiting the link. Hence, this order is a clear indication of the user's relative preference and it can be directly taken as the relative preference order $r_o{}^{(s)}$ for the session. The implicit assumption here is that the user has reviewed all the results, even though typically only 10 limits are presented in one page. Studies have shown that 62% of users abandon after the first page, and nearly 90% after the first 3 pages [iProspect Search Engine User Behavior Study, April 11, 2006, White paper].

3.2   Preference Orders Based on Time Measure

The click sequence gives only the ranking given by the user based on the abstract accompanying the URLs in the search results page. This inference is based on incomplete/partial information. Since the relevance judgment is based only on the text accompanying the result, the user may feel it irrelevant immediately after seeing its content and may return back to the search results page to visit another link. Hence, the click sequence alone is not very reliable feedback measure. This necessitates the search for additional measures which can give relative relevance judgment of the user given after reading the document. The user will read a document for a longer time, if he/she feels it's contents to be relevant. Hence, time spent on a document will be a useful relevance judgment measure.

The time spent in a particular document can be calculated from the click time of that document and the click time of the next immediate click in the same session. It is assumed that the user is not opening more than one window/tab at a time and he/she is reading the selected page till he find, it irrelevant. Further it is assumed that the user is not distracted by other activities while reading the document and hence the time measures are assumed to be unbiased. Let T_l be the time of selection of URL l and T_{l+1} be the time of selection of the URL which is immediate next to l in the log file, then the time spent on URL l is given by:

$$t_l = T_{l+1} - T_l \tag{1}$$

Here the time spent is only an indicator to establish the partial ordering over the document set D. Hence, it can be directly used. Otherwise normalization, inter and intra session comparability issues need to be addressed [21]. The visited documents can be sorted based on the time spent, to get the partial ordering $r_t^{(s)}$.

4   Preference Aggregation

The rank orders $r^{(s)}$ given in all sessions $s \in S_q$ need to be combined to get a single rank order for the query q. The combined ranking should order the documents with a minimum deviation from the multiple rank orders sourced from the different sessions. Such an aggregation can be done indirectly by deriving a function which will assign weights to all the documents, so that most of the rank ordered pairs gets satisfied. The combined ranking can be obtained by ordering the documents based on these weights. Let F be the family of functions that for each query 'q' assigns a weight to each document in D_q. A function $f \in \{F : q \longrightarrow \mathbb{R}^{|D_q|}\}$ has to be chosen such that most of the ordered pairs across the sessions with the same query are satisfied. Let $r_{f(q)}$ be the ranking over D_q established by the weights from $f(q)$, then it implies:

$$(d_i, d_k) \in r_{f(q)} \Longleftrightarrow w_i > w_k \tag{2}$$

If $(d_i, d_k) \in r_{f(q)}$, then it implies that d_i is preferred over d_k as per the ranking $r_{f(q)}$ (i.e., $d_i \prec r_{f(q)} d_k$). So, the function $f(q)$ has to assign weights to the documents in such a way that $w_i > w_k \ \forall i < k \ (d_i, d_k) \in r_{f(q)}$. Such a function f can be trained by using the margin maximization technique. The following notations are used in the model:

Indices/Sets:
Q : Set of queries in the training sample
D : Set of documents in the training sample
D_q : Set of documents visited by the users for the query q
D_s : Set of documents visited by in the session s
S_q : Set of sessions having the same query q
q : Index for the queries
i, k : Indices for the documents in D_q
j : Rank index
s : Index for the sessions

Parameters:
$r^{(s)}$: Rank order given in session s
R_s : Function that maps the document i to the rank j given by the rank order $r^{(s)}$. $R_s(i) = \{j : j$ is the rank of document $d_i \in D_q$ as per $r^{(s)}\}$
ϕ_s : Function that maps the rank j from $r^{(s)}$ to the index i of the document in the set D_q. $\phi_s(j) = \{i : d_i \in D_q, R_s(i) = j\})$
C : Control parameter to tradeoff between the training error and margin size
ξ_{iks} : Non-negative slack variables (deviations) for the document pair (d_i, d_k) in the session s

Decision Variables:
w_i : weight of the document d_i

The weights of the documents for a query q can be calculated by considering each of the preference relation given by the user as a constraint and the basic formulation is as follows:

$$\text{Minimize} : \frac{1}{2} \boldsymbol{w}^T \boldsymbol{w}$$

Subject to :

$$w_i > w_k \quad \forall (d_i, d_k) \in r^{(s)}, \ \forall s \in S_q$$

$$w_i \geq 0 \quad \forall i$$

This formulation will be feasible only when the preference order given by the users are consistent. But most of the times the users' preference orders will be conflicting, so, while aggregating there is a possibility that the above formulation might lead to an infeasible solution. So, a non-negative deviation variable ξ_{ijs} is added while considering each of the documents pair (d_i, d_k) with respect to session s. But still there is a possibility to get 0 weights for all the documents except any one of the document. In order to force the weights to move from this trivial solution $1 - \xi_{iks}$ is added for each pairwise comparisons in the above model.

$$\text{Minimize} : \frac{1}{2} w^T w + C \sum_{iks} \xi_{iks}$$

Subject to :

$$w_i > w_k + 1 - \xi_{iks} \qquad \forall (d_i, d_k) \in r^{(s)}, \quad \forall s \in S_q$$
$$\xi_{iks} \geq 0 \qquad \forall i, k, s$$
$$w_i \geq 0 \qquad \forall i$$

In general a user does not visit all the documents in D_q and hence the ranking $r^{(s)}$ given in the session s will not be complete. If the rankings given are over a subset of D_q then the weights obtained from the above model might be misleading. For instance, let $D_q = \{d_1, d_2, d_3, d_4, d_5\}$ and there are 3 sessions (3 rankings). Let $r^{(1)} \implies d_2 \prec d_5 \prec d_3$, $r^{(2)} \implies d_2 \prec d_4$ and $r^{(3)} \implies d_1 \prec d_2$. In the first two sessions the document d_2 is preferred than other documents and in the third session the document d_1 is preferred than d_2. The document d_2 has been selected in all the 3 sessions and out of those three, twice it was ranked 1, hence, d_2 is expected to get more weigtage. But the above model will rate d_1 as the best because the document d_1 has got selected only once and in that session it was ranked better than d_2 which is the best document in the other two sessions. Hence, the above model needs to be tuned to overcome this drawback.

Let $D_s \subseteq D_q$ be the set of documents visited in the session s, then the ranking $r^{(s)}$ will establish an ordering over the set D_s rather than in D_q. In this article it is assumed that all the documents in the set D_s are preferred to the documents in $D_q - D_s$ and all the documents in $D_q - D_s$ are equally ranked. Establishing the order given over D_s is more important than the comparisons given over $D_q - D_s$. So those comparisons given over D_s should be given more consideration than the others. The weights can be assigned as follows.

Consider the session 1 $(r^{(1)} \implies d_2 \prec d_5 \prec d_3)$ in the above example. If the relations are consistent and taken together then the relation $d_2 \prec d_5$ is not only explaining the relation $d_2 \prec d_5$ but it also explains all the other relations of d_2 like $d_2 \prec d_3$, $d_2 \prec d_1$ and $d_2 \prec d_4$, hence, it should be given a weight of 4. Similarly the relation $d_5 \prec d_3$ should be given a weight of 3. Since the documents d_1 and d_4 were not visited in session s they both are considered to be equally preferred and hence the relations $d_3 \prec d_1$ and $d_3 \prec d_4$ doesn't explain any thing more than what it is, therefore, they should be given a weight of 1 each. The weights to these comparisons (constraints) can be incorporated in the model by giving the respective weights to the deviational variable ξ_{is} in the objective function. The modified model is given below:

$$\text{Minimize} : \frac{1}{2} \boldsymbol{w}^T \boldsymbol{w} + C \sum_{s=1}^{|S_q|} \left[\sum_{j=1}^{|D_s|-1} (|D_q| - j)\, \xi_{\phi_s(j)s} + \sum_{d_i \in |D_q - D_s|} \xi_{is} \right]$$

Subject to :

$$w_{\phi_s(j)} > w_{\phi_s(j+1)} + 1 - \xi_{\phi_s(j)s} \qquad \forall (d_{\phi_s(j)}, d_{\phi_s(j+1)}) \in D_s,\ \ \forall s \in S_q$$

$$w_{\phi_s(|D_s|)} > w_i + 1 - \xi_{is} \qquad \forall d_i \in D_q - D_s,\ \ \forall s \in S_q$$

$$\xi_{is} \geq 0 \qquad \forall i, s$$

$$w_i \geq 0 \qquad \forall i$$

The above model will assign weight to all the documents in D_q such that the rank ordering implied by those weights will satisfy most of the preference orders given by the users of the individual sessions. We denote this optimal preference order by $r_{f(q)}$ and it can be taken as $r_o{}^{(q)}$ or $r_t{}^{(q)}$ depending on the type of feedback used in the model.

Next we use this model on an actual search engine log file to test the information content in the two types of feedback measures under consideration.

5   Experimental Results

In this experiment a 24 hour (i.e., one day) log data recorded on **6th February 2001** by **AlltheWeb.com** has been used. This data set has been previously used to study the emerging trends in web searching by Jansen et.al [22]. A small description of the dataset is given in the Table 1.

A snapshot of this dataset is given in Table 2. Each tuple in the dataset corresponds to a click event made by a user. The log contains the userID (masked IP), clickTime (i.e., the time at which the click has been made), the query posed and the URL on which the click has been made. In the experiment the entries that have same userID, same query and the clickTime within 30 minutes are considered to belong to the same session. For example in the Table 2 the user 4.16.116.98 has posted two queries "free pics" and "free download mp3" one after the other and in this experiment this will be split into two sessions. In this experiment, 30 non-trivial queries (omitting queries like "google") were chosen that had sufficient number of sessions (≥ 8). These 30 queries will form the query

Table 1. Descriptive statistics of the log data

Variables	Count	Percentage
Queries	451,551	
Mean terms per query	2.4	
Mean pages viewed per query	2.2	
Terms per query		
1 term	113,447	25%
2 terms	161,541	36%
3+ terms	176,563	39%

Table 2. A snapshot from the log file

UserID	clickTime	query	clickURL
4.16.103.153	14:47:19	mp3 to wave	www.mp3towave.com/
4.16.103.153	15:00:23	cd to mp3	www.zy2000.com/
4.16.103.153	15:06:29	cd to mp3	www.birdcagesoft.com/
4.16.116.98	22:19:03	free pics	kingdomcum.com/
4.16.116.98	22:19:14	free pics	kingdomcum.com/
4.16.116.98	22:19:46	free pics	www.adult-worx.com/pics/
4.16.116.98	22:22:23	free download mp3	www.mp3dd.net/

Table 3. Order-based and time-based rank frequency distribution of the documents for query **"waitangi day"**

URLs	Order-based 1 2 3 4 5 6 7	Time-based 1 2 3 4 5 6 7
www.hpl.govt.nz/Waitangi.html	1 4 3 2 0 0 0	4 3 2 1 0 0 0
www.kidlink.org/KIDPROJ/MCC/mcc0361.html	1 0 1 1 1 0 0	2 0 1 1 0 0 0
www.maaori.com/develop/waitangi99.html	0 1 0 0 0 0 0	1 0 0 0 0 0 0
www.mtcarmelchurch.org/waitangi/	15 1 3 0 1 0 0	9 5 3 2 1 0 0
www.muaupoko.iwi.nz/claims.htm	0 1 0 0 0 0 0	0 1 0 0 0 0 0
www.nzhistory.net.nz/gallery/treaty/	2 8 2 2 1 1 1	1 6 5 2 1 1 1
www.pasifika.net/	0 0 0 0 1 0 0	0 0 0 0 1 0 0

set Q and the documents visited through these queries put together will form the set D.

For each query, the order in which the user has clicked the URL is considered as the judgment made by the user before reading the document. This judgment is based only on the short text accompanying the link. It is the order in which the URL gets logged in the log file. But the time based preference order is not explicit. The time spent by the user in a document is calculated as described in Section 3. The documents were then sorted based on the relative time spent by the user within the same session. This sorted order is used as the user's preference order given after glancing the document. For example the click order based and time order based ranking of the documents with respect to the query "baby names" is given in the Table 4.

The order based information is ordinal and the time based information is continuous, so, the time based ranking will be more sensitive and hence less skewed, but, the order based ranking will be skewed. It is highly probable that a page which has been selected first by many of the users to get a different ranks in the time based ranking. This fact is very much seen in the distribution of these two ranks. One such instance is explained in the Table 3. In Table 3 the order based rank frequency distribution for the page www.mtcarmelchurch.org/waitangi/ is highly skewed when compared to the time based rank frequency distribution.

Table 4. A Few sessions for query **"baby names"**

session ID	click Time	Time Spent(s)	click URL	Order Rank	Time Rank
1	5:22:54	37	www.kabalarians.com/	1	3
1	5:23:31	53	www.babynamer.com/	2	2
1	5:24:24	122	www.babycenter.com/babyname/	3	1
1	5:26:26	0	www.heptune.com/names/nameinde.html	-	-
2	18:24:32	5	www.babynames.com/	1	3
2	18:24:37	24	www.kabalarians.com/	2	2
2	18:25:01	57	www.babycenter.com/babyname/	3	1
2	18:25:58	0	www.babycenter.com/babyname/	-	-

If the user visits n documents in a session, there won't be sufficient information to calculate the time spent by the user in the nth (last) document. So the time spent on the nth document is assumed to be 0. But this information loss is not very predominent in the dataset, since in most of the sessions the last visited pages is not the same. It is evident in the example given in Table 4.

These session based rankings were combined using the quadratic program mentioned Section 4. The quadratic program will give the weight of each pages with respect to a particular query. The pages were arranged based on their weights to get the combined ranking. But these rankings are partial rankings (with ties). The partial rankings got from the quadrating programming for the query **"baby names"** is given in Table 5.

For all the 30 queries the partial rankings were calculated as explained above and they were compared using the Kendall τ rank correlation coefficient[23]. In this article pairs that are tied in both the rankings are considered to be concordant. The distribution of Kendall's τ for the 30 queries is shown in Figure 6.

Table 5. Combined Ranking for query **"baby names"**

URLs	Order Rank	Time Rank
www.babynames.com/	1	3
www.babynamer.com/	2	1
www.babyuniversity.com/	2	6
www.kabalarians.com/	2	3
www.babycenter.com/babyname/	5	1
www.indiaexpress.com/specials/babynames/	6	6
www.heptune.com/names/nameinde.html	6	6
www.girlbabynames.com/	6	6
bnf.parentsoup.com/	6	3
callmenames.com/	6	6
www.zelo.com/firstnames/	11	6
babyzone.com/babynames	12	12
www.thinkupnames.com/	12	12
www.4babynames.com/	12	12

Table 6. Kendall's τ distribution for the test data set

Kendall's measure	<0.2	0.2-0.4	0.4-0.6	0.6-0.8	>0.8
Frequency	3	13	10	4	0

Table 6 shows the distribution of τ to be skewed towards 0. For around 86% of the queries the Kendall's τ measure is less than 0.6 and for nearly 55% of the queries the Kendall's τ measure is less than 0.4. This indicates a strong tendency of the two ranking mechanisms to be independent. This observation is quite interesting because it indicates that the information content in the two feedback measures are not in complete conformity with each other and hence some incremental information can be gathered from them.

6 Conclusion

The need to overcome spamming attempts by page authors highlights the importance of incorporating the user relevance judgments on the quality of a page. However, users can rarely be motivated to give direct/explicit feedback on page quality. In this article we evaluated two implicit feedback measures that can be collected from search engine log files. Specifically we considered the click sequence and the time spent by a user in reading a document as measures of document importance for a query. Initially, we developed a mathematical programming model to collate the feedback from different session and provide an overall partial ordering of the documents. These partial ordering were then evaluated to determine if they conveyed the same information or not by Kendall's τ measure. Further experiments were conducted on an actual search engine log, which indicated that the partial ordering of documents are different and hence there is incremental information content in the two implicit feedback measures.

References

1. http://news.netcraft.com/archives/web_server_survey.html (May 2007)
2. Yuwono, B., Lee, D.L.: Search and ranking algorithms for locating resources on the world wide web. In: ICDE, pp. 164–171 (1996)
3. Arasu, A., Cho, J., Garcia-Molina, H., Paepcke, A., Raghavan, S.: Searching the web. ACM Trans. Inter. Tech. 1(1), 2–43 (2001)
4. Brin, S., Page, L.: The anatomy of a large-scale hypertextual web search engine. Computer Networks and ISDN Systems 30(1-7), 107–117 (1998)
5. Kleinberg, J.M.: Authoritative sources in a hyperlinked environment. J. ACM 46(5), 604–632 (1999)
6. Salton, G., McGill, M.J.: Introduction to Modern Information Retrieval. McGraw-Hill, New York (1983)
7. Frakes, W.B., Baeza-Yates, R.: Information retrieval: Data structures and algorithms. Prentice-Hall, Englewood Cliffs (1992)

8. Goecks, J., Shavlik, J.: Learning users' interests by unobtrusively observing their normal behavior. In: Proceedings of the 5th international conference on Intelligent user interfaces, pp. 129–132 (2000)

9. White, R., Ruthven, I., Jose, J.M.: The use of implicit evidence for relevance feedback in web retrieval. In: Proceedings of the 24th BCS-IRSG European Colloquium on IR Research, pp. 93–109. Springer, Heidelberg (2002)

10. Claypool, M., Le, P., Wased, M., Brown, D.: Implicit interest indicators. In: IUI 2001. Proceedings of the 6th international conference on Intelligent user interfaces, pp. 33–40 (2001)

11. Kelly, D., Teevan, J.: Implicit feedback for inferring user preference: a bibliography. SIGIR Forum 37(2), 18–28 (2003)

12. Sriram, S., Shen, X., Zhai, C.: A session-based search engine. In: ACM SIGIR, pp. 492–493 (2004)

13. Beeferman, D., Berger, A.: Agglomerative clustering of a search engine query log. In: ACM SIGKDD, pp. 407–416 (2000)

14. Jones, R., Rey, B., Madani, O., Greiner, W.: Generating query substitutions. In: WWW '06: Proceedings of the 15th international conference on World Wide Web. pp. 387–396 (2006)

15. Radlinski, F., Joachims, T.: Query chains: learning to rank from implicit feedback. In: ACM SIGKDD, pp. 239–248 (2005)

16. Kim, J., Oard, D., Romanik, K.: Using implicit feedback for user modeling in internet and intranet searching. Technical report, College of Library and Information Services, University of Maryland at College Park (2000)

17. Kelly, D., Belkin, N.J.: Display time as implicit feedback: understanding task effects. In: ACM SIGIR, pp. 377–384 (2004)

18. Joachims, T.: Optimizing search engines using clickthrough data. In: ACM SIGKDD, pp. 133–142 (2002)

19. Tan, Q., Chai, X., Ng, W., Lee, D.L.: Applying co-training to clickthrough data for search engine adaptation, pp. 519–532 (2004)

20. Ding, C., Chi, C.H.: Towards an adaptive and task-specific ranking mechanism in web searching (poster session). In: ACM SIGIR, pp. 375–376 (2000)

21. Ramachandran, P.: Discovering user preferences by using time entries in clickthrough data to improve search engine results. In: Hoffmann, A., Motoda, H., Scheffer, T. (eds.) DS 2005. LNCS (LNAI), vol. 3735, pp. 383–385. Springer, Heidelberg (2005)

22. Jansen, B.J., Spink, A.: An analysis of web searching by european alltheweb.com users. Inf. Process. Manage. 41(2), 361–381 (2005)

23. Kendall, M.: A new measure of rank correlation. Biometrika 30(1/2), 81–93 (1938)

Pharmacophore Knowledge Refinement Method in the Chemical Structure Space

Satoshi Fujishima[1], Yoshimasa Takahashi[1], and Takashi Okada[2]

[1] Department of Knowledge-based Information Engineering,
Toyohashi University of Technology,
1-1 Hibarigaoka, Tempaku-cho, Toyohashi, Aichi 441-8580, Japan
{fujisima, taka}@mis.tutkie.tut.ac.jp
[2] Department of Informatics, School of Science and Technology,
Kwansei Gakuin University,
2-1 Gakuen, Sanda, Hyogo 669-1337, Japan
okada-office@ksc.kwansei.ac.jp

Abstract. Studies on the structure–activity relationship of drugs essentially require a relational learning scheme in order to extract meaningful chemical subgraphs; however, most relational learning systems suffer from a vast search space. On the other hand, some propositional logic mining methods use the presence or absence of chemical fragments as features, but rules so obtained give only crude knowledge about part of the pharmacophore structure. This paper proposes a knowledge refinement method in the chemical structure space for the latter approach. A simple hill-climbing approach was shown to be very useful if the seed fragment contains the essential characteristic of the pharmacophore. An application to the analysis of dopamine D1 agonists is discussed as an illustrative example.

Keywords: Knowledge refinement, Chemical structure space, Structure activity relationship, Pharmacophore.

1 Introduction

It is important to establish relationships between the structures of chemical compounds and their physiological activities. After King et al. succeeded in applying the inductive logic programming method to mutagenicity analysis [3], several techniques were proposed for extracting characteristic substructures from chemical compounds with a variety of structures [7]. In one of these approaches, a mining technique is confined to propositional logic, and numerous substructures are generated initially and used as features. Then, a combination of a few features is expected to explain the pharmacophore. The problem with this approach is that the description of the pharmacophore is limited by the space expressed in the features. Therefore, the classifiers do not always show a substructure that medicinal chemists can understand easily, even if the classifiers possess strong discriminating power.

We applied this approach using linear fragments and the cascade model [5], and succeeded in determining new pharmacophore structures. However, chemists must

V. Corruble, M. Takeda, and E. Suzuki (Eds.): DS 2007, LNAI 4755, pp. 243–247, 2007.
© Springer-Verlag Berlin Heidelberg 2007

inspect all of the rule conditions and collateral correlations carefully, referring to the supporting chemical structures in translating rules to the pharmacophore structure, and the time and effort involved in this work is truly prohibitive. The purpose of this study was to find an easier way to obtain the pharmacophore.

In this paper, we start from obtained rules, and place the essential fragment of the rule into the chemical structural formula space. The seed fragment is then refined to give the pharmacophore structure using a simple hill-climbing procedure. We expect this method to provide information that chemists can understand easily.

2   Refinement Method

The knowledge refinement system accepts a seed and searches its neighborhood to reach a pertinent piece of information. The pharmacophore knowledge is a substructure pattern that affects an activity, and we use the SMILES/SMARTS language as a way to specify chemical structures and the pattern. The Simplified Molecular Input Line Entry System (SMILES) is a line notation developed to enter and represent molecules and reactions using short ASCII strings [8]. SMARTS is a straightforward extension of SMILES language that describes molecular patterns [1].

We can easily retrieve molecules with a specified pattern when we apply a chemical structure database system using SMILES/SMARTS expressions. In addition, a chemist can enter a seed pattern based on an idea and start the refinement procedure. Once molecules matching the pattern are retrieved, their adequacy as the pharmacophore can be judged using the BSS values, and we can apply a hill-climbing approach in the search. It shares the idea of using SMILES string in the detection of the pharmacophore with SMIREP [2], but its building blocks and the optimization criterion is completely different from the refinement process proposed here.

The refinement process necessitates a seed SMARTS pattern, a database of specific structures written in SMILES, and lists of atom and bond types (Table 1) to be inserted into the pattern. The following algorithm describes the refinement process:

Algorithm 1. Refinement algorithm from a SMARTS pattern.

(1) Set *Alist* to atom types and *Blist* to bond types.
(2) Set *ptrn* to initial seed fragment written in SMARTS.
(3) Compute *BSS* from the instances retrieved by *ptrn*.
(4) Set *maxBSS* = 0.0.
(5) For each combination of *atom* in *Alist* and *bond* in *Blist*,
(6)   For each *position* in *ptrn*,
(7)   Generate *newptrn* by inserting *atom* and *bond* at the *position* in *ptrn*.
(8)   Set *tuples* to the instances retrieved by *newptrn*.
(9)   Compute *newBSS* from *tuples*.
(10)   If *newBSS* > *maxBSS*,
                set *maxBSS* = *newBSS*; *maxptrn* = *newptrn*.
(11) If *maxBSS* < *BSS* then return *ptrn*.
(12) Set *ptrn* = *maxptrn*; *BSS* = *maxBSS*
(13) Goto step (4).

Table 1. Default atom and bond types lists that can be inserted into the seed fragment in the system. Lowercase letters indicate atoms in a conjugated system, and [ ] implies that one of the atoms in parentheses is to be chosen.

List	Elements
Atom list:	C, O, N, S, c, o, n, s, [I, Br, Cl, F]
Bond list:	-, =, :

We employ the between-groups sum of squares (BSS), computed in step (9), to indicate the strength of the pattern [5], which is calculated by the following formula:

$$\text{BSS}^g = \frac{n^g}{2} \sum_\alpha (p^g(\alpha) - p(\alpha))^2 \tag{1}$$

where $p(\alpha)$ and $p^g(\alpha)$ are the probabilities of active/inactive compounds retrieved using the initial seed fragment and the *newptrn*, respectively, and n^g is the number of compounds selected by *newptrn*.

3 Results and Discussion

The MDDR database by MDL Inc. was used as the data source [4]. It contains 369 records that describe dopamine (D1, D2, and autoreceptor) agonist activity, with 63, 143, and 186 active compounds possessing D1, D2, and autoreceptor activity, respectively. Some of the compounds affected multiple receptors.

The knowledge refinement system is applied to the pharmacophore of the dopamine D1 agonist activity. Previously we analyzed this activity using the cascade model and obtained 16 rules, from which chemists selected the four rules. Here, we examine the strongest rule R1 shown in Table 2. The linear fragment in this rules is given to the refinement system as a seed pattern, and the resulting pharmacophore is evaluated comparing to the hand-carved pharmacophore shown in Fig. 1.

Rule R1 has [O2H-c3:c3-O2H: y] as main condition and no preconditions exist. The change in the activity ratio is very sharp, from 17% in 369 compounds to 96% in 52 compounds, and this might be detected using any mining method. However, this is not the final pharmacophore that chemists expect.

Table 2. Rule suggesting the strongest characteristic of dopamine D1 agonist activity

Rule ID	#compounds	Main condition	Preconditions	Distribution change
R1	369 → 52	[O2H-c3:c3-O2H: y]	none	17% →96%

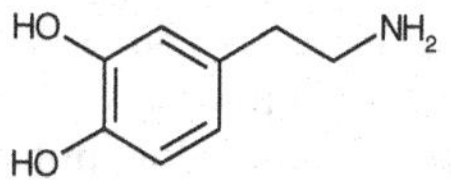

Fig. 1. Pharmacophore structure derived from the strongest rule for dopamine D1 agonist activity

Table 3. Structural formula of the seed fragment and five patterns generated from the seed in the refinement procedure

	Step 0	Step 1	Step 2	Step 3	Step 4	Step 8
BSS	12.2	12.2	14.0	14.0	22.3	22.3
Support (Y/N)	57 / 58	57 / 58	35 / 15	35 / 15	35 / 2	35 / 2
Structural Formula	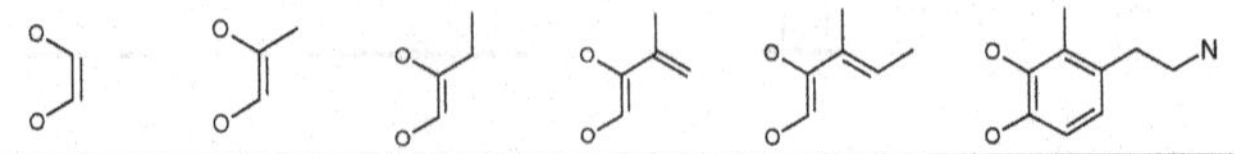					

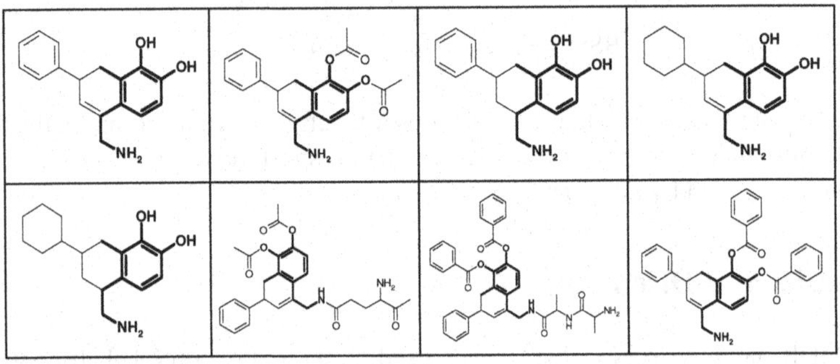

Fig. 2. Example of structures supporting the refined pharmacophore at step 8 in Table 3. The bold line part indicates the final pattern obtained. Red is used instead of the bold line in the actual application.

Let us examine the refinement results starting from the initial pattern OccO, which is a simplified interpretation of the main condition. Table 3 summarizes the process at steps 0–4 and 8. We can see that the BSS value increases at steps 2 and 4, while the value does not change at the other steps. Few alternative pattern candidates occur with the same BSS at every refinement step, but we reach the same final pattern even if we use other pattern expressions in the refinement process. The BSS increases at steps 2 and 4 suggest that the attached components at these steps have significant meaning for understanding the pharmacophore. The decreases in the number of supporting negative instances are the main reason of these BSS increase, and chemists can recognize the group of compounds excluded by growth of the pattern.

The resulting pharmacophore shown in step 8 of Table 3 is very similar to the hand-carved one illustrated in Fig. 1. Some supporting molecules that form the final pattern are shown in Fig. 2, where the bold line indicates the pattern. Active and inactive molecules are shown at the top and bottom of the browser window. This illustration makes it very easy for chemists to evaluate the adequacy of the proposed pharmacophore.

Another refinement starting from the initial pattern: [O;H1]cc[O;H1] has reached the identical structure to that in Fig. 1. This seed has attached hydrogen atoms at the oxygen, and it is the strict interpretation of the rule fragment. The seed had 52 supporting compounds, of which 50 had dopamine D1 agonist activity. Only one inactive compound was deleted from the supporting compounds during the refinement process. This result shows that the fragment that appeared under the main condition of R1 has captured the essential point of the pharmacophore, and the refinement process has succeeded to capture the total pharmacophore structure.

4 Conclusions

We have shown that knowledge refinement is very effective at determining valid pharmacophore hypotheses. The method used was a simple hill-climbing approach based on optimizing the BSS value. We cannot attain a valid conclusion if we start from an invalid initial pattern. Therefore, the success of our approach can be attributed to two reasons. First, we can start the refinement process from a fragment with good characteristics. Second, the BSS criterion used is an effective guide leading to a satisfactory hypothesis.

Similar approaches are expected to provide high-quality knowledge in other domains. That is, a good starting hypothesis is obtained using the standard mining method, and then refinement proceeds in the original problem space. Some examples involve mining from sentences in natural language and from plant operations in which the quality of knowledge should be judged in the problem-specific representation space.

From the viewpoint of applications in chemistry, the burden of rule interpretation has been reduced greatly. Moreover, the system allows chemists to incorporate their own ideas in the initial hypothesis, which can then give a somewhat unconstrained feeling to users. Our system is now one of the principal software components in the pharmacophore knowledge-base project being developed in our laboratory.

Acknowledgments. The authors thank Dr. Masumi Yamakawa for his work on dopamine activity analysis as a medicinal chemist. Part of this research is supported by Grant-in-Aid for Scientific Research (A) 14208032.

References

1. Daylight Chemical Information Systems: Daylight Theory: SMARTS Manual(2004), http://www.daylight.com/dayhtml/doc/theory/theory.smarts.html
2. Karwath, A., De Raedt, L.: SMIREP: Predicting Chemical Activity from SMILES. Journal of Chemical Information and Modeling 46(6), 2432–2444 (2006)
3. King, R.D., Muggleton, S.H., Srinivasan, A., Sternberg, M.J.E.: Structure–activity relationships derived by machine learning: The use of atoms and their bond connectivities to predict mutagenicity by inductive logic programming. Proceedings of the National Academy of Sciences 93(1), 438–442 (1996)
4. MDL: Drug Data Report, MDL, 2001.1 (2001)
5. Okada, T.: Rule induction in cascade model based on sum of squares decomposition. In: Żytkow, J.M., Rauch, J. (eds.) PKDD 1999. LNCS (LNAI), vol. 1704, pp. 468–474. Springer, Heidelberg (1999)
6. Okada, T., Yamakawa, M., Niitsuma, H.: Spiral mining using attributes from 3D molecular structures. In: Tsumoto, S., Yamaguchi, T., Numao, M., Motoda, H. (eds.) AM 2003. LNCS (LNAI), vol. 3430, pp. 287–302. Springer, Heidelberg (2005)
7. Okada, T.: Mining from chemical graphs. In: Holder, L.B., Cook, D.J. (eds.) Mining Graph Data, pp. 347–379. Wiley–Interscience, Chichester (2006)
8. Weininger, D.: SMILES, a chemical language and information system. 1. Introduction and encoding rules. Journal of Chemical Information and Computer Science 28(1), 31–36 (1988)

An Attempt to Rebuild C. Bernard's Scientific Steps

Jean-Gabriel Ganascia and Bassel Habib

Université Pierre and Marie Curie, Laboratoire d'informatique de Paris VI
104, avenue du Président Kennedy, F-75016 Paris, France

Abstract. Our aim is to reconstruct Claude Bernard's empirical investigations with a computational model. We suppose that he had in mind what we call "kernel models" that provide simplified views of physiology, which allowed him to make hypotheses and to draw out their logical consequences. We show how those "kernel models" can be specified using both description logics and multi-agent systems. Then, the paper will explain how it is possible to build a virtual experiment laboratory, which lets us construct and conduct virtual experiments.

1 Introduction

During the past, there have been many attempts to rationally reconstruct scientific discoveries with Artificial Intelligence techniques [5,2]. In a way, the science of discovery results from those attempts. Nevertheless, a question remains concerning the logical status of the discovery: is it mainly an inductive, a deductive on an abductive process? Philosophers do not agree in this point; but whatever their underlying theories, it appears that inferences involved in discovery are many in number and various. Nevertheless, up to now, most of the simulations of scientific discovery processes that have been achieved in Artificial Intelligence correspond to the simulation of inductive processes. This paper constitutes an attempt to reconstruct some of the Claude Bernard's scientific steps that are mainly abductive. It explores with the help of Knowledge Representation and Multi-Agent techniques, some aspects of the discovery science that are not directly related to inductive processes.

Let us recall that Claude Bernard (1813–1878) was not only one of the most eminent 19th century physiologists, but also a theoretician who generalized his experimental method in his famous book, "Experimental Medicine" [1], which is nowadays a classic that all young students in medicine are supposed to have read. The goal of our project [3] is mainly to clarify and to generalize this experimental method by formalizing it with artificial intelligence techniques and by simulating it on computers.

More precisely, Claude Bernard had in mind an ontology of the physiology which he used to express scientific hypotheses concerning both the organ functions and the activity of toxic and/or medicinal substances. He also used this ontology to design experiments that were intended to discriminate among the different scientific hypotheses. Our first aim here is to rebuild the ontology described in the Claude Bernard's works with modern knowledge representations techniques. Then, we want to construct, on the top of this ontology, "kernel models", which simulate the experiments that Claude Bernard's had in mind when he investigated the effects of toxic substances, e.g. carbon monoxide and curare.

V. Corruble, M. Takeda, and E. Suzuki (Eds.): DS 2007, LNAI 4755, pp. 248–252, 2007.

The first part formalizes the Bernard's medical ontology. The second is dedicated to the description of a two level model built to simulate his experimental method. The third describes the notion of "kernel model"; the fourth, the virtual laboratory on the top of which virtual experiments may be done. The final and last part envisages the hypothesis generation module and other possible generalizations.

2   The Claude Bernard's Ontology

In his writings, Claude Bernard presumes that organisms are composed of organs, themselves analogous to organisms since each of them has its own aliments, poisons, excitations, actions etc. Organs are categorized into three classes – skeleton, tissues (e.g. epithelium, glandular tissue or mucous membrane) and fibers (i.e. muscles and nerves) – that are recursively sub-categorized into subclasses, sub-subclasses etc. Each class and subclass has its own characteristics, which can easily be formulated, according to Claude Bernard's explanations.

The internal environment – i.e. the "milieu intérieur" –, mainly the blood, carries organ poisons and aliments, while the organ actions may have different effects on other organs and, consequently, on the whole organism. More precisely, for Claude Bernard, life is synonymous of exchanges. The organisms exchange through the external medium that is the air for outside animals or the water for fish. The external medium may also carry aliments, poisons etc. Similarly, organs can be viewed as some sorts of organisms living in the body and participating to its life. Their life is also governed by exchanges; but the medium that supports exchanges is not air or water; it is the so-called "milieu intérieur", which mainly corresponds to blood.

The Claude Bernard's ontology may simply be derived from these considerations. It is then easy to formulate it in an ontology description language similar to those that are nowadays used in the life sciences to represent biological and medical knowledge [7]. Note that most of the ontologies used in the biomedical community, for instance the OBO – the Open Biological Ontologies http://obofoundry.org/ – refer to three levels: one for the organs and the anatomy, the second for the cells and the third for molecules. For obvious reasons the Claude Bernard's ontology refers mainly to the first, i.e. to organs and anatomy. However, it would possible to extend our model to a three level ontology that is more appropriate in contemporaneous medicine. For instance, below are some of the previous assertions expressed with description logics [6]:

- The organs are parts of the organism: $Organ \sqsubseteq \exists PART.Organism.$
- The organs are tissues, skeleton or fibers: $Organ \equiv Tissue \sqcup Skeleton \sqcup Fiber$
- Fibers may be nerves or muscles: $Fiber \equiv Nerve \sqcup Muscle$
- Nerves may be sensitive or motor: $Nerve \equiv Sensitive\_Nerve \sqcup Motor\_Nerve$
- Epithelium, glandular tissue, mucous membrane etc. are tissues: $Tissue \sqsupseteq Epithelium \sqcup Glandular\_Tissue \sqcup Mucous\_Membrane \sqcup \cdots$

3   Two-Level Model

As previously stated, abduction played a crucial role in Claude Bernard's investigations. More precisely, he always considered an initial hypothesis, which he called an "idea" or a "theory". He then tried to test it by designing *in vivo* experiments. According to the observational results of his experiments, he changed his hypotheses, until he reached a satisfying theoretical explanation of empirical phenomena.

To design a computational model that simulates the intellectual pathways leading Claude Bernard to his discoveries, we have supposed that he had in mind what we call "kernel models". Those "kernel models" contain basic physiological concepts — such as internal environment, organ names etc. — upon which he builds his "ideas". More precisely, "ideas" correspond to hypothetical organ functions that Claude Bernard wanted to elucidate, while "kernel models" describe the physical architecture of the simplified organisms on the top of which his experiments were designed. Claude Bernard assumed that one can use toxic substances as tools of investigation — he evoked the idea of "chemical scalpel" — to dissociate and identify the functions of different organs. He presupposed, as an underlying principle, that each toxic substance neutralizes one organ first. The simulation of a "kernel model" makes explicit the consequences of each working hypothesis. All his "ideas", i.e. all his working hypotheses, were then evaluated by the confrontation of their potential consequences, i.e. the consequences derived from "kernel models" simulation, to the consequences observed through empirical experiments.

Our aim, in this paper, is to build and to simulate those "kernel models" using multi-agent architectures. Such simulations have to show, on a simplified view, both the normal behavior of the organism and the consequences of an organ dysfunction.

Nevertheless, other questions need to be solved when we want to rationally reconstruct the discovery process: how are "ideas", i.e. working hypotheses, generated and how are validating experiments designed? In order to answer the first question, we add to the "kernel model" a "working hypothesis management" module that has both to guide working hypothesis generation and to design experiments. The second is out of the scope of our study.

4 "Kernel Model" Simulation

The "kernel models" contain organs and connections between organs through the internal environment, mainly the blood. Both organs — e.g. muscles, hart, lung, nerves etc. — and connections between organs are represented using agents that communicate with other organs and evolves in the "milieu intérieur" viewed as the internal environment. The agents correspond to the concepts of the previously described ontology. It is possible, for the internal environment, to lose or gain some substance, for instance oxygen, and some pressure when passing by an organ. In the usual case, e.g. for muscles, the input internal environment corresponds to arterial blood while the output corresponds to venous blood. The organism, which is a set of connected organs, is modeled as a synchronous multi-agent system, where each agent has its own inputs, transfer function and states. The organ activation cycle follows the blood circulation. The time is supposed to be discrete and after each period of time, the states of the different agents belonging to the "kernel model" and their outputs are modified.

The implementation makes use of object oriented programming techniques. It helps both to simulate the "kernel model" evolutions and to conduct virtual experimentations (see next section) on those "kernel models", which fully validates our first ideas concerning the viability of the notion of "kernel model". Within this implementation, organs, i.e. instantiations of concepts of the initial ontology, and connections between organs are associated to objects that implement agents. The inheritance and instantiation mechanisms of object oriented programming facilitate the implementation of those agents. However, since our ultimate goal is to simulate the hypothesis generation and especially the abductive reasoning on which relies the discovery process, we chose to build "kernel models" using logic programming techniques on which it is easy to simulate logical inferences, whatever they are, either deductive or abductive.

The logic programming implementation is programmed in SWI Prolog[1]. It makes use of modules to emulate object oriented programing techniques, i.e. mainly the instantiation, inheritance and message sending mechanisms.

5 Virtual "Thought Experiments"

Once the "kernel model" is built, it is not only possible to simulate normal organism behavior, but also to introduce pathologies (i.e. organ deficiencies) in the multi-agent system that models the organism and then emulate its evolution. In a way, these abnormal behavior simulations can be viewed as virtual experiments: they help to draw consequences of virtual situations under a working hypothesis, i.e. a supposition concerning both the effect of a substance on some organs and the function of the implied organs. In order to complete the range of virtual experiments, we introduce, according to Claude Bernard's practices, some virtual experimental operators, such as injection and ingestion of substances, application of tourniquet on members, excitations, etc. For instance, if one wants to understand the effects of a substance A, one can hypothesize that its concentration in the blood may affect such or such organ subclass, which has such or such function in the organism. Under these hypotheses, it is possible with the "kernel model" simulation to predict the consequences of a direct injection of A combined with any combination of experimental operations (applying a tourniquet on a member and/or exciting another part of the organism before or after injecting the substance A etc.). In other words, it is possible to specify virtual experiments and to anticipate the subsequent model behavior under a working hypothesis.

To be concrete, take a simple example of intoxication with curare that is presented in Claude Bernard's personal writings. In this experiment, Claude Bernard poisons an animal. The voluntary movements are the first to be paralyzed. This is only when respiratory disorders appear, due to the paralysis of lung muscles, that the animal is asphyxiated. To simulate such an evolution, we introduced a virtual organism with a voluntary muscle, a kidney that is progressively evacuating the curare and a muscle that control the lung movements. We supposed that curare affects the muscles. We injected a dose of curare in the virtual organism and we obtained the following evolution: if the curare dose is sufficient, after 5 steps, the voluntary muscle is progressively paralyzed, but it takes more than 30 steps to see the lung paralyzed and the animal asphyxiated. If the curare dose is very low, the muscle is paralyzed, but there is no asphyxia, and the curare is evacuated. etc.

---

[1] See http://www.swi-prolog.org/ for more details.

6   Conclusion

A virtual laboratory has been programmed in PROLOG. It allows to build virtual experiments associated with different working hypotheses about the toxic effects of carbon monoxide and curare. It was then possible to correlate those virtual experiments to actual experiments done by Claude Bernard, and then to corroborate or refute working hypotheses according to the observations. As a consequence, we are able to computationally reconstruct part of Claude Bernard's intellectual pathway. As it was previously suggested, the virtual experiments are achieved under working hypotheses that assume, for instance, that a substance A affects such or such a function of such or such an organ class. Being given a toxic substance, one has to explore all the possible hypotheses and suggest, for each, experiments that could corroborate or refute them by showing observable consequences. It is the role of the working hypothesis management module to investigate all these hypotheses. Nevertheless, the goal is neither to achieve, nor to generate experiments, as would be the case with a robot scientist (see for instance [4]). The next step is to build such an hypothesis management module.

We also investigate the possibility to build multi-scale "kernel models" in which physiological behaviors can be studied at different scales — organ, cell, molecule etc. —. It should open new perspectives to modern clinical medicine. As a matter of fact, principles on which lay down Claude Bernard's empirical method are always valid, even if the ontology on which are built the "kernel models" considerably changed with time. Today, the effect of new substances is usually studied at the cell or molecule scale, while the organ scale was dominant at Claude Bernard's epoch. A model that could help to simulate the consequences of physiological dysfunctions at different levels would be of great help to determine the effects of new substances by recording different experiments and by ensuring that all the plausible hypotheses have already been explored.

References

1. Bernard, C.: An Introduction to the Study of Experimental Medicine. In: First English translation by Henry Copley Greene, Macmillan & Co., Ltd (1927)
2. Davis, R., Lenat, D.: AM: Discovery in Mathematics as Heuristic Search in Knowledge-Based System in Artificial Intelligence, Part One. McGraw-Hill, New York (1982)
3. Ganascia, J.-G., Debru, C.: CYBERNARD: A Computational Reconstruction of Claude Bernardś Scientific Discoveries. In: Studies in Computational Intelligence (SCI), vol. 64, pp. 497–510. Springer-Verlag, Heidelberg (2007)
4. King, R., Whelan, K., Jones, F., Reiser, P., Bryant, C., Muggleton, S.: Functional genomic hypothesis generation and experimentation by a robot scientist. Nature 427, 247–252 (2005)
5. Kulkarni, D., Simon, H.: Experimentation in Machine Discovery. In: Shrager, J., Langley, P. (eds.) Computational Models of Scientific Discovery and Theory Formation, Morgan Kaufmann, San Francisco (1990)
6. Nebel, B., Smolka, G.: Attributive description formalisms... and the rest of the world. In: Herzog, O., Rollinger, C -R. (eds.) Text Understanding in LILOG. LNCS, vol. 546, pp. 439–452. Springer, Heidelberg (1991)
7. Smith, B., Ceusters, W.: Ontology as the core Discipline of Biomedical Informatics, Legacies of the Past and Recommendations for the Future Direction of Research. In: Crnkovic, G.D., Stuart, S. (eds.) Forthcoming in computing, philosophy, and cognitive science, Cambridge Scholar press, Cambridge (2006)

Semantic Annotation of Data Tables Using a Domain Ontology

Gaëlle Hignette[1], Patrice Buche[1], Juliette Dibie-Barthélemy[1],
and Ollivier Haemmerlé[2]

[1] UMR AgroParisTech/INRA MIA - INRA Unité Mét@risk,
AgroParisTech, 16 rue Claude Bernard, F-75231 Paris Cedex 5, France
{hignette, buche, dibie}@agroparistech.fr
[2] IRIT - Université Toulouse le Mirail, Dpt. Mathématiques-Informatique,
5 allées Antonio Machado, F-31058 Toulouse Cedex
ollivier.haemmerle@univ-tlse2.fr

Abstract. In this paper, we show the different steps of an annotation process that allows one to annotate data tables with the relations of a domain ontology. The columns of a table are first segregated according to whether they represent numeric or symbolic data. Then, we annotate the numeric columns with their corresponding numeric type, and the symbolic columns with their corresponding symbolic type, combining different evidences from the ontology. The relations represented by a table are recognized using both the table title and the types of the columns. We give experimental results for our annotation method.

1 Introduction

In the scientific world, many experimental data are produced and continually published on the web; a lot of these data are presented synthetically in the form of tables. Our aim is to create an XML data warehouse in which tables are annotated with a domain ontology for subsequent querying. Our work is applied to food microbiology and is integrated within an existing system called MIEL [1] in which data are manually entered into a relational database, indexed with a domain ontology. In this ontology, users select the food products, microorganisms and relations that they are interested in: the database is then queried to find data corresponding to or close to the users selection criteria. Our XML data warehouse is designed to be queried simultaneously with the relational database: the data in the tables from the web are annotated with the same ontology as the one used in the MIEL system. The ontology is structured as follows:

- numeric types are described by the type name, the units for the type and the interval of possible values for the type;
- symbolic types are described by the type name and the type hierarchy (the possible values for the type, partially ordered by the subsumption relation);
- relations are described by the relation name and the relation signature, compounded of a result type (the measure the experiment was set for) and access types (controlled factors that vary according to the experimental plan);

V. Corruble, M. Takeda, and E. Suzuki (Eds.): DS 2007, LNAI 4755, pp. 253–258, 2007.

- the "no-result indicators" are terms used to represent the absence of data, for example "No Result" or "NS" (for Not Specified);
- the stop-list lists words with low semantic content (articles, conjunctions,...).

There has been a lot of research on table recognition [2], but works on table annotation using an ontology are rare. Techniques of wrapper induction based on the structure, such as presented in Lixto [3], are not adapted to our problem since each author structures his tables differently. The work in [4] is more similar to ours: frames are constructed from tables using the table signature which is deduced from an ontology. However, their ontology is generic and relations have no *a priori* meaning as they are constructed directly from the table signatures, whereas we want to recognize predefined relations in an ontology specific to the targetted domain. In [5], relations from an ontology are instanciated using various HTML structures including tables. They identify binary concept-role relations between instances that are assumed to be recognized by a pre-treatment which is not part of their process. Our work differs as we focus on the recognition of n-ary relations and we propose a step-by-step algorithm including the recognition of element types.

This paper is a following work of [6] with several enhancements: (1) we use a different similarity measure between the terms from the web and the terms from the ontology; (2) numeric columns are treated specifically; (3) column annotation is more robust as it takes into account both column title and column content; (4) scoring relations helps choosing which relations to use to annotate the table.

The first step of our annotation process consists in classifying columns as numeric or symbolic (Sect. 2). The similarity measure between the terms from the web and the terms from the ontology that we use through our annotation process is presented in Sect. 3. Then we present the way of finding the type of a numeric column (Sect. 4) and the type of a symbolic column (Sect. 5). We then explain how we find the relations represented in the table (Sect. 6).

2 Distinction Between Numeric and Symbolic Columns

Let *col* be a column of the table we want to annotate. We search *col* for all occurrences of numbers (in decimal or scientific format) and of units of numeric types described in the ontology. We also count the words occuring in *col*: a word is an alphabetic character sequence that is neither a unit nor a "no result indicator". Let c be a cell of the column *col*. We apply the following rules:

- if c contains a number immediately followed by a unit, or a number in scientific format, then c is numeric;
- else, if c contains more numbers and units than words, then c is numeric;
- else, if c contains more words than numbers and units, then c is symbolic;
- if words are as numerous as numbers+units, the status of c is unknown.

Once all cells of the column *col* are classified using those rules, *col* is classified as symbolic if it contains more symbolic cells than numeric cells: it is then further annotated according to Sect.5. Else, *col* is classified as numeric and further annotated according to Sect.4.

3 Similarity Measure Between a Term from the Web and a Term from the Ontology

Throughout our whole annotation process, we will be using a similarity measure that allows to compare a term from the web with a term from the ontology, a term being a set of consecutive words. The initial similarity measure used in [6] was the proportion of words in common between the two terms. We go further by manually weighting the words in terms from the ontology, according to their importance in the meaning of the term. The different weights are: 0 for non-informative words listed in the stop-list, 1 for the most informative word(s) of the term, 0.2 for secondary words. Weighted terms are represented as vectors, in which the coordinates represent the different possible lemmatised words and their weight in the term (0 if the word does not belong to the term). The similarity between a term from the web and a term from the ontology is then the cosine similarity measure between the two vectors. Let w be a term from the web, represented as the weighted vector $\boldsymbol{w} = (w_1, \ldots, w_n)$ and o a term from the ontology, represented as the weighted vector $\boldsymbol{o} = (o_1, \ldots, o_n)$.

$$sim(w, o) = \frac{\sum_{k=1}^{n} w_k \times o_k}{\sqrt{\sum_{k=1}^{n} w_k^2 \times \sum_{k=1}^{n} o_k^2}} \tag{1}$$

We will call score of a type t for a column *col*, noted $score_{title}(t, col)$, the similarity between the type name and the column title. We will as well call score of a relation *rel* for a table *tab* according to the table title, noted $score_{title}(rel, tab)$, the similarity between the relation name and the table title.

4 Numeric Column Annotation

Let u be a unit and T_u the set of numeric types that can be expressed in this unit, the score of a type t for the unit u is $score(t, u) = 0$ if $t \notin T_u$, and $score(t, u) = \frac{1}{|T_u|}$ if $t \in T_u$.

The score of a numeric type t for the column *col* according to the units in the column is computed as the maximum of the scores of t for each unit present in *col*. If no unit was identified in the column *col*, then *col* is considered as presenting the unit "no unit", which is a treated as a normal unit in the ontology.

If all numbers in the column *col* are compatible with the range of the numeric type t, the final score of t for *col* is $score_{final}(t, col) = 1 - (1 - score_{title}(t, col))(1 - score_{unit}(t, col))$. Else $score_{final}(t, col) = 0$.

Numeric types of the ontology are ordered according to their final score for the column: let *best* be the type with the best score and *secondBest* be the type with the second best score. We compute the proportional advantage of *best* over *secondBest* on the column *col*:

$$advantage(best, col) = \frac{score_{final}(best, col) - score_{final}(secondBest, col)}{score_{final}(best, col)} \quad (2)$$

If $advantage(best, col)$ is greater than a threshold θ_{numCol}, then the column col is annotated with the type $best$. Else the type of col is considered as unknown.

5   Symbolic Column Annotation

Let c be the term in a cell of a symbolic column col, let t be a symbolic type and $hier(t)$ the set of all terms in the hierarchy of the type t. The score of the type t for the cell c is $score(t, c) = \sum_{x \in hier(t)} sim(c, x)$. Let $best$ be the type having the best score for the cell c. If the proportional advantage of $best$ is higher than a threshold $\theta_{symbCell}$, then the c is annotated with the type $best$, else c has an unknown type. Let n be the number of cells in the column col and n_t the number of cells having the type t, then the score of t for the column col according to the column contents is $score_{contents}(t, col) = \frac{n_t}{n}$.

The final score of a symbolic type t for the symbolic column col is computed as $score_{final}(t, col) = 1 - (1 - score_{title}(t, col))(1 - score_{contents}(t, col))$. If the proportional advantage of the best type for the column is greater than a threshold $\theta_{symbCol}$, then the column is annotated with this best type, else the type of the column is considered as unknown.

6   Finding the Semantic Relations Represented by the Table

Let $Sign_{rel}$ be the set of types in the signature of a relation rel (i.e. the access types and the result type), and $Sign_{tab}$ the set of types that were recognized for the columns of table tab. If the result type of rel was recognized among the columns of tab then $score_{signature}(rel, tab) = \frac{|Sign_{rel} \cap Sign_{tab}|}{|Sign_{rel}|}$. Else $score_{signature}(rel, tab) = 0$.

The final score of a relation rel for the table tab is computed as
$score_{final}(rel, tab) = 1 - (1 - score_{title}(rel, tab))(1 - score_{signature}(rel, tab))$

Two relations are called *concurrent* if they have the same result type. If a relation has a non-zero score for the table and has no *concurrent relation*, this relation is used to annotate the table. If there are several *concurrent relations* with non-zero scores for the table, then we only keep the one(s) with the highest score for the annotation of the table.

We have experimented this annotation method on 60 tables from publications on food microbiology. Those tables were manually annotated with the 16 relations of the ontology: each table was annotated with 1 to 5 relations, which gives a total of 123 relations. 117 relations were correctly annotated with our annotation system, only 6 were not recognized but 52 relations were recognized while not present in the manual annotations.

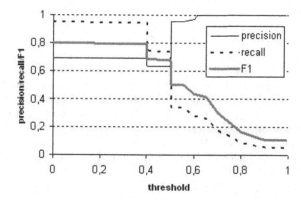

Fig. 1. Quality measures of relation annotation given the value of the threshold θ_{rel}

In order to improve precision, we have tried to apply a score threshold on the relations: only relations with score greater than a threshold θ_{rel} are used for the annotation. The variation of precision, recall and $F1$-value with the value of θ_{rel} is shown in Fig. 1.

7 Conclusion and Perspectives

In this paper, we have shown the different steps of the annotation of data tables with a domain ontology. The columns of a table are first classified as numeric or symbolic. Then, we annotate the columns using both the column title and the column contents (units for numeric columns, terms inside cells for symbolic columns). The relations represented by a table are recognized using both the table title and the types of the columns. When all steps are run one after the other, we obtain a high recall on relation recognition, with an acceptable precision level.

Our future works will aim at instanciating the relations according to the content of the cells in the table. Fuzzy sets [7] will be used to represent similarities between symbolic cells and terms from the ontology, and to represent imprecise data for numeric cells.Then we will focus on the querying of the annotated data tables: the querying system must be integrated to the one already used in the MIEL system, which means that we will have to deal with user preferences, also expressed as fuzzy sets. Our querying system will have to take into account the different scores that we have computed during the annotation, that give hints about how sure we are of these annotations.

Acknowledgements

Financial support from the French National Research Agency (ANR) for the project WebContent in the framework of the National Network for Software Technology (RNTL) is gratefully acknowledged.

References

1. Buche, P., Dervin, C., Haemmerlé, O., Thomopoulos, R.: Fuzzy querying of incomplete, imprecise, and heterogeneously structured data in the relational model using ontologies and rules. IEEE T. Fuzzy Systems 13(3), 373–383 (2005)
2. Zanibbi, R., Blostein, D., Cordy, J.R.: A survey of table recognition: Models, observations, transformations, and inferences. IJDAR 7, 1–16 (2004)
3. Baumgartner, R., Flesca, S., Gottlob, G.: Visual web information extraction with Lixto. In: VLDB, pp. 119–128 (2001)
4. Pivk, A., Cimiano, P., Sure, Y.: From tables to frames. In: McIlraith, S.A., Plexousakis, D., van Harmelen, F. (eds.) ISWC 2004. LNCS, vol. 3298, pp. 166–181. Springer, Heidelberg (2004)
5. Tenier, S., Toussaint, Y., Napoli, A., Polanco, X.: Instantiation of relations for semantic annotation. In: WI, pp. 463–472 (2006)
6. Gagliardi, H., Haemmerlé, O., Pernelle, N., Saïs, F.: An automatic ontology-based approach to enrich tables semantically. In: AAAI Context and Ontologies Workshop (2005)
7. Zadeh, L.: Fuzzy sets. Information and control 8, 338–353 (1965)

Model Selection and Estimation Via Subjective User Preferences

Jaakko Hollmén

Helsinki Institute of Information Technology,
Helsinki University of Technology, Laboratory of Computer and
Information Science, P.O. Box 5400, FI-02015 TKK, Espoo, Finland
Jaakko.Hollmen@tkk.fi

Abstract. Subjective opinions of domain experts are often encountered in data analysis projects. Often, it is difficult to express the experts' opinions in model form or integrate their professional knowledge in the analysis. In this paper, we approach the problem directly in the context of model selection and estimation: we ask the expert for subjective preferences between readily computed model solutions, and compute an optimal solution based on the recorded opinions. We consider the pre-computed models as graph nodes, and calculate the preferential relations between the nodes based on the recorded opinions as conditional probabilities. Using a random surfer model from the Web analysis community, we compute the stationary distribution of the preferences. The stationary distribution can be used in model selection by selecting the most probable model or in model estimation by averaging over the models according to their posterior probabilities. We present a real-life application in a regression problem of tree-ring width series data.

1 Introduction

The standard solution to estimating a regression function $y_i = f(x_i; \theta) + \varepsilon_i$ from a given data set (x_i, y_i), $i = 1, \ldots, N$, is to minimize the sum of squared residuals as a function of the parameters θ [5]. This approach uses the data only and depending on the optimization framework used and the difficulty of the problem, solution of differing quality can be achieved. Sometimes this approach may be flawed in the sense that it is desirable for the residual signal to contain information that can be related to other data. In the application of our interest in this paper, we focus on modeling non-climatic component in the tree-ring width series related to the age-related growth and wish to leave the climatic signal as the residual. In this case, the standard solution is not satisfactory. A qualified domain expert — a senior researcher involved in quantitative modeling of forest growth and qualified to judge regression models — was asked to express his subjective opinions in terms of preferential choices. From a controlled experiment collecting a relatively large number of preferential choices between two randomized pairs of pre-computed models, we compute an optimal model based on a preference graph using a random surfer model [1].

V. Corruble, M. Takeda, and E. Suzuki (Eds.): DS 2007, LNAI 4755, pp. 259–263, 2007.
© Springer-Verlag Berlin Heidelberg 2007

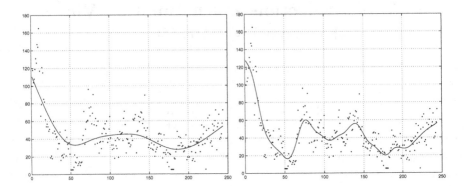

Fig. 1. Two precomputed solutions to the regression problem are illustrated. The left panel presents a simple smooth regression function, whereas the right function models more detailed variation in the data. The comparison between two given models is the basis of the controlled experiment: the expert is forced to select a better, or preferred, model out of the two presented models.

2   Model Selection and Estimation Problem

The basis of our proposed solution is to compute a set of solutions to the modeling problem and to let the expert express his subjective opinion during a controlled experiment. The model selection and estimation problem is illustrated in Figure 2. Two pre-computed solutions to modeling the phenomenon with different complexities are presented to the domain expert. The merit of the model is not, however, calculated based on the fit to the data but rather defined how well it satisfies experts' interest in the modeling task. In our application of modeling tree growth, the task was defined as to "model the age-related growth trend while leaving most of the long-term variability related to climatic information as a residual signal". This judgment was left to the domain expert.

For the purpose of recording the experts' opinions, we implemented a Web based tool that presents two solutions of different complexity simultaneously. The task of the expert is to select the model which better suits his purpose. The pre-computed models were selected in random order and the choices were recorded. The experimental setting corresponds to the two-alternatives, forced choice (2AFC) experiment [4].

In the experimental setting, the expert is shown two examples out of the k possible solutions, and a forced choice is recorded. As a result, the logged choices will be of the form (i, j, k), where i and j represent the indices of the presented stimuli and k the preferred one. In our application, the indices range from 1 to 9. Because of human error, inconsistencies, and other factors, the result of an individual trial may be different in different repeats. Therefore, we turn to probabilistic modeling of choices and represent the probabilities of preference.

3   Experiments

Now we will turn our attention to the application, for which we developed the methodology. The problem is to model the tree-ring width series and to model the non-climatic variation (assumed, hypothetical normal tree growth) so that the climatic variation would be left as a residual signal. Standard regression analysis would attempt to model as much of the signal as possible, leaving only a non-explanatory random signal. Our data consists of the age of the tree as the x variable and the annual growth recorded from the tree-rings as the y variable. From a set of trees, the task is to estimate a model for normal growth as a regression function: $y = f(x)$.

3.1   Smoothing Splines as a Model Class

We will use smoothing splines as our model class for removing the non-climatic variance of forest interior tree-ring width series [3,2]. The smoothing splines [6] are regression functions built on local polynomials that are connected together in a way that ensures a smooth function. In estimation, the error function to be minimized consists of a squared error term summarizing the discrepancy between the measurements and the model and a term penalizing for large curvatures. The trade-off between selecting the accuracy of the function and the smoothness is controlled by the parameter p; large values of p will give importance to the model fit, small values of p will underline the smoothness of the mapping. The selection of the parameter p through recorded subjective choices of the expert is the focus of this paper. The optimization problem for estimating the smoothing spline for the interval $[a, b]$ is

$$\min_{p} \left\{ p \times \sum_{i=1}^{n} (y_i - f(x_i))^2 + (1 - p) \times \int_{a}^{b} (f'')^2 \right\}$$

The model selection and estimation problem is to find an optimal p that would reflect the subjective opinions of the expert. The domain expert is presented with two instances of model solutions simultaneously and is told to select the preferred solution. In all, 2395 pairs of random pairs modeling the growths of the same tree (same data) were presented, and the results were logged. 34 different trees were included in the set of stimuli.

The sufficient statistics for the conditional probability distributions representing the preferential relations are the counts n_{ij} where the model i is preferred over the model j. By normalizing the rows of the matrix containing the recorded preferences, the matrix becomes a stochastic matrix with non-negative entries. In essence, the entries p_{ij} can be given an interpretation of preferring j over i when a stimulus pair (i, j) has been presented. The stochastic matrix defines a fully connected graph, in which the arcs and the associated probabilities imply the preference relations between the models. Assimilating the model with a Web sites and links between the sites, we can use techniques for scoring the sites by using

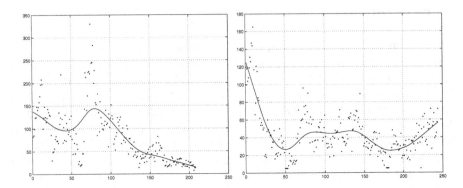

Fig. 2. The optimal models computed from the expert preferences

the random surfer model [1]. Initializing a state of preference randomly and surfing the connected preference graph repeatedly, we approach the stationary distribution, which reflects the posterior probability of preferences. The stationary distribution under the random surfer model for the normalized stochastic matrix is $s_j = (0.0482, 0.0514, 0.1312, 0.2325, 0.2280, 0.2098, 0.0776, 0.0214, 0)$. Stationary distribution can be used in selecting the most probable model; also, all the information in the distribution can be used in model averaging: parameter values for the smoothing parameter p can be calculated as the weighted average of the possible values as

$$\hat{p} = \sum_{j=1}^{9} s_j \times p_j$$

and we get the estimate $\hat{p} = 10^{-4.4129}$. This solution, is approximately of the same complexity as the models perceived to be satisfactory in the article [3]. However, the authors only comment the complexity of the models qualitatively. The optimal solution is illustrated for two example trees in Figure 2.

In the experiment, the expert was presented with the interpolating model, which goes through all the data points. The interpolating function was never judged to be superior to any other model. In the subsequent versions, the interpolating function will be removed from the experiment. Interestingly, during the initial testing with a non-expert, an interpolating function was sometimes judged better, underlying different perceptions in assessing model quality. In the future, we will repeat the experiment with a number of experts as it would be interesting to compare the perceptions of different experts and also to compare the individual opinions with the pooled consensus opinion of all experts together.

4   Summary and Conclusions

We presented a solution to a problem, where the model complexity is determined with the help of expert opinion, rather than data analytic considerations such as generalization ability measured with the cross-validation error. For this purpose, a two-alternatives, forced choice experiment was conducted and the individual choices were summarized as preferential relations. Furthermore, we can summarize the preferences with conditional probability tables. Given the probabilistic preferential relations, a random surfer model is applied to yield the equilibrium distribution. The posterior probabilities of the equilibrium distribution is used to yield the final model. Most probable model may be chosen for the purpose of model selection and model averaging may be used for model estimation.

Acknowledgments

The author thanks Harri Mäkinen at the Finnish Forest Research Center for the expert opinion used in the model selection experiment.

References

1. Baldi, P., Frasconi, P., Smyth, P.: Modeling the Internet and the Web — Probabilistic Methods and Algorithms. John Wiley & Sons, Chichester (2003)
2. Cook, E., Kairiukstis, L. (eds.): Methods of Dendrochronology — Applications in the Environmental Sciences. Kluwer Academic Publishers, Dordrecht (1989)
3. Cook, E.R., Peters, K.: The smoothing spline: A new approach to standardizing forest interior tree-ring width series for dendroclimatic studies. Tree-ring bulletin 41, 45–55 (1981)
4. Egan, J.: Signal Detection Theory and ROC Analysis. Academic Press, London (1975)
5. Hastie, T., Tibshirani, R., Friedman, J.: The Elements of Statistical Learning — Data Mining, Inference and Prediction. In: Springer Series in Statistics, Springer, Heidelberg (2001)
6. Wahba, G.: Spline models for observational data. Society of Industrial and Applied Mathematics (SIAM) (1990)

Detecting Concept Drift Using Statistical Testing*

Kyosuke Nishida and Koichiro Yamauchi

Graduate School of Information Science and Technology, Hokkaido University,
Kita 14 Nishi 9, Kita, Sapporo, 060-0814, Japan
{knishida, yamauchi}@complex.eng.hokudai.ac.jp

Abstract. Detecting concept drift is important for dealing with real-world online learning problems. To detect concept drift in a small number of examples, methods that have an online classifier and monitor its prediction errors during the learning have been developed. We have developed such a detection method that uses a statistical test of equal proportions. Experimental results showed that our method performed well in detecting the concept drift in five synthetic datasets that contained various types of concept drift.

1 Introduction

A difficult problem in learning scenarios is that the underlying distribution of the target concept may change over time. This is generally known as "concept drift" [1]. We have developed a method to detect concept drift in an online learning scenario in which a classifier is sequentially presented with training examples. The classifier outputs a class prediction for the given input, x_t, at each time step and then updates its hypothesis based on the true class label, y_t. Each example is independently drawn from the current distribution of the target concept, $\Pr_t(x, y)$. If concept drift occurs at time t, $\Pr_t(x, y)$ differs from $\Pr_{t-1}(x, y)$. The task of the method is to detect changes quickly and accurately to enable the classifier to minimize cumulative prediction errors during online learning.

The detection of changes is one way to respond to concept drift. Examples of real problems where change detection is relevant include user modeling, monitoring in biomedicine and industrial processes, fault detection and diagnosis [2]. There has been much work on detecting changes in online data streams [2,3,4]; however, most of it is based on estimating the underlying distribution of examples, which requires a large number of examples.

Detection methods that monitor classification errors in an online classifier during online learning have been proposed recently [5,6,7]. These methods do not depend on the type of input attribute. Moreover, they are able to detect concept drift from a small number of examples and thus have low computational costs.

---

* This study was partly supported by a Grant-in-Aid for JSPS Fellows (18-4475) from the Japan Society for the Promotion of Science.

V. Corruble, M. Takeda, and E. Suzuki (Eds.): DS 2007, LNAI 4755, pp. 264–269, 2007.

We have proposed such a drift detection method that uses a statistical test of equal proportions (STEPD) to detect various types of concept drift quickly and accurately. We demonstrated experimentally the performance of the method using five synthetic datasets that contain concept drift.

2   Related Drift Detection Methods

Gama et al. proposed a drift detection method with an online classifier (DDM) [5]. For each time, t, the error rate is the probability of misclassifying, p_t, with standard deviation, $s_t = \sqrt{p_t(1 - p_t)/t}$. It is assumed that p_t decreases as time advances if the target concept is stationary, and any significant increase of p_t suggests that the concept is changing. If the concept is unchanged, then the $1-\alpha$ confidence interval for p_t with $n > 30$ examples is approximately $p_t \pm z_{\alpha/2}s_t$, where $z_{\alpha/2}$ denotes the $(1 - \alpha/2)$th percentile of the standard normal distribution. DDM stores the values of p_t and s_t when $p_t + s_t$ reaches its minimum value (obtaining $p_{\min}$ and $s_{\min}$) and stores examples in short-term memory while $p_t + s_t \geq p_{\min} + 2s_{\min}$ is satisfied. DDM then rebuilds the classifier from the stored examples and resets all variables if $p_t + s_t \geq p_{\min} + 3s_{\min}$. DDM performs well for sudden changes; however, it has difficulties detecting gradual changes.

To improve the detection of gradual changes, Baena-García et al. developed the early drift detection method (EDDM) [6]. Their key idea is to consider the time interval (distance) between two occurrences of classification errors. They assume that any significant decrease in the distance suggests that the concept is changing. Thus, EDDM calculates the average distance between two errors, p'_t, and its standard deviation, s'_t, and stores these values when $p'_t + 2s'_t$ reaches its maximum value (obtaining $p'_{\max}$ and $s'_{\max}$). EDDM stores examples in short-term memory while $v_t \ (= (p'_t + 2s'_t)/(p'_{\max} + 2s'_{\max})) < \alpha$ is satisfied. It then rebuilds the classifier from the stored examples and resets all variables if $v_t < \beta$. Note that it starts detecting drift after 30 errors have occurred. EDDM performs well for gradual changes; however, it is not good at detecting drift in noisy examples.

We previously developed a drift detection method in a multiple classifier system [7]. We have now simplified it. This simplified method (ACED) uses only an online classifier. ACED observes the predictive accuracy of the online classifier for recent W examples, q_t, and calculates the $1 - \alpha_d$ confidence interval for q_t at every time t. Our key idea is that q_t will not fall below the lower endpoint of the interval at time $t - W$, q_{t-W}^l, if the target concept is stationary. Thus, it initializes the classifier if $q_t < q_{t-W}^l$. Note that it starts detecting drift after receiving $2W$ examples. ACED is able to detect concept drift quickly when W is small; however, such small windows often cause misdetection.

3   STEPD: Detection Method Using Statistical Testing

STEPD has been developed to achieve quick and accurate detection. The basic principle is to consider two accuracies: the recent one and the overall one. We assume two things: the accuracy of a classifier for recent W examples will be equal to the overall accuracy from the beginning of the learning if the target concept is stationary; and a significant decrease of recent accuracy suggests that the concept is changing. The test is performed by calculating the following statistic,

$$T(r_o, r_r, n_o, n_r) = \frac{|r_o/n_o - r_r/n_r| - 0.5(1/n_o + 1/n_r)}{\sqrt{\hat{p}(1 - \hat{p})(1/n_o + 1/n_r)}}, \tag{1}$$

and comparing its value to the percentile of the standard normal distribution to obtain the observed significance level (P-value)[1]. r_o is the number of correct classifications among the overall n_o examples except for recent W examples, r_r is the number of correct classifications among the $W (= n_r)$ examples, and $\hat{p} = (r_o + r_r)/(n_o + n_r)$. If the P-value, P, is less than a significance level, then the null hypothesis $(r_o/n_o = r_r/n_r)$ is rejected and the alternative hypothesis $(r_o/n_o > r_r/n_r)$ is accepted, namely concept drift has been detected. STEPD uses two significance levels: α_w and α_d. It stores examples in short-term memory while $P < \alpha_w$ is satisfied. It then rebuilds the classifier from the stored examples and resets all variables if $P < \alpha_d$. Note that it starts detecting drift after satisfying $n_o + n_r \geq 2W$ and the stored examples are removed if $P \geq \alpha_w$.

4   Experiment and Results

We used five synthetic datasets based on sets used in other papers concerning concept drift [5,6,8]. All the datasets have two classes. Each concept has 1000 examples. The number of training examples is 4000, except for STAGGER, which has 3000. The number of test examples is 100. The training and test examples were generated randomly according to the current concept.

- STAGGER (1S). sudden. The dataset has three nominal attributes: size (*small, medium, large*), color (*red, blue, green*), and shape (*circle, square, triangle*), and has three concepts: 1) [size = *small* and color = *red*], 2) [color = *green* or shape = *circle*], and 3) [size = *medium* or *large*].
- GAUSS (2G). sudden, noisy. The examples are labeled according to two different but overlapped Gaussian, $N([0, 0], 1])$ and $N([2, 0], 4)$. The overlapping can be considered as noise. After each change, the classification is reversed.
- MIXED2 (3M). sudden, noisy. The dataset has two boolean attributes (v, w) and two continuous attributes (x, y) from $[0, 1]$. The examples are classified as positive if at least two of the three following conditions are satisfied: v, w, $y < 0.5 + 0.3\sin(3\pi x)$. After each change, the classification is reversed. Noise is introduced by switching the labels of 10% of the examples.
- CIRCLES (4C). gradual. The examples are labeled according to the condition: if an example is inside the circle, then its label is positive. The change is achieved by displacing the center of the circle $((0.2, 0.5) \rightarrow (0.4, 0.5) \rightarrow (0.6, 0.5) \rightarrow (0.8, 0.5))$ and growing its radius $(0.15 \rightarrow 0.2 \rightarrow 0.25 \rightarrow 0.3)$.

---

[1] We should use the Fisher's exact test where sample sizes are small; however, we did not use it due to its high computational costs. The statistic in Eq. (1) is equivalent to the chi-square test with Yates's continuity correction.

Table 1. Cumulative prediction error rate with 95% confidence interval, number of drift detection (N_d), and number of required examples to detect drift correctly (N_e)

Data set	Method	IB1			Naive Bayes (NB)		
		Error Rate	N_d	N_e	Error Rate	N_d	N_e
1S	STEPD	**.0059**±.0001	**2.000−.000**	**4.29**	**.0076**±.0002	1.998 − .010	**4.89**
	DDM	.0064 ±.0001	**2.000−**.106	7.35	.0087 ±.0002	**2.000−**.370	8.94
	EDDM	.0214 ±.0000	**2.000−.000**	47.3	.0208 ±.0000	**2.000−.000**	42.0
	ACED	.0085 ±.0001	**2.000−.000**	11.4	.0100 ±.0002	**2.000−**.008	11.4
	Not Use	.3134 ±.0007			.3351 ±.0006		
2G	STEPD	**.1676**±.0007	**2.966−1.102**	**10.6**	**.1109**±.0004	2.964 − 1.180	**7.89**
	DDM	.2039 ±.0044	2.766 − **.892**	35.2	.1347 ±.0029	**2.970−1.008**	26.0
	EDDM	.1898 ±.0016	2.918 −10.56	26.4	.1250 ±.0009	2.934 − 7.682	18.0
	ACED	.1749 ±.0008	2.880 −5.306	12.0	.1156 ±.0005	2.910 − 3.434	9.68
	Not Use	.4456 ±.0006			.4737 ±.0007		
3M	STEPD	**.2143**±.0009	**2.968−**.932	**12.8**	**.1885**±.0006	**2.976−** .586	**11.2**
	DDM	.2439 ±.0036	2.672 − **.748**	43.9	.2008 ±.0013	2.942 − **.364**	36.8
	EDDM	.2443 ±.0014	2.884 −13.20	33.6	.2175 ±.0009	2.952 − 8.704	33.5
	ACED	.2262 ±.0009	2.866 −6.690	14.1	.2043 ±.0008	2.850 − 5.604	12.8
	Not Use	.4534 ±.0007			.4864 ±.0007		
4C	STEPD	**.0286**±.0002	**2.952−**.190	**26.8**	.0956 ±.0007	1.584 − **2.292**	42.5
	DDM	.0320 ±.0003	2.318 −1.490	58.9	.1072 ±.0010	.686 − 3.450	60.9
	EDDM	.0318 ±.0002	2.618 − .462	49.5	**.0920**±.0004	**1.588−** 7.934	50.0
	ACED	.0529 ±.0002	1.498 − .908	31.6	.1046 ±.0009	.786 − 2.952	**37.5**
	Not Use	.1365 ±.0004			.1536 ±.0005		
5H	STEPD	**.2254**±.0012	1.406		.1182 ±.0014	2.000	
	DDM	.2361 ±.0016	.048		.1278 ±.0017	1.518	
	EDDM	.2327 ±.0013	6.834		**.1110**±.0011	4.800	
	ACED	.2326 ±.0009	7.486		.1176 ±.0011	3.398	
	Not Use	.2465 ±.0021			.1590 ±.0028		

Notes: The prediction error rate is only calculated from the error on training data. The form of the N_d column (ex. $n-m$) means that n is the number of detection within 100 examples after each change and m is otherwise one (corresponding to the number of misdetection). We excluded misdetection in the calculation of N_e.

- HYPERP (5H). very gradual. The examples uniformly distributed in multi-dimensional space $[0, 1]^{10}$ are labeled satisfying $\sum_{i=1}^{10} a_i x_i \geq a_0$ as positive. The weights of the moving hyperplane, $\{a_i\}$, which are initialized to $[-1, 1]$ randomly, are updated as $a_i \leftarrow a_i + 0.001 s_i$ at each time, where $s_i \in \{-1, 1\}$ is the direction of change for each weight. The threshold a_0 is calculated as $a_0 = \frac{1}{2}\sum_{i=1}^{10} a_i$ at each time. $\{s_i\}$ is reset randomly every 1000 examples.

We compared STEPD with DDM, EDDM, ACED, and classifiers that did not use any methods (Not Use). The parameters of STEPD and ACED were $W = 30$, $\alpha_d = 0.003$, and $\alpha_w = 0.05$. Those of EDDM were $\alpha = 0.95$ and $\beta = 0.90$. We used two distinct classifiers with the methods: the Weka implementations of IB1 and Naive Bayes (NB) [9]. All results were averaged over 500 trials.

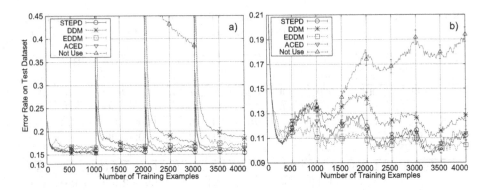

Fig. 1. Test error rate with 95% confidence intervals for a) 2G–IB1 and b) 5H–NB

Figure 1 and Table 1 show that all the detection methods improved the performance of the two classifiers in all the datasets. STEPD performed the best for sudden changes. Moreover, its performance was comparable to EDDM for gradual changes. ACED and EDDM were able to detect gradual changes well, whereas much misdetection occurred while the target concept was static because they were too sensitive to errors and noise (see N_d values for 2G and 3M). DDM detected sudden changes correctly; however, its detection speed was very slow. STEPD performed well in the presence of sudden and gradual changes and noise.

5   Conclusions

Our proposed drift detection method, STEPD, uses the statistical test of equal proportions. Experiments showed the test enables STEPD to detect various types of concept drift quickly and accurately. Future work will involve reducing misdetection and improving drift detection when changes are gradual.

References

1. Widmer, G., Kubat, M.: Learning in the presence of concept drift and hidden contexts. Machine Learning 23(1), 69–101 (1996)
2. Basseville, M., Nikiforov, I.V.: Detection of Abrupt Changes: Theory and Application. Prentice-Hall, Englewood Cliffs (1993)
3. Markou, M., Singh, S.: Novelty detection: a review — part 1: Statistical approaches. Signal Processing 83(12), 2481–2497 (2003)
4. Markou, M., Singh, S.: Novelty detection: a review — part 2: Neural network based approaches. Signal Processing 83(12), 2499–2521 (2003)
5. Gama, J., Medas, P., Castillo, G., Rodrigues, P.: Learning with drift detection. In: Proc. 17th Brazilian Symp. Artificial Intelligence, pp. 285–295 (2004)
6. Baena-García, M., del Campo-Ávila, J., Fidalgo, R., Bifet, A., Gavaldà, R., Morales-Bueno, R.: Early drift detection method. In: Proc. ECML/PKDD 2006, Work. Knowledge Discovery from Data Streams, pp. 77–86 (2006)

7. Nishida, K., Yamauchi, K., Omori, T.: ACE: Adaptive classifiers-ensemble system for concept-drifting environments. In: Proc. 6th Int. Work. Multiple Classifier Systems, pp. 176–185 (2005)
8. Wang, H., Fan, W., Yu, P.S., Han, J.: Mining concept-drifting data streams using ensemble classifiers. In: Proc. 9th ACM SIGKDD Int. Conf. Knowledge Discovery and Data Mining, pp. 226–235 (2003)
9. Witten, I.H., Frank, E.: Data Mining: Practical machine learning tools and techniques, 2nd edn. Morgan Kaufmann, San Francisco (2005)

Towards Future Technology Projection: A Method for Extracting Capability Phrases from Documents

Risa Nishiyama, Hironori Takeuchi, and Hideo Watanabe

Tokyo Research Laboratory, IBM Research
1623-14 Shimotsuruma, Yamato-shi, Kanagawa, 242-8502 Japan
{lisa, hironori, hiwat}@jp.ibm.com

Abstract. This paper deals with novel approaches for discovering phrases expressing technical capabilities in technical literature (such as patents), intended to support strategic consultants introducing new technologies and their capabilities to their clients. An extracted capability phrase is scored based on its expected business impact, which can also be considered as unexpectedness of the capability in a specified technology field. The proposed capability extraction method and unexpectedness estimation method are implemented in a "Future Technology Projection tool." The tool will be utilized by the consultants to provide lists of capability phrases related to a technology field of interest to the consultants.

1 Introduction

Many corporate executives pay attention to "innovation" these days [1], and they need business insights as well as new technologies for innovation. A strategic consultation is a key activity for such companies, and the consultants need to present new technologies that can be utilized for innovative business ideas. Currently, the consultants manually survey websites and books to search for the seeds of future new technologies which may have large impacts on business, but such manual processes may not be able to collect sufficient amounts of information. The insufficiency of manual survey would also decrease reliability of the resulted technology lists.

In such situations, we found that natural language processing technologies are able to support consultants by analyzing technical articles [2]. In this paper, we solve two types of problem: First, a method to extract capability phrases from technical literature such as patent documents is introduced. A capability phrase is a phrase mentioning a technical capability such as "enables high performance computing," or "providing a new experience for a speech-recognized kitchen." Second, a scoring method for the extracted capability phrases is also proposed. Our method estimates potential business impact of the technical capabilities and use the estimated business impact for the scoring (High/Low). We particularly focus on unexpectedness of the technical capability, since it is one of the important factors to determine the business impact.

V. Corruble, M. Takeda, and E. Suzuki (Eds.): DS 2007, LNAI 4755, pp. 270–274, 2007.
© Springer-Verlag Berlin Heidelberg 2007

Phrasal Patterns	Capability Phrase Examples
～を向上する (improves X)	ユーザの使い勝手を向上する (improves usability)
～を高める (improves X)	光の利用効率を高める (improves light use efficiency)
～を改善する (improves X)	初期の劣化を改善する (improves initial degradation)
～に優れる (excels over X)	冷熱サイクル性に優れる (excels over refrigeration cyclic performance)
～の向上 (improvement of X)	処理速度の向上 (improvement of processing speed)
～の … 化 (-ed X)	全体の薄型化 (thinned body)
～の改善 (improvement of X)	コントラストの改善 (improvement of contrast)
～を防止する (prevent X)	画像の劣化を防止する (prevents image quality degradation)
～を抑制する (control X)	変動による影響を抑制する (controls fluctuation effects)
～を低減する (reduce X)	消費電力を低減する (reduce power consumption)
～を抑える (control X)	半導体素子の電流を抑える (controls electric current in the chip)
～を可能にする (enables X)	強度を確保することを可能にする (allows maintaining strength)
～することができる (can X)	カウントすることで把握することができる (can determine by counting)
～を容易にする (make X easy)	チャンネルを選択することを容易にする (make channel selection easier)

Fig. 1. The phrasal patterns utilized to extract capability phrases from Japanese patent documents and examples of the extracted phrases

2   Capability Phrase Extraction

Before the capability phrase extraction method is introduced, we describe the data collection used in this work. The original document collection was first morphologically analyzed and parsed as preprocessing. In this paper, 290,889 Japanese patent documents that were filed from 2004 to 2006 are used as the original document set. A Japanese patent document consists of an abstract, claims, a background section, the embodiment, and an effect section. We extracted the abstract sections, which consist of issue and solution subsections, and the effect sections from the original patent journals, since these two sections describe important information for future technology projection, such as what problem is solved by the technology and how.

We found that the phrases in Japanese patents mentioning certain technical capabilities are able to be extracted by using phrasal patterns in the left column in Figure 1. By using those patterns, phrases such as those in the right column in Figure 1 are extracted from the parsing results.

In each runtime process, an arbitrary technology field of interest to the user is input as a query. The query is a string standing for the field such as "voice recognition" or "robot" in this work. By this query input, documents containing the input string are selected, since they are expected to be related to the field. Then the capability phrases in the collected documents are extracted and scored in the following process.

3   Scoring of Extracted Capability Phrases

This process gives capability phrases high or low score based on estimated business impact. Actual business impact is given by many factors, but we first focus on the impact due to the unexpectedness of the product or solution in the specified technology field. The unexpectedness is estimated by whether the capability phrase contains nouns that are normally unrelated to the technology field. In the case of robotics, for example, words such as "houses" or "intentions" would be more unexpected than words such as "throughput" or "hands." Difference of the given score is based on the different degrees of unexpectedness among the nouns used in the capability phrases. If a phrase contains unexpected nouns for that technology field, then high business impact is associated with the phrase, on otherwise low business impact is assigned.

The unexpected nouns W_u in the capability phrases are determined with Eq. (1). W_n denotes all nouns in the extracted capability phrases, and a noun is considered as "unexpected" in the field if it is not in any of W_c, W_t, or W_s. W_c denotes common nouns in the given technical document collection and determined with Eq. (2), where $N(w)$ denotes the number of documents containing the word w. If the noun w appears in more than n documents, the noun is regarded as common[1]. For Japanese patent documents, words such as "Hatsu-mei" (invention) and "Souchi" (device) are in W_c. W_t denotes nouns typical of the specified technology field and is determined with Eq. (3), where $RRF(w)$ denotes the ratio of relative frequency, $N(w \wedge q)$ denotes the number of documents in D_q having the word w, and D_q denotes the documents collected by the input query q. A noun w is considered as typical of the technology field if the ratio of its relative frequency is higher than x, so the noun is relatively common in the selected document set compared with its appearances in all of the documents[2]. W_s denotes a stop list and contains manually selected uninformative words.

$$W_u = W_n \cap \overline{W_c \cup W_t \cup W_s} \tag{1}$$

$$W_c = \{w | N(w) > n\} \tag{2}$$

$$W_t = \{w | RRF(w) > x\}, \quad RRF(w) = \frac{N(w \wedge q)/N_{D_q}}{N(w)/N_{D_{all}}} \tag{3}$$

The capability phrase extraction method described in the previous section and the scoring method for the extracted phrases are implemented in a "Future Technology Projection tool," and the tool outputs a list of capability phrases that are expected to be related to the technology field input by the user. Samples of the results will be shown in the next section.

4   Case Study

In this case study, the input query to specify a technology field was "robot," and 601 patent documents containing the string were collected. By using the capability phrase extraction method, 1,236 capability phrases were obtained from the selected documents, and 406 phrases out of the 1,236 were given high business impact score and other phrases were given low business impact score using the

---

[1] The value of n was set to 600 in the following case study. The least frequent nouns in W_c appear in approximately one-tenth of the document collection (290,889 documents).

[2] The value of x was set to 1.0 in the following case study.

High	Low
家屋内を掃除することができる (can clean inside of the house)	撮影装置の発行手段から放射される光を検知することができる (can detect light emitted from the flash of cameras)
基板搬送時間の短縮を図ることのできる (win reduce the conveyance time of boards)	本数不足等を回避することができる (can avoid cases such as number insufficiency)
保守サービス提供を実現する (can enable maintenance service provision)	処理液で処理することができる (can process with treatment liquid)
作動音を把握することができる (can comprehend loudness of the operation noise)	省力化を図ることができる (will save power)
飽きさせない話題を提供することのできる (can provide topics which do not make users bored)	省スペース化を図ることができる (can save space)
膜質の安定化を図る (can stabilize membranes)	人手を介することなく変更することのできる (can modify without assistance)
非接触で行うことができる (enables contact-free actions)	載置することができる (can place)
感情や雰囲気を詳細に表現することができる (can express emotions and atmospheres)	高スループット化を実現できる (can enable high throughput)
ホームページであることを確認することができる (can confirm that it is a webpage)	簡単にして確実に回避できる (can avoid easily and certainly)
使用者が利用することができる (can be used by users)	検査プロセスのスループットを増大させることができる (can speed inspection process)
直感的な遠隔操作を可能とすることができる (can enable intuitive remote control)	満遍なく洗浄し乾燥することができる (can throughly wash and dry)
自由な使い勝手を実現することができる (can enable unrestrictive usability)	2つのアームを含むロボットに切り替えることができる (can switch to robots with two arms)
長期間のバックアップを可能にする (enables long term backup)	基板が汚染されない様にする (avoids contamination of the boards)

Fig. 2. Example capability phrases extracted for the field of "robot" with scores (high or low)

unexpectedness estimation method. Figure 2 shows selected examples of the 1,236 phrases with their scores.

In the output list, the phrases given low business impact contain words showing normally anticipated improvements given by robot technology, e.g. "Kou-throughput-ka wo jitsugen-dekiru" (enables high throughput) and "Kouritsu-yoku hansousuru-koto ga dekiru" (can efficiently convey). On the other hand, several phrases given high business impact showed capabilities expected to make major impressions on the market in coming years, e.g. "Kaoku-nai wo soujisuru-koto ga dekiru" (can clean inside of the house), "Akisasenai wadai wo teikyousuru-koto ga dekiru" (can provide topics which do not make users bored).

In fact, Toshiba Corporation announced its new housecleaning robot, which was the first such consumer product in Japan, in 2002[3]. In 2005, NEC Corporation also developed a robot companion that can talk with the user to make a good impression[4]. Those real business cases support the validity of our proposed estimation method for business impact related to the unexpectedness of the application or product as described in the capability phrases.

5 Discussion and Conclusion

In this paper, we introduced a new technical document analysis method to support strategic consultants introducing new technologies and their capabilities

---

[3] http://www.toshiba.co.jp/about/press/2002_09/pr_j0501.htm (in Japanese)
[4] http://www.incx.nec.co.jp/robot/english/papero2005/index.html

to corporate executives. The tool implementing the proposed phrase extraction method and the scoring method is able to list the phrases mentioning technical capabilities with their business impact scores given by their estimated unexpectedness in the technology field. A case study showed examples of the extracted phrases and their estimated business impact scores. Several high-scored phrases supported by real business cases were also introduced.

There are few methods aiming at extracting phrases mentioning technical capabilities from patents and other technical documents. Most conventional analysis methods for patents aimed at summarizing the documents by classification [3,4], and several other methods focus on the occurrences of noun phrases in the documents to estimate trends in the technology field [5,6]. Those conventional methods could support the strategic consultation for innovation, but they require professional knowledge of the technology field to match the information to the customer-specific business cases. That is why our tool allows consultants and client executives to perceive how they can utilize the technology for their situations.

Future study will involve extension of the capability phrase extraction method for English documents and other types of documents such as Web news or academic papers. The evaluation of the unexpectedness estimation method will also be done with consultants, who are the expected users of the tool.

Acknowledgement. The authors would like to thank Tetsuya Nasukawa in IBM Research and Junji Maeda, Toshiyuki Kuramochi, Akihiro Kuroda, and Eiji Hayashiguchi in IBM Business Consulting Services for their valuable comments on our research.

References

1. Council on Competitiveness: Innovate America (National Innovation Initiative Report) (2004), http://www.compete.org/pdf/NII_Interim_Report.pdf (Date Accessed: 2007.07.19)
2. Nishiyama, R., Takeuchi, H., Watanabe, H., Nasukawa, T., Maeda, J., Kuramochi, T., Hayashiguchi, E.: Technical Document Mining Method for Future Technology Projections. In: Proc. of JSAI 2007(in japanese) (2007)
3. Murata, M., Kanamaru, T., Shirado, T., Isahara, H.: Using the K Nearest Neighbor Method and BM25 in the Patent Document Categorization Subtask at NTCIR-5. In: Proc. of NTCIR-5 Workshop Meeting (2005)
4. Kim, J.H., Huang, J.X., Jung, H.Y., Choi, K.S.: Patent Document Retrieval and Classification at KAIST. In: Proc. of NTCIR-5 Workshop Meeting (2005)
5. Lent, B., Agrawal, R., Srikant, R.: Discovering Trends in Text Databases. In: Proc. of KDD 1997, pp. 227–230 (1997)
6. Porter, A.L., Newman, N.C.: Patent Profiling for Competitive Advantage: Deducing Who Is Doing What, Where, and When. In: Handbook of Quantitative Science and Technology Research, pp. 587–612. Kluwer Academic Publishers, Dordrecht (2005)

Efficient Incremental Mining of
Top-K Frequent Closed Itemsets*

Andrea Pietracaprina and Fabio Vandin

Dipartimento di Ingegneria dell'Informazione, Università di Padova, Via Gradenigo
6/B, 35131, Padova, Italy
{capri,vandinfa}@dei.unipd.it

Abstract. In this work we study the mining of top-K frequent closed
itemsets, a recently proposed variant of the classical problem of mining frequent closed itemsets where the support threshold is chosen as
the maximum value sufficient to guarantee that the itemsets returned
in output be at least K. We discuss the effectiveness of parameter K in
controlling the output size and develop an efficient algorithm for mining
top-K frequent closed itemsets in order of decreasing support, which exhibits consistently better performance than the best previously known
one, attaining substantial improvements in some cases. A distinctive feature of our algorithm is that it allows the user to dynamically raise the
value K with no need to restart the computation from scratch.

1 Introduction

The discovery of frequent (closed) itemsets is a fundamental primitive in data
mining. Let $\mathcal{I}$ be a set of *items*, and $\mathcal{D}$ a (multi)set of *transactions*, where each
transaction $t \in \mathcal{D}$ is a subset of $\mathcal{I}$. For an *itemset* $X \subseteq \mathcal{I}$ we define its *conditional
dataset* $\mathcal{D}_X \subseteq \mathcal{D}$ as the (multi)set of transactions $t \in \mathcal{D}$ that contain X, and
define the *support of X w.r.t. $\mathcal{D}$*, $\text{supp}_{\mathcal{D}}(X)$ for short, as the number of transactions in $\mathcal{D}_X$. An itemset X is *closed w.r.t.* $\mathcal{D}$ if there exists no itemset Y, with
$X \subset Y \subseteq \mathcal{I}$, such that $\text{supp}_{\mathcal{D}}(Y) = \text{supp}_{\mathcal{D}}(X)$. The standard formulation of the
problem requires to discover, for a given support threshold σ, the set $\mathcal{F}(\mathcal{D}, \sigma)$
of all itemsets with support at least σ, which are called *frequent itemsets* [1].
In order to avoid the redundancy inherent in $\mathcal{F}(\mathcal{D}, \sigma)$, it was proposed in [5] to
restrict the discovery to the subset $\mathcal{FC}(\mathcal{D}, \sigma) \subseteq \mathcal{F}(\mathcal{D}, \sigma)$ of all closed itemsets
with support at least σ, called *frequent closed itemsets*.

A challenging aspect regarding the above formulation of the problem is related
to the difficulty of predicting the actual number of frequent (closed) itemsets for
a given dataset $\mathcal{D}$ and support threshold σ. Indeed, in some cases setting σ too
small could yield a number of frequent itemsets impractically large, possibly
exponential in the dataset size [11], while setting σ too big could yield very few
or no frequent itemsets.

---

* This work was supported in part by MIUR of Italy under project MAINSTREAM,
and by the EU under the EU/IST Project 15964 *AEOLUS*.

V. Corruble, M. Takeda, and E. Suzuki (Eds.): DS 2007, LNAI 4755, pp. 275–280, 2007.

In [10], an elegant variant of the problem has been proposed which, for a given "desired" output size $K \geq 1$, requires to discover the set $\mathcal{FC}_K(\mathcal{D})$ of top-K *frequent closed itemsets* (top-K f.c.i., for short), defined as the set $\mathcal{FC}(\mathcal{D}, \sigma_K)$ where σ_K is the maximum support threshold that such that $|\mathcal{FC}(\mathcal{D}, \sigma_K)| \geq K$. Although one is not guaranteed that $|\mathcal{FC}_K(\mathcal{D})| = K$, it is conceivable that parameter K be more effective than an independently fixed support threshold σ in controlling the output size. In the same paper, the authors present an efficient algorithm, called TFP, for mining the top-K f.c.i. TFP discovers frequent itemsets starting with a low support threshold which is progressively increased, as the execution proceeds, by means of several heuristics, until the final value σ_K is reached. TFP allows also the user to specify a minimum length $\min_\ell$ for the closed itemsets to be returned. The main drawbacks of TFP are that no bound is given on the number of non-closed or infrequent itemsets that the algorithm must process, and that an involved itemset closure checking scheme is required. Moreover, TFP does not appear to be able to handle efficiently a dynamic scenario where the user is allowed to raise the value K. A number of results concerning somewhat related problems can be found in [3,4,8].

The mining of top-K f.c.i. is the focus of this paper. In Section 2, we study the effectiveness of parameter K in controlling the output size by proving tight bounds on $|\mathcal{FC}_K(\mathcal{D})|$. In Section 3 we present a new algorithm, TopKMiner, for mining top-K f.c.i. in order of decreasing support. Unlike algorithm TFP, TopKMiner features a provable bound on the number of itemsets touched during the mining process, and, moreover, it allows the user to dynamically raise the value K efficiently without restarting the computation from scratch. Section 4 reports some results from extensive experiments which show that TopKMiner consistently exhibits better performance than TFP, with substantial improvements in some cases (more than two orders of magnitude). The efficiency of TopKMiner becomes considerably higher when used in the dynamic scenario[1].

2   Effectiveness of K in Controlling the Output Size

Let $\Delta(n)$ be the family of all datasets $\mathcal{D}$ whose transactions comprise n distinct items. Define $\rho(n, K) = \max_{\mathcal{D} \in \Delta(n)} (|\mathcal{FC}_K(\mathcal{D})|/K)$, which provides a worst-case estimation of the deviation of the output size from K when mining top-K closed frequent itemsets. Building on a result by [2] we can prove the following theorem.

Theorem 1. *For every $n \geq 1$ and $K \geq 1$, we have $\rho(n, K) \leq n$. Moreover, for every $n \geq 1$ and every constant c, there are $\Omega(n^c)$ distinct values of K such that $\rho(n, K) \in \Omega(n)$.*

We note that in several tests we performed on real and synthetic datasets with values of K between 100 and 10000, the ratio $|\mathcal{FC}_K(\mathcal{D})|/K$ turned out to be much smaller than n and actually very close to 1. In fact, it can be shown that

---

[1] For lack of space many details have been omitted from the paper and can be found in the companion technical report [6].

$\rho(n, K) = n$ is attained only for $K = 1$, and we conjecture that $\rho(n, K)$ is a decreasing function of K.

3 TopKMiner

In this section, we briefly describe our algorithm TopKMiner for mining the top-K f.c.i. for a dataset $\mathcal{D}$ defined over the set of items $\mathcal{I} = \{a_1, a_2, \ldots, a_n\}$ (the indexing of the items is fixed but arbitrary). The algorithm crucially relies on the notion of ppc-extension introduced in [9] and recalled below. For an itemset $X \subseteq \mathcal{I}$, we define its j-th prefix as $X(j) = X \cap \{a_i : 1 \le i \le j\}$, with $1 \le j \le n$, and define $\mathrm{Clo}_{\mathcal{D}}(X) = \bigcap_{t \in \mathcal{D}_X} t$, which is the smallest closed itemset that contains X. The core index of a closed itemset X, denoted by $\mathrm{core}(X)$, is defined as the minimum j such that $\mathcal{D}_X = \mathcal{D}_{X(j)}$. A closed itemset X is a called a prefix-preserving closure extension (ppc-extension) of a closed itemset Y if: (1) $X = \mathrm{Clo}_{\mathcal{D}}(Y \cup \{a_j\})$, for some $a_j \notin Y$ with $j > \mathrm{core}(Y)$; and (2) $X(j-1) = Y(j-1)$. Clearly, if X is a ppc-extension of Y, then $\mathrm{supp}(X) < \mathrm{supp}(Y)$. Let $\perp = \mathrm{Clo}_{\mathcal{D}}(\emptyset)$, which is the possibly empty closed itemset consisting of the items occurring in all transactions. It is shown in [9] that any closed itemset $X \ne \perp$ is the ppc-extension of exactly one closed itemset Y. Hence, all closed itemsets can be conceptually organized in a tree whose root is $\perp$, and where the children of a closed itemset Y are its ppc-extensions.

TopKMiner generates the frequent closed itemsets in order of decreasing support by performing a best-first (i.e., highest-support-first) exploration of the nodes of the tree defined above. Specifically, the algorithm receives in input the dataset $\mathcal{D}$, the value K for which the top-K f.c.i. are sought, and a value $K^* \ge K$. During the course of the algorithm, the user is allowed to dynamically raise K up to K^*. The algorithm makes use of a priority queue Q (implemented as a max heap), whose entries correspond to closed itemsets. Let $\mathrm{E}(Y, s)$ denote an entry of Q relative to a closed itemset Y with support s. The value s is the key for the entry. Q is initialized with entries corresponding to the ppc-extension of $\perp$. Then, a main loop is executed where in each iteration the entry $\mathrm{E}(Y, s)$ with maximum s is extracted from Q, the corresponding itemset Y is generated and returned in output, and entries for all ppc-extensions of Y are inserted into Q.

A variable σ is used to maintain an approximation from below to the support σ_{K^*} of the K^*-th most frequent closed itemset. This variable is initialized by suitable heuristics similar to those employed in [10], and it is updated in each iteration of the main loop to reflect the support of the K^*-th most frequent closed itemset seen so far. The value σ is used to avoid inserting entries with support smaller that σ into Q. Also, after each update of σ, entries with support smaller than σ, previously inserted into Q, can be removed from the queue.

After the K-th closed itemset is generated, its support is stored in a variable σ'. The main loop ends when the last closed itemset of support σ' is generated. At this point the user may decide to raise K to a new value K_{new}. In this case, the main loop is reactivated and the termination condition will depend now on

K_{new}. It can be shown that TopKMiner processes (i.e., inserts into Q) at most nK^* closed itemsets[2], while one such bound is not known for TFP [10].

Implementation. The efficient implementation of TopKMiner is a challenging task. For lack of space, we will limit ourself to briefly mention the key ingredients of our implementation. More details can be found in [6]. As in [7] the dataset $\mathcal{D}$ is represented through a Patricia trie $T_{\mathcal{D}}$ built on the set of transactions with items sorted by decreasing support. An entry $E(Y, s)$ of Q, corresponding to some closed itemset Y, is represented by the quadruple $(\mathcal{D}_Y, s, i, Y(i-1))$, where i is the core index of Y. For space and time efficiency, we represent $\mathcal{D}_Y$ through the list $L_{\mathcal{D}}(Y)$ of nodes of $T_{\mathcal{D}}$ which contain the core index item a_i of Y and belong to paths associated with transactions in $\mathcal{D}_Y$. We observe that $(\mathcal{D}_Y, s, i, Y(i-1))$ contains only a prefix, $Y(i-1)$, of Y. The actual generation of Y is delayed to the time when the entry $(\mathcal{D}_Y, s, i, Y(i-1))$ is extracted from Q. At this point a clever traversal of the subtrie of $T_{\mathcal{D}}$ whose leaves are the nodes of $L_{\mathcal{D}}(Y)$, is employed to generate Y and the quadruples for all of its ppc-extensions to be inserted into Q.

4   Experimental Evaluation

We experimentally compared the performance of TopKMiner and TFP [10] on all real and synthetic datasets from the FIMI repository (http://fimi.cs.helsinki.fi). The experiments have been conducted on a HP Proliant, using one AMD Opteron 2.2GHz processor, with 8GB main memory, 64KB L1 cache and 1MB L2 cache. Both TopKMiner and TFP have been coded in C++ and the source code for TFP has been provided to us by its authors. Due to lack of space, we report only a few representative results relative to datasets kosarac and accidents. The results for the other datasets are consistent with those reported here. The characteristics of the datasets are reported in the following table, including the values $\sigma_K/|\mathcal{D}|$ for $K = 1000$ and $K = 10000$:

Dataset	#Items	Avg. Trans. Length	# Transactions	$\sigma_{1000}/|\mathcal{D}|$	$\sigma_{10000}/|\mathcal{D}|$
accidents	468	33.8	340,183	0.656	0.483
kosarac	41,270	8.1	990,002	0.006	0.002

Figure 1.(a) and 1.(b) show the relative running times of TopKMiner and TFP on kosarac and accidents, respectively, for values of K ranging from 1000 to 10000 with step 1000. For TopKMiner, we imposed $K = K^*$ so to assess the relative performance of the two algorithms when focused on the basic task of mining top-K f.c.i. In another experiment, we tested the effectiveness of the TopKMiner's feature which allows the user to dynamically raise the value K up to a maximum value K^*. To this purpose we simulated a scenario where K is raised from 1000 to 10000 with step 1000 and run TopKMiner with $K^* = 10000$ measuring the

---

[2] In fact, with a slight modification of the algorithm the bound can be lowered to nK.

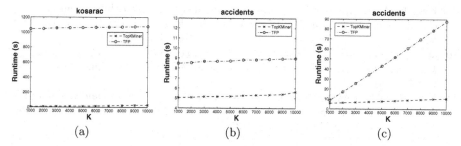

Fig. 1. Running times of TopKMiner and TFP for: (a) kosarac without dynamic update of K; (b) accidents without dynamic update of K; and (c) accidents with dynamic update of K

running time after the computation for each value K ended. We compared these running times with those attained by executing TFP from scratch for each value K and accumulating the running times of previous executions. The results are shown in Figure 1.(c) for dataset accidents.

We also analyzed the memory usage of TFP and TopKMiner on all datasets, for K between 1000 and 10000. TopKMiner requires less memory than TFP in almost all cases and, in the worst case, requires a factor 1.5 more memory.

References

1. Agrawal, R., Imielinski, T., Swami, A.: Mining association rules between sets of items in large databases. In: Proc. of the ACM SIGMOD Intl. Conference on Management of Data, pp. 207–216 (1993)
2. Boros, E., Gurvich, V., Khachiyan, L., Makino, K.: On maximal frequent and minimal infrequent sets in binary matrices. Annals of Mathematics and Artificial Intelligence 39, 211–221 (2003)
3. Cheung, Y., Fu, A.: Mining frequent itemsets without support threshold: with and without item constraints. IEEE Trans. on Knowledge and Data Engineering 16(9), 1052–1069 (2004)
4. Fu, A., Kwong, R., Tang, J.: Mining n-most interesting itemsets. In: Proc. of the Intl. Symp. on Methodologies for Intelligent Systems, pp. 59–67 (2000)
5. Pasquier, N., Bastide, Y., Taouil, R., Lakhal, L.: Discovering frequent closed itemsets for association rules. In: Proc. of the 7th Int. Conference on Database Theory, pp. 398–416 (1999)
6. Pietracaprina, A., Vandin, F.: Efficient Incremental Mining of Top-K Frequent Closed Itemsets. Technical Report, available at http://www.dei.unipd.it/c~apri/PietracaprinaVTR07.pdf
7. Pietracaprina, A., Zandolin, D.: Mining frequent itemsets using Patricia tries. In: Proc. of the Workshop on Frequent Itemset Mining Implementations, FIMI 2003. CEUR-WS Workshop On-line Proceedings. Melbourne, USA, vol. 90 (2003)
8. Seppanen, J., Mannila, H.: Dense itemsets. In: Proc. of the 10th ACM SIGKDD Intl. Conference on Knowledge Discovery and Data Mining, pp. 683–688 (2004)

9. Uno, T., Asai, T., Uchida, Y., Arimura, H.: An efficient algorithm for enumerating closed patterns in transaction databases. In: Suzuki, E., Arikawa, S. (eds.) DS 2004. LNCS (LNAI), vol. 3245, pp. 16–31. Springer, Heidelberg (2004)

10. Wang, J., Han, J., Lu, Y., Tzvetkov, P.: TFP: An efficient algorithm for mining top-k frequent closed itemsets. IEEE Trans. on Knowledge and Data Engineering 17(5), 652–664 (2005)

11. Yang, G.: The complexity of mining maximal frequent itemsets and maximal frequent patterns. In: Proc. of the 10th ACM SIGKDD Intl. Conference on Knowledge Discovery and Data Mining, pp. 344–353 (2004)

An Intentional Kernel Function
for RNA Classification

Hiroshi Sankoh, Koichiro Doi, and Akihiro Yamamoto

Graduate School of Informatics, Kyoto University
Yoshida-Honmachi, Sakyo-ku, Kyoto 606-8501, Japan
sankoh@mbox.kudpc.kyoto-u.ac.jp,
{doi, akihiro}@i.kyoto-u.ac.jp

Abstract. This paper presents a kernel function class K_{RNA} which is based on the concept of the intentional kernel (Doi et al., 2006) as opposed to that of the convolution kernel (Haussler, 1999). A kernel function in K_{RNA} computes the similarity between two RNA sequences from the viewpoint of secondary structures. As an instance of K_{RNA}, we give the definition and the algorithm of K_{RNA}^N which takes a pair of RNA sequences as its inputs, and facilitates Support Vector Machine (SVM) classifying RNA sequences in a higher dimension space. Our experimental results show a high performance of K_{RNA}^N, compared with the string kernel which is a convolution kernel.

1 Introduction

Much attention has been paid to computational analysis and prediction of noncoding RNA sequences [1]. The greatest characteristic of RNA sequences is that every RNA sequence has a secondary structure (Fig. 1) determined by interactions between Watson-Crick complementary base pairs (a-u, c-g). Various attempts have so far been made at modeling an RNA secondary structure by using Stochastic Context Free Grammars (SCFGs) [2]. Because few great advance with these methods are expected in decreasing time-complexity or in improving classification accuracy, the Support Vector Machine (SVM) technique combined with kernel functions is now attracting much attention in this field. Several kernel functions have been proposed to compute the similarity between two RNA sequences [3].

The purpose of this paper is to devise a new kernel function class K_{RNA} based on the intentional kernel [4], and to show that the kernel function K_{RNA}^N, which

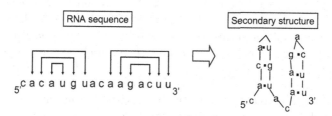

Fig. 1. An RNA sequence and its secondary structure

V. Corruble, M. Takeda, and E. Suzuki (Eds.): DS 2007, LNAI 4755, pp. 281–285, 2007.
© Springer-Verlag Berlin Heidelberg 2007

is an instance of K_{RNA} is better suitable to RNA classification than the string kernel [5], which is a convolution kernel [6].

2   Designing a New Kernel Function

2.1   Convolution Kernel and Intentional Kernel for Structured Data

SVM is a high-powered classifier which constructs a linear separating function in an input space. If input data cannot be linearly separated, by using some kernel functions, SVM maps any data to a point in a high dimensional feature space so that the mapped data can be linearly separated [7]. Recently, kernel functions for not only numerical vectors but also non-numerical structured data have been proposed in many scientific areas, for example, bioinformatics.

Such kernel functions for structured data are usually instances of the convolution kernel. A *convolution kernel* $K(x, y)$ computes its value by summing up the kernel values of every pair of substructure data from x and y, that is, $K(x, y) = \sum_{s \in s(x)} \sum_{t \in s(y)} K^s(s, t)$, where $s(x)$ represents the set of substructures in x.

On the other hand, the intentional kernel is a class radically contrasted with the convolution kernel. We assume that all data are in a partially ordered set $\mathcal{P}$, and for elements x and y in $\mathcal{P}$, an *intentional kernel* $K(x, y)$ is defined with the set $E(x, y) = \{z \mid z \succeq x, z \succeq y\}$. The name is from the point that $a \succeq b$ can be interpreted as "a subsumes b". A typical intentional kernel K_{TERM} [4], which takes a pair of first-order terms, is defined as $K_{TERM}(x, y) = \#(E(x, y))$, where $\#$ represents the cardinality.

2.2   Definition of K_{RNA}

The intentional kernel K_{RNA} for RNA sequences is designed by representing an RNA secondary structure with a simple Context Free Grammar G. The grammar G has non-terminal symbols P, L, R, S, E: P for emitting a canonical base pair, L and R for emitting only one base to the left and right respectively, and S and E for representing the start and the end of a derivation. The grammar G consists of the following production rules: $S \rightarrow P \mid L \mid R \mid E$, $P \rightarrow xPy \mid xLy \mid xRy \mid xEy$, $L \rightarrow xP \mid xL \mid xR \mid xE$, $R \rightarrow Px \mid Lx \mid Rx \mid Ex$, where x is in $\{a, u, c, g\}$, and (x, y) is a pair in $\{(a, u), (u, a), (c, g), (g, c)\}$.

The partial ordered set $\mathcal{P}$ for K_{RNA} is defined as the set of congruence classes of derivations starting with S. Two derivations are equivalent if they become same after (1) removing terminals, (2) removing Ls, Rs and an E following the last P, and (3) sorting subsequences consisting of Ls and Rs. For two congruence classes x and y, we define $x \succeq y$ if a shortest derivation in x is the substring of one in y. For example, a derivation $S \rightarrow R \rightarrow Lu \rightarrow aPu \rightarrow aaEuu \rightarrow aauu$ is equivalent to a derivation $S \rightarrow L \rightarrow aR \rightarrow aPu \rightarrow auEau \rightarrow auau$, and they are the elements in the congruence class $[S \rightarrow L \rightarrow R \rightarrow P]$. We define K_{RNA} by counting the number of common congruence classes for two input sequences.

Computing the value of K_{RNA} is equivalent to counting the number of common candidates of the secondary structures for two input sequences. For

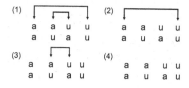

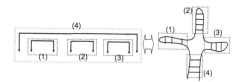

Fig. 2. The common candidates of the secondary structures for *aauu* and *auau*

Fig. 3. The structure of a tRNA sequence

example, we consider RNA sequences *aauu* and *auau*. The common congruence classes of derivations for the two sequences are $(1)[S \to P \to P]$, $(2)[S \to P]$, $(3)[S \to L \to R \to P]$, and $(4)[S]$. These four congruence classes are correspond to the secondary structures in Fig. 2, and every common derivation to *aauu* and *auau* represents exactly one of the secondary structures in Fig. 2. For example, the derivation $S \to R \to L \to P$ represents the same secondary structure of (3). So the number of congruence classes for *aauu* and *auau* is 4.

We introduce a kernel function K_{RNA}^N as an instance of K_{RNA} for treating input pairs of which a sequence is much larger than another. For two RNA sequences x and y of length n and m respectively, $K_{RNA}^N(x, y)$ is defined with an N-length window (Fig. 4) which extracts the two subsequences $x^{(i)} = x_i \cdots x_{i+N-1}$ $(i = 0, \ldots, n - N)$, and $y^{(j)} = y_j \cdots y_{j+N-1}(j = 0, \ldots, m - N)$ of equal length. We define the number of secondary structures shared by $x^{(i)}$ and $y^{(j)}$ as $K_{common}(x^{(i)}, y^{(j)})$.

The number N can take any integer which is no more than the least length of all training input sequences. So the definition of K_{RNA}^N is

$$K_{RNA}^N(x, y) = \sum_{i=0}^{n-N} \sum_{j=0}^{m-N} K_{common}(x^{(i)}, y^{(j)}).$$

The feature space of K_{RNA}^N has coordinates each of which corresponds to a secondary structure that RNA sequences of N-length can take, and the value of each coordinate is 1 if the input RNA sequence has the corresponding secondary structure, and is 0, otherwise.

This model is so simple that it cannot represent the complicated structures such as pseudoknot-structures and bifurcations. We outlive this problem by taking known structures of RNA families into account, and we apply K_{RNA}^N to each substructure of input RNA sequences. For example, tRNA sequences have four substructures (Fig. 3), and so we calculate the value of K_{RNA}^N for each of the four substructures and sum them up.

2.3 Algorithm for K_{RNA}^N

We represent the computing of $K_{common}(s, t)$ in a recursive expression, where $s = s_0 s_1 \cdots s_{N-1}$ and $t = t_0 t_1 \cdots t_{N-1}$ are N-length sequences. We let $K(i, j) =$

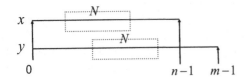
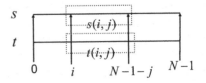

Fig. 4. N-length windows   **Fig. 5.** Two subsequences of $K(i,j)$

$K_{common}(s(i,j), t(i,j))$, where $s(i,j) = s_i s_{i+1} \cdots s_{N-1-j}$, and $t(i,j) = t_i t_{i+1} \cdots t_{N-1-j}$ (Fig. 5). It holds that $K_{common}(s,t) = K(0,0)$. Now, we consider computing the value of $K(i, j-1)$ from the value of $K(i,j)$. We let $C(i,j) = \{l \mid s_l = \bar{s}_{N-j}, t_l = \bar{t}_{N-j}, i \leq l \leq N-1-j\}$, where $\bar{x}$ represents the complementary base pair of x, and get the following recursive expression:

$$K(i, j-1) = K(i,j) + \sum_{l \in C(i,j)} K(l+1, j). \tag{1}$$

We have to give the initial values for computing $K(i,j)$ with (1). There are two cases where a pair of sequences of length 2 are both forming the complementary base pair or not, and the value of $K(i,j)$ is 2 or 1, so we can get the following expression by sliding the 2-length window from the position 0 to $N-2$:

$$K(i, N-2-i) = \begin{cases} 2 & \text{if } s_i = \bar{s}_{i+1} \text{ and } t_i = \bar{t}_{i+1}, \\ 1 & \text{otherwise,} \end{cases} \quad (i = 0, \ldots, N-2),$$

and, for sequences of length 1 and 0, we let $K(i,j) = 1$. Thus we can compute the values of $K_{common}(s,t) = K(0,0)$ with (1) and the default values of $K(i,j)$ in $O(N^2)$, and $K_{RNA}^N(x,y)$ in $O(mnN)$ with a little ingenuity, omitting the details due to the lack of space.

3   Experimental Results

In order to evaluate the performance of K_{RNA}^N, we used three RNA families; tRNA, miRNA, and 5SrRNA taken from the Rfam database (http://www.sanger.ac.uk/Software/Rfam/), extracted 2500, 347, and 1955 sequences respectively from each family as positive examples, and generated a same number of negative examples by shuffling the bases in the sequences. We trained SVM classifiers combined with K_{RNA}^N and the string kernel [5] for comparison of performance, with these examples in each family, and conducted 10-fold cross-validation. We implemented K_{RNA}^N and the string kernel as the original kernel function for SVM^{light} [8]. For tRNA family, we applied K_{RNA}^N to its four substructures, for 5SrRNA family, we applied K_{RNA}^N to its five substructures. After trying classifications with several values of N, we decide the value of N which is for the best accuracy. We also used the weighted subsequences version of the string kernel.

Table 1. Comparison of K_{RNA}^N and the string kernel

family	length	# sub-parts N	K_{RNA}^N accuracy	precision	recall	string kernel accuracy	precision	recall
tRNA	$65 - 93$	4	16 98.1	98.6	97.6	57.2	61.2	73.2
miRNA	$60 - 187$	1	60 65.1	65.0	65.2	55.0	54.0	68.0
5SrRNA	$81 - 130$	5	18 74.1	74.3	73.8	58.3	65.9	57.2

We show the result in Table 1. This result shows the high performance of K_{RNA}^N, on the point that it gives the highest classification accuracy of tRNA family. This is because K_{RNA}^N fully takes secondary structures of tRNA into account. On the other hand, the accuracy of miRNA is not high. We conjecture that this is caused by the large difference between the length of the longest (sub)sequence and the window length N.

4   Conclusions and Future Work

In this paper, we devised an intentional kernel class K_{RNA} and showed that the kernel function K_{RNA}^N which is an instance of K_{RNA} has a high performance, compared with the string kernel which is a convolution kernel. The window length N is an important parameter for performance, so we have to select the value carefully in experiments. In the future, we will make an improvement to K_{RNA}^N by weighting parameters on the secondary structures.

References

1. Eddy, S.: Non-coding RNA genes and the modern RNA world. Nature Reviews Genetics 2(12), 919–929 (2001)
2. Sakakibara, Y., Brown, M., Hughey, R., Mian, I., Sjolander, K., Underwood, R., Haussler, D.: Stochastic context-free grammars for tRNA modeling. Nucleic Acids Research 22(23), 5112–5120 (1994)
3. Kin, T., Tsuda, K., Asai, K.: Marginalized kernels for RNA sequence data analysis. Genome Informatics 13, 112–122 (2002)
4. Doi, K., Yamashita, T., Yamamoto, A.: An efficient algorithm for computing kernel function defined with anti-unification. In: ILP 2006. LNCS (LNAI), vol. 4455, pp. 139–153. Springer, Heidelberg (2007)
5. Lodhi, H., Saunders, C., Shawe-Taylor, J., Cristianini, N., Watkins, C.: Text classification using string kernels. The Journal of Machine Learning Research 2, 419–444 (2002)
6. Haussler, D.: Convolution kernels on discrete structures. UC Santa Cruz Technical Report UCS-CRL-99-10 (1999)
7. Vapnik, V.: The Nature of Statistical Learning Theory. Springer, Heidelberg (2000)
8. Joachims, T.: Making large-scale SVM learning practical. In: Schölkopf, B., Burges, C., Smola, A. (eds.) Advances in Kernel Methods-Support Vector Learning (1999)

Mining Subtrees with Frequent Occurrence of Similar Subtrees

Hisashi Tosaka, Atsuyoshi Nakamura, and Mineichi Kudo

Graduate School of Information Science and Technology Hokkaido University
Kita 14, Nishi 9, Kita-ku, Sapporo, 060–0814 Japan

Abstract. We study a novel problem of mining subtrees with frequent occurrence of similar subtrees, and propose an algorithm for this problem. In our problem setting, frequency of a subtree is counted not only for equivalent subtrees but also for similar subtrees. According to our experiment using tag trees of web pages, this problem can be solved fast enough for practical use. An encouraging result was obtained in a preliminary experiment for data record extraction from web pages using our mining method.

1 Introduction

Mining frequent subtrees from labeled trees is now a popular research field in the area of data mining and has a lot of application areas such as computer networks, web mining, bioinformatics, XML document mining, etc. [1]. In this field, various types of subtrees are mined. The most simple one is an *ordered bottom-up subtree* [2]. However, ordered bottom-up subtrees are too simple to catch frequent substructures of ordered trees. Therefore, richer classes of subtrees such as the class of *ordered induced subtrees* [3] and the class of *ordered embedded subtrees* [4] have been targeted.

In this paper, we study the problem of mining ordered bottom-up subtrees with frequent occurrence of similar subtrees. The difference from previously studied problems is the way of frequency counting. In our problem setting, not only equivalent subtrees but also similar subtrees are counted for a given similarity measure. Using previous frequent mining methods for induced or embedded subtrees, common substructures among similar subtrees might be found, however, such substructures do not always exist. Furthermore, the fact that two bottom-up subtrees have a common substructures does not imply that those are similar with respect to our similarity measure.

In this paper, we give a definition of our mining problem and show an algorithm to solve the problem. In order not to doubly count essentially the same parts of a tree, only maximal ones among similar subtrees are counted. As a similarity measure, we use measures based on the tree alignment/edit distance [5]/[6]. An algorithm for the problem can be easily created using the algorithm of calculating tree alignment/edit distance between two trees [5]/[6] because it

V. Corruble, M. Takeda, and E. Suzuki (Eds.): DS 2007, LNAI 4755, pp. 286–290, 2007.

calculates distance for all pairs of subtrees on the way. We propose tree conversion to the compact form before distance calculation to speed up the algorithm for trees containing many equivalent subtrees.

By virtue of the conversion, our algorithm runs in time practically usable according to our experiment for web pages of e-commerce search engine. Encouraging result was also obtained for an application of our subtree mining to extracting data records from those web pages.

2   Problem Setting

In this section, we define the problem of finding subtrees with frequent occurrence of similar subtrees in a labeled ordered tree. Note that the term "trees" are always used to refer to labeled ordered trees throughout this paper.

Since an ordered tree is a kind of a rooted tree, all branches are supposed to be directed from parent nodes to child nodes. $|T|$ denotes the number of nodes in T. If there is a path from node u to node v, node u is called an *ancestor* of node v (node v is called a *descendant* of node u). A *(bottom-up) subtree* of tree T rooted by node u, denoted by T_u, is a tree that is composed of all descendants of node u. We define $\mathscr{S}(T)$ as the set of all subtrees of tree T. For $\mathscr{U} \subseteq \mathscr{S}(T)$, we say that T_u is *maximal* in $\mathscr{U}$ if $T_u \in \mathscr{U}$ and no node $v(\neq u)$ is an ancestor of node u for all node v with $T_v \in \mathscr{U}$. Let sim be a similarity function on labeled ordered trees. For a given *similarity threshold* $k \geq 0$, a tree S is said to be *similar* to a tree T if $\mathrm{sim}(T, S) \geq k$. Let $\mathscr{S}_u(T)$ denote the set of all subtrees in $\mathscr{S}(T)$ similar to T_u. For a given *minimum support* $\sigma \geq 0$, a subtree T_u is said to be a *subtree with frequent occurrence of similar subtrees*, an *SFOSS* for short, if $\mathscr{S}_u(T)$ has at least σ maximal subtrees.

Let us now define our mining problem.

Problem. Given a tree T, a similarity threshold $k \geq 0$ and a minimum support $\sigma \geq 0$, find all maximal subtrees of T in the set of SFOSSs.

Remark. Frequency of similar subtrees should be counted by counting the number of independent occurrences. Thus, in the above problem setting, we take account of only maximal ones in frequency counting. Only maximal SFOSSs are enumerated in our problem setting for reducing result complexity.

As a similarity function $\mathrm{sim}(T, S)$,

$$\max\{1 - \frac{d(T, S)}{d(T, \theta)}, 0\}, 1 - \frac{d(T, S)}{\max\limits_{|T'|=|T|, |S'|=|S|} d(T', S')}$$

and so on are considered, where d is the tree alignment/edit distance [5]/[6] and θ is the empty tree.

Here, we only consider similarity functions defined using the tree alignment/edit distance.

3   Algorithm

Our algorithm consists of three steps: conversion to the compact form, distance table calculation and enumeration of maximal SFOSSs.

First step of our algorithm is conversion to the compact form. The *compact form R* of a tree T [7] is a factored representation that is an edge-ordered rooted DAG (directed acyclic graph) in which all common subtrees are shared being represented only once. Although this conversion takes time $O(|T|)$ [7], it reduces the computation time of the following steps when T contains many equivalent subtrees.

Next, the algorithm calculates the distance table that contains distance for all the pairs of subtrees in the compact form R. Fortunately, this can be done by applying a conventional algorithm calculating the tree alignment/edit distance [5]/[6] to the same two DAG R, because distance for all the pairs of subtrees are calculated by the algorithm on the way. Note that the algorithm is applicable to DAGs without modification though it is not designed for DAGs but for trees. Also notice that redundant calculation for the same pairs of subtrees can be avoided using the compact form instead of the tree itself.

The last step of our algorithm is enumeration of maximal SFOSSs. In this step, frequency of maximal subtrees similar to each subtree represented by a node in R is calculated. Similarity between two subtrees are obtained using the distance table. Counting only maximal ones can be realized by returning to the parent right after the subtree represented by the current node becomes known to be a similar tree in the preorder traversal of DAG R. Enumerating only maximal SFOSSs can be realized similarly.

Computation time needed for making distance table is known to be $O(|T|^2 \cdot \deg^2(T))$ for the tree alignment distance, which is the rate-determining step of our algorithm. Here, $\deg(T)$ is the maximum number of children of any node in the tree T. By converting to the compact form first, our algorithm for that distance runs in time $O(|T| + \text{edge}^2(R) \cdot \deg^2(T))$, where $\text{edge}(R)$ is the number of edges in R.

4   Experiments

We conducted two experiments, a running time performance experiment and a preliminary experiment for data record extraction from web pages. The data set we used in our experiments is data set 3 used in [8]. The data set is composed of 50 web pages collected from 50 different websites, and each page was randomly selected from the pages of the same site included in Omini testbed [9], which contains more than 2000 pages. The data set mainly consists of result pages of e-commerce search engines.

Running time of our algorithms for the above data set is shown in Table 1. In this experiment, we used the similarity function $\text{sim}(S, T) = \max\{1 - \frac{d(T,S)}{d(T,\theta)}, 0\}$, where d is the tree alignment distance, and set a similarity threshold k to 0.5 and a minimum support σ to 5. We implemented our algorithms by Java and

Table 1. Running Time($k = 0.5, \sigma = 5$)

Algorithm	Mean [s]	Max [s]	Min [s]
without CF	2.635	19.886	0.034
with CF	0.786	5.282	0.024

CF: compact form conversion

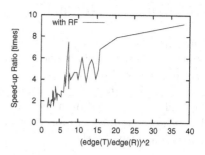

Fig. 1. Speed-up ratio for each data

conducted this experiment using a Linux computer with 3.4GHz CPU and 2GB memory. Average running time of our algorithm with CF was 0.786 seconds, which indicates practical usefulness of the algorithm. Conversion to the compact form was actually effective. The algorithm with CF is 3 times faster than that without CF on average. Speed-up ratio for each data is shown in Fig. 1. The larger the edge reduction ratio, the larger the speed-up ratio.

The second experiment is on data record extraction from web pages. Since data records of the same type in a web page have similar tag structure, they must be frequent. Furthermore, main data records such as search result records (SRRs) have the largest structure among those with high probability. We considered a simple algorithm using this idea, and conducted experiments using a part of the data set.

The algorithm we used is the following. First, we fixed minimum support σ and similarity threshold k to a certain number ($\sigma = 5, k = 0.7$ in our experiment). This means that we gave up extraction of data records with occurrence less than σ. Next, our subtree mining algorithm is executed and SFOSS $T_{\max}$ with the maximum number of nodes is picked up. Since $T_{\max}$ might be too far from the smallest SFOSS of the same-type data records with respect to alignment distance, the algorithm picks up one maximal SFOSS T_{freq} with the largest frequency among the maximal similar subtrees of $T_{\max}$. The final estimated data records of our algorithm are those corresponding to all maximal similar subtrees of T_{freq}.

Our subtree mining algorithm cannot find frequent subforests, so the algorithm above cannot extract data records with forest tag structure. Thus, we used a smaller data set that is made by removing 14 pages that have SRRs with forest tag structure or occurrence less than $\sigma(= 5)$.

We compared performance of our method with those of ViNTs [8] and MDR [10] calculated from the detailed result table[1]. Recalls for our method, ViNTs and MDR were 91.9%, 98.9% and 50.4%, respectively. Precisions for the three methods were 99.3%, 99.2% and 98.0%, respectively. Considering that ViNTs and MDR are complicated methods that make use of domain knowledge and heuristics, this result looks encouraging.

---

[1] http://www.data.binghamton.edu:8080/vints/mdrCompare.html

5 Concluding Remarks

In our experiment, we used the tree alignment distance instead of the tree edit distance, but which distance should be used depends on applications. As for data record extraction from web pages, the tree alignment distance looks suitable because the next step of the data record extraction is alignment of component item/fields in extracted data records [11].

Extension to mining sub-forests would make our mining method more effective and it is our future work.

References

1. Chi, Y., Muntz, R.R., Nijssen, S., Kok, J.N.: Frequent Subtree Mining-An Overview. Fundamenta Informaticae 66(1), 161–198 (2005)
2. Luccio, F., Enriquez, A.M., Rieumont, P.O., Pagli, L.: Exact rooted subtree matching in sublinear time. Technical Report TR-01-14, Universita Di Pisa (2001)
3. Asai, T., Abe, K., Kawasoe, S., Arimura, H., Sakamoto, H., Arikawa, S.: Efficient substructure discovery from large semi-structured data. In: Proceedings of the 2nd SIAM international conference on Data Mining, pp. 158–174 (2002)
4. Zaki, M.: Efficiently mining frequent trees in a forest: algorithms and applications. IEEE Transactions on Knowledge and Data Engineering 17(8), 1021–1035 (2005)
5. Jiang, T., Wang, L., Zhang, K.: Alignment of Trees-An Alternative to Tree Edit. Theoretical Computer Science (143), 1021–1035 (1994)
6. Zhang, K., Shasha, D.: Simple fast algorithms for the editing distance between trees and related problems. SIAM Journal on Computing 18(6), 1245–1262 (1989)
7. Flajolet, P., Sipala, P., Steyaert, J.M.: Analytic variations on the common subexpression problem. Automata, Languages and Programming, 220–234 (1990)
8. Zhao, H., Meng, W., Wu, Z., Raghavan, V., Yu, C.: Fully automatic wrapper generation for search engines. In: Proceedings of the 14th international conference on World Wide Web, pp. 66–75 (2005)
9. Buttler, D., Liu, L., Pu, C.: A Fully Automated Object Extraction System for the World Wide Web. In: Proceedings of the International Conference on Distributed Computing Systems (2001)
10. Liu, B., Grossman, R., Zhai, Y.: Mining data records in Web pages. In: Proceedings of the ninth ACM SIGKDD international conference on Knowledge discovery and data mining, pp. 601–606 (2003)
11. Zhai, Y., Liu, B.: Web data extraction based on partial tree alignment. In: Proceedings of the 14th international conference on World Wide Web, pp. 76–85 (2005)

Semantic Based Real-Time Clustering for PubMed Literatures

Ruey-Ling Yeh[1], Ching Liu[1], Ben-Chang Shia[2], I-Jen Chiang[3,4],Wen-Wen Yang[5], and Hsiang-Chun Tsai[4]

[1] Division of Biometrics, Graduate Institute of Agronomy, National Taiwan University, Taipei, Taiwan
[2] Department of Statistics and Information Science, Fu Jen Catholic University, Taipei, Taiwan
[3] Graduate Institute of Biomedical Informatics, Taipei Medical University, Taipei, Taiwan
[4] Institute of Biomedical Engineering, National Taiwan University, Taipei, Taiwan
[5] Graduate Institute of Medical Sciences, Taipei Medical University, Taipei, Taiwan
{d90621202, m485} @ntu.edu.tw, stat1001@mails.fju.edu.tw, {ijchiang, d102094022}@tmu.edu.tw, and ginnitsai@gmail.com

Abstract. This paper addresses to use the latent semantic topology to real-time cluster the literatures retrieved by PubMed in response to clinical queries and evaluates its performance by professional experts. The result shows that semantic clusters properly offer an exploratory view on the returned search results, which saves users' time to understand them. Besides, most experts conceive that the documents assigned to the identical cluster are similar and the concepts of clusters are appropriate.

Keywords: real-time, semantic clustering, combinatorial topology, Web Mining.

1 Introduction

An overwhelming amount of biomedical literatures stored in PubMed grows rapidly and becomes quickly diverse. Online Mendelian Inheritance in Man (OMIM) classifies varied PubMed literatures base on a biomedical taxonomy ontology. XplorMed [1] and GoPubMed [2] use the predefined classes from the MeSH or GeneOntology to classify biomedical literatures. However, the taxonomy needs a pile of laborious maintenance work and is unable to satisfy medical specialists' requests. It is necessary to classify the immensely retrieved literatures from PubMed immediately. Therefore, latent semantic clustering is considered to be one predominant approach [3] to automatically cluster data into meaningful groups.

Document clustering has been contemplated as one of the most pivotal techniques for dealing with the diverse and enormous amount of information on the World Wide Web. Traditional methods based on k-means, hierarchical clustering, and nearest neighbor clustering select a set of key terms or phrases to organize the feature vectors corresponding to different documents. Zamir *et al.* [4] presented a suffix-tree clustering (STC), which identified the sets of documents that share common phrases and formed

V. Corruble, M. Takeda, and E. Suzuki (Eds.): DS 2007, LNAI 4755, pp. 291–295, 2007.
© Springer-Verlag Berlin Heidelberg 2007

document clusters depending on the similarity between documents. Ertoz *et al.*[5] proposed a clustering approach that found out the nearest neighbors of each data point and then identified core points and then built clusters.

The semantic topology-based method [3,6] yielded better results than the k-means, AutoClass, HCA, and PDDP [7] on classifying the high-dimensional data, such as the Web pages from [7], the newswire articles from the Reuters-21578, and so forth. In this paper, we apply it to produce a real-time clustering on a vast amount of biomedical literatures retrieved by PubMed in response to clinical queries. In our framework, documents are represented as a topology of features, e.g., keywords. An agglomerative clustering algorithm to construct a semantic hierarchy based on the combination of those features is in use to discover latent semantics behind those documents.

This paper is organized as follows. We briefly review feature selections and the latent semantic topology method in the next section. Section 3 discusses some experimental results and evaluations, followed by the conclusion.

2 The Method

We use Hidden Markov Models (HMMs) to generate a part-of-speech tagger [8] for biomedical literatures to extract noun phrases in a sentence. All the collections of noun phases are considered to be the key features in a document. Then *tfidf* indexing is applied to weight features. Those features with higher *tfidf* values are selected and put in the feature list of the document collection. We believe that the set of co-occurred key features (within in a short distance, e.g., a sentence or a paragraph; in our paper, we use a paragraph) reflects latent semantics in the collection. According to topological property it naturally organized a hierarchical lattice of the co-occurred feature sets, which is called semantic topology. The upper level hierarchy filters the verbose terms contained in the lower lever hierarchy, therefore, it induces more concrete and kernel information brings from some lower sets in the topology.

A latent semantic topology illustrates a hierarchy of concept disciplines associated with the extracting features. Basically, the algorithm is divided into three main parts (referring [6] in detail): first, to construct an undirected connected graph, i.e., a skeleton S_0^1 of the simplicial complex, from a data set; second, to generate the concepts from graph recursively; third, to cluster the data based on generated concepts.

We built a Web-based clustering search engine, it consists three layers as follows.

1. Presentation layer: This layer presents the search results and their hierarchical semantic structures.
2. Business layer: This layer contains the main processing logic of clustering. The statistical mechanism, Hidden Markov Model (HMM), is used for feature extraction. The features are extracted from the returned search results using the HMM-based part-of-speech tagger [8] to generate simplical complex to make an agglomerative clustering of the search results.
3. Data layer: The data layer stores metadata of the returned PubMed literatures, such as title, abstract, authors, and so on. Different parts of PubMed documents will be assigned different weights for document clustering.

3   PubMed Experiments and Results

We conducted an expert evaluation of total twenty-six volunteers who are pharmacists, medical engineers, physicians and public health experts. Three types of test queries, ambiguous, entity, and general terms showed in Table 1 were selected by them. Ambiguous terms yield multiple interpretations in biomedical fields. General terms cause common concepts in biomedical fields. Besides, they provided three entity terms in their respective fields for the evaluation. Some qualitative parameters [9-10] are chosen to evaluate the comprehension of auto-generated hierarchies.

Table 1. Three types of query terms

Type	Query terms
Ambiguous terms	cure, drug, NEC, peer, neglect, response, channel, quality control, prime, order
Entity terms	NTG, celecoxib, metformin, vaccine, hospital, addiction, parkinson's disease, schizophrenia, spinal cord
General terms	NSAIDs, antibiotics, sex hormones, POCT, video game, medical devices, Speech Recognition, patient safety, X-ray, immunity

The definition of each parameter used in the experiment is described in the following:

1.Summarization: whether the clusters at top level are enough summarization.
2.Missing concepts: whether the clusters at top level have any missing concepts.
3.Redundancy: whether the concepts of clusters at top level have redundancy.
4.Cohesiveness:whether the documents assigned to the identical cluster are similar.
5.Isolation: whether the clusters at the same level are discriminating and their concepts do not subsume one another.
6.General to specific concept: whether the generated concept hierarchy is traversed from broader concepts at the higher levels to narrower concepts at the lower levels.
7.Navigation balance: whether the fan-out at each level of the hierarchy is appropriate.
8.Readability:whether the concepts of clusters are appropriate.
9.Search Time: whether our online system compared with PubMed really helps in reducing time to locate information.

For each parameter, the evaluators were asked to rate each type on a scale of 1-10: a higher value indicates a higher agreement. Table 2 shows the results, median and quartile deviation for all terms. To evaluate reliability, internal consistency methods are widely used and Cronbach α is used to evaluate it. The 0.67 of Cronbach α implies a good reliability and credibility [11]. We find that the expert's opinions have a wide discrepancy even though they have the same discipline such as Physician. The results show a high diversity of "redundancy at top level" in ambiguous terms and entity terms for physician. One of the reasons is just like the feedback from some evaluators that the ambiguous terms have their innate equivocation, especially in biomedical domain. Besides, 3 evaluators replied in support of the cluster-based information retrieval that the clusters from the retrieved search results aroused them to concretize their information needs. The results (table 2 and fig.1) show that almost all evaluators agree our clustering search engine can reduce search time as compared with PubMed (V9).In addition, most experts conceive that the documents assigned to the identical cluster are similar (V4) and the concepts of clusters are appropriate (V8).

Table 2. Expert study for evaluating the concept hierarchies

Item	Ambiguous terms		Entity terms for pharmacist		Entity terms for physician		Entity terms for public health		General terms	
	Me	Q.D.	Me	Q.D.	Me	Q.D.	Me	Q.D.	Me	Q.D.
Summarization at top level(V1)	7.0	1.5	7.0	0.5	4.0	2.1	7.0	0.5	7.0	0.6
Missing concepts at top level(V2)	7.0	1.0	8.0	1.5	4.0	1.4	6.0	0.5	7.0	0.6
Redundancy at top level(V3)	3.0	2.5	6.0	2.0	6.5	3.4	5.0	0.0	6.0	1.4
Overall cohesiveness(V4)	8.0	1.0	7.0	1.5	6.5	1.3	8.0	0.0	7.0	1.0
Overall isolation(V5)	5.0	2.0	7.0	0.5	6.0	1.8	7.0	0.5	6.0	0.5
Overall general to specific concept(V6)	7.0	2.5	7.0	0.5	6.0	1.8	6.0	0.0	6.0	0.6
Overall navigation balance(V7)	5.0	1.0	6.0	1.0	6.0	1.8	7.0	0.5	6.5	0.6
Overall readability(V8)	6.0	0.5	9.0	1.0	6.0	1.0	7.0	0.5	7.0	1.0
Overall search Time(V9)	7.0	2.0	7.0	1.0	7.0	1.1	7.0	0.0	7.0	0.1

Note:1. Me indicates Median; 2. Q.D. indicates quartile deviation.

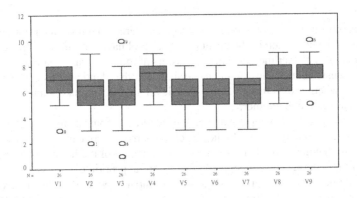

Fig. 1. The boxplots of items response

4   Conclusion

This paper applies the combinatorial topology-based semantic clustering method to real-time cluster search results from PubMed. Although the real-time clustering is not easy to be objectively evaluated, we attempt to built several measures as a tool of overall appraisal. The results demonstrate that building meaningful clustering search results from PubMed is useful for health professionals to save their time. Besides, most experts are in agreement on that the documents assigned to the identical cluster are similar and the concepts of clusters are appropriate.

Acknowledgments. We wish to thank the anonymous referees for their valuable comments which helped us to improve the paper.

References

1. Perez-Iratxeta, C., Perez, A.J., Bork, P., Andrade, M.A.: Update on XplorMed: a web server for exploring scientific literature. Nucleic Acids Research 31, 3866–3868 (2003)
2. Doms, A., Schroeder, M.: GoPubMed: exploring PubMed with the Gene Ontology. Nucleic Acids Research 33, W783–W786 (2005)
3. Chiang, I.-J.: Discover the Semantic Topology in High- Dimensional Data. Expert Systems with Applications 33, 256–262 (2007)
4. Zamir, O., Etzioni, O.: Web document clustering: a feasibility demonstration. In: Proc. 19th International ACM SIGIR Conference on Research and Development in Information Retrieval (SIGIR 98), pp. 46–54 (1998)
5. Ertoz, L., Steinbach, M., Kumar, V.: Finding Clusters of Different Sizes, Shapes, and Densities in Noisy, High Dimensional Data. In: Proc. 2003 SIAM International Conference on Data Mining (SDM'03), San Francisco. CA, pp. 59–70 (2003)
6. Lin, T.Y., Chiang, I.-J.: A simplicial complex, a hypergraph, structure in the latent semantic space of document clustering. International Journal of Approximate Reasoning 40, 55–80 (2005)
7. Boley, D., Gini, M., Gross, R., Han, E.-H., Hastings, K., Karypis, G., et al.: Document categorization and query generation on the world wide web using webace. Artificial Intelligence Review 13(5–6), 365–391 (1999)
8. Dias Guilloré, S., Lopes, J.G.P.: Extracting Textual Associations from Part-Of-Speech Tagged Corpora. In: European Association for Machine Translation Workshop on Harvesting Existing Resources, Ljubljana, Slovenia (2000)
9. Chuang, S.-L., Chien, L.-F.: A practical web-based approach to generating topic hierarchy for text segments. In: Proc. ACM Conference on Information and Knowledge Management (CIKM'04), pp. 127–136 (2004)
10. Zheng, H.-J., He, Q.-C., Chen, Z., Ma, W.-Y., Ma, J.: Learning to cluster Web search results. In: Proc. SIGIR 2004, pp. 210–217 (2004)
11. Cronbach, L.J.: Coefficient Alpha and the Internal Structure of tests. Psychometrika 16(3), 297–334 (1951)

Author Index

Arimura, Hiroki   42, 219
Azevedo, Paulo J.   56

Bannai, Hideo   161
Blum, Avrim   39
Boley, Mario   68
Buche, Patrice   253

Cai, Zhihua   104
Chiang, I-Jen   291

Dartnell, Christopher   91
Deslandres, Véronique   78
Dibie-Barthélemy, Juliette   253
Dieterich, Thomas G.   9
Doi, Koichiro   281
Dussauchoy, Alain   78

Elghazel, Haytham   78

Fujishima, Satoshi   243

Ganascia, Jean-Gabriel   248
Gretton, Arthur   40

Ha, Inay   116
Habib, Bassel   248
Haemmerlé, Ollivier   253
Hagège, Hélène   91
Hatano, Kohei   127, 161
Hignette, Gaëlle   253
Hollmén, Jaakko   259

Inoue, Katsumi   173
Ishino, Akira   127

Jiang, Liangxiao   104
Jo, Geun-Sik   116
Jorge, Alípio M.   56
Jung, Jin-Guk   116

Kheddouci, Hamamache   78
Kim, Heung-Nam   116
Kitsuregawa, Masaru   1

Kobayashi, Hayato   127
Kudo, Mineichi   286

Liu, Ching   291

Madhavan, C.E. Veni   197
Minato, Shin-ichi   139
Mononen, Tommi   151
Mostafa, Javed   185
Myllymäki, Petri   151

Nakamura, Atsuyoshi   286
Narisawa, Kazuyuki   161
Nishida, Kyosuke   264
Nishiyama, Risa   270

Okada, Takashi   243

Pietracaprina, Andrea   275
Pryczek, Michal   209

Raghupathy, Narayanan   197
Ramachandran, Parthasarathy   231
Ray, Oliver   173

Sallantin, Jean   91
Sankoh, Hiroshi   281
Schmidhuber, Jürgen   26
Schölkopf, Bernhard   40
Seki, Kazuhiro   185
Shekar, B.   197
Shia, Ben-Chang   291
Shimozono, Shinichi   42
Shinohara, Ayumi   127
Smola, Alex   40
Song, Le   40
Suresh, V.   197
Szczepaniak, Piotr S.   209

Takahashi, Yoshimasa   243
Takeda, Masayuki   161
Takeuchi, Hironori   270
Tomczyk, Arkadiusz   209
Tosaka, Hisashi   286
Tsai, Hsiang-Chun   291

Uno, Takeaki   42, 219

Vandin, Fabio   275
Veilumuthu, Ashok   231

Wang, Dianhong   104
Watanabe, Hideo   270

Yamamoto, Akihiro   281
Yamauchi, Koichiro   264
Yang, Wen-Wen   291
Yeh, Ruey-Ling   291

Zhang, Harry   104

Lecture Notes in Artificial Intelligence (LNAI)

Vol. 4755: V. Corruble, M. Takeda, E. Suzuki (Eds.), Discovery Science. XI, 298 pages. 2007.

Vol. 4754: M. Hutter, R.A. Servedio, E. Takimoto (Eds.), Algorithmic Learning Theory. XI, 403 pages. 2007.

Vol. 4737: B. Berendt, A. Hotho, D. Mladenic, G. Semeraro (Eds.), From Web to Social Web: Discovering and Deploying User and Content Profiles. XI, 161 pages. 2007.

Vol. 4733: R. Basili, M.T. Pazienza (Eds.), AI*IA 2007: Artificial Intelligence and Human-Oriented Computing. XVII, 858 pages. 2007.

Vol. 4724: K. Mellouli (Ed.), Symbolic and Quantitative Approaches to Reasoning with Uncertainty. XV, 914 pages. 2007.

Vol. 4722: C. Pelachaud, J.-C. Martin, E. André, G. Chollet, K. Karpouzis, D. Pelé (Eds.), Intelligent Virtual Agents. XV, 425 pages. 2007.

Vol. 4720: B. Konev, F. Wolter (Eds.), Frontiers of Combining Systems. X, 283 pages. 2007.

Vol. 4702: J.N. Kok, J. Koronacki, R. Lopez de Mantaras, S. Matwin, D. Mladenič, A. Skowron (Eds.), Knowledge Discovery in Databases: PKDD 2007. XXIV, 640 pages. 2007.

Vol. 4701: J.N. Kok, J. Koronacki, R.L.d. Mantaras, S. Matwin, D. Mladenič, A. Skowron (Eds.), Machine Learning: ECML 2007. XXII, 809 pages. 2007.

Vol. 4696: H.-D. Burkhard, G. Lindemann, R. Verbrugge, L.Z. Varga (Eds.), Multi-Agent Systems and Applications V. XIII, 350 pages. 2007.

Vol. 4694: B. Apolloni, R.J. Howlett, L. Jain (Eds.), Knowledge-Based Intelligent Information and Engineering Systems, Part III. XXIX, 1126 pages. 2007.

Vol. 4693: B. Apolloni, R.J. Howlett, L. Jain (Eds.), Knowledge-Based Intelligent Information and Engineering Systems, Part II. XXXII, 1380 pages. 2007.

Vol. 4692: B. Apolloni, R.J. Howlett, L. Jain (Eds.), Knowledge-Based Intelligent Information and Engineering Systems, Part I. LV, 882 pages. 2007.

Vol. 4687: P. Petta, J.P. Müller, M. Klusch, M. Georgeff (Eds.), Multiagent System Technologies. X, 207 pages. 2007.

Vol. 4682: D.-S. Huang, L. Heutte, M. Loog (Eds.), Advanced Intelligent Computing Theories and Applications. XXVII, 1373 pages. 2007.

Vol. 4676: M. Klusch, K.V. Hindriks, M.P. Papazoglou, L. Sterling (Eds.), Cooperative Information Agents XI. XI, 361 pages. 2007.

Vol. 4667: J. Hertzberg, M. Beetz, R. Englert (Eds.), KI 2007: Advances in Artificial Intelligence. IX, 516 pages. 2007.

Vol. 4660: S. Džeroski, J. Todorovski (Eds.), Computational Discovery of Scientific Knowledge. X, 327 pages. 2007.

Vol. 4659: V. Mařík, V. Vyatkin, A.W. Colombo (Eds.), Holonic and Multi-Agent Systems for Manufacturing. VIII, 456 pages. 2007.

Vol. 4651: F. Azevedo, P. Barahona, F. Fages, F. Rossi (Eds.), Recent Advances in Constraints. VIII, 185 pages. 2007.

Vol. 4648: F. Almeida e Costa, L.M. Rocha, E. Costa, I. Harvey, A. Coutinho (Eds.), Advances in Artificial Life. XVIII, 1215 pages. 2007.

Vol. 4635: B. Kokinov, D.C. Richardson, T.R. Roth-Berghofer, L. Vieu (Eds.), Modeling and Using Context. XIV, 574 pages. 2007.

Vol. 4632: R. Alhajj, H. Gao, X. Li, J. Li, O.R. Zaïane (Eds.), Advanced Data Mining and Applications. XV, 634 pages. 2007.

Vol. 4629: V. Matoušek, P. Mautner (Eds.), Text, Speech and Dialogue. XVII, 663 pages. 2007.

Vol. 4626: R.O. Weber, M.M. Richter (Eds.), Case-Based Reasoning Research and Development. XIII, 534 pages. 2007.

Vol. 4617: V. Torra, Y. Narukawa, Y. Yoshida (Eds.), Modeling Decisions for Artificial Intelligence. XII, 502 pages. 2007.

Vol. 4612: I. Miguel, W. Ruml (Eds.), Abstraction, Reformulation, and Approximation. XI, 418 pages. 2007.

Vol. 4604: U. Priss, S. Polovina, R. Hill (Eds.), Conceptual Structures: Knowledge Architectures for Smart Applications. XII, 514 pages. 2007.

Vol. 4603: F. Pfenning (Ed.), Automated Deduction – CADE-21. XII, 522 pages. 2007.

Vol. 4597: P. Perner (Ed.), Advances in Data Mining. XI, 353 pages. 2007.

Vol. 4594: R. Bellazzi, A. Abu-Hanna, J. Hunter (Eds.), Artificial Intelligence in Medicine. XVI, 509 pages. 2007.

Vol. 4585: M. Kryszkiewicz, J.F. Peters, H. Rybinski, A. Skowron (Eds.), Rough Sets and Intelligent Systems Paradigms. XIX, 836 pages. 2007.

Vol. 4578: F. Masulli, S. Mitra, G. Pasi (Eds.), Applications of Fuzzy Sets Theory. XVIII, 693 pages. 2007.

Vol. 4573: M. Kauers, M. Kerber, R. Miner, W. Windsteiger (Eds.), Towards Mechanized Mathematical Assistants. XIII, 407 pages. 2007.

Vol. 4571: P. Perner (Ed.), Machine Learning and Data Mining in Pattern Recognition. XIV, 913 pages. 2007.

Vol. 4570: H.G. Okuno, M. Ali (Eds.), New Trends in Applied Artificial Intelligence. XXI, 1194 pages. 2007.

Vol. 4565: D.D. Schmorrow, L.M. Reeves (Eds.), Foundations of Augmented Cognition. XIX, 450 pages. 2007.

Vol. 4562: D. Harris (Ed.), Engineering Psychology and Cognitive Ergonomics. XXIII, 879 pages. 2007.

Vol. 4548: N. Olivetti (Ed.), Automated Reasoning with Analytic Tableaux and Related Methods. X, 245 pages. 2007.

Vol. 4539: N.H. Bshouty, C. Gentile (Eds.), Learning Theory. XII, 634 pages. 2007.

Vol. 4529: P. Melin, O. Castillo, L.T. Aguilar, J. Kacprzyk, W. Pedrycz (Eds.), Foundations of Fuzzy Logic and Soft Computing. XIX, 830 pages. 2007.

Vol. 4520: M.V. Butz, O. Sigaud, G. Pezzulo, G. Baldassarre (Eds.), Anticipatory Behavior in Adaptive Learning Systems. X, 379 pages. 2007.

Vol. 4511: C. Conati, K. McCoy, G. Paliouras (Eds.), User Modeling 2007. XVI, 487 pages. 2007.

Vol. 4509: Z. Kobti, D. Wu (Eds.), Advances in Artificial Intelligence. XII, 552 pages. 2007.

Vol. 4496: N.T. Nguyen, A. Grzech, R.J. Howlett, L.C. Jain (Eds.), Agent and Multi-Agent Systems: Technologies and Applications. XXI, 1046 pages. 2007.

Vol. 4483: C. Baral, G. Brewka, J. Schlipf (Eds.), Logic Programming and Nonmonotonic Reasoning. IX, 327 pages. 2007.

Vol. 4482: A. An, J. Stefanowski, S. Ramanna, C.J. Butz, W. Pedrycz, G. Wang (Eds.), Rough Sets, Fuzzy Sets, Data Mining and Granular Computing. XIV, 585 pages. 2007.

Vol. 4481: J. Yao, P. Lingras, W.-Z. Wu, M. Szczuka, N.J. Cercone, D. Ślęzak (Eds.), Rough Sets and Knowledge Technology. XIV, 576 pages. 2007.

Vol. 4476: V. Gorodetsky, C. Zhang, V.A. Skormin, L. Cao (Eds.), Autonomous Intelligent Systems: Multi-Agents and Data Mining. XIII, 323 pages. 2007.

Vol. 4456: Y. Wang, Y.-m. Cheung, H. Liu (Eds.), Computational Intelligence and Security. XXIII, 1118 pages. 2007.

Vol. 4455: S. Muggleton, R. Otero, A. Tamaddoni-Nezhad (Eds.), Inductive Logic Programming. XII, 456 pages. 2007.

Vol. 4452: M. Fasli, O. Shehory (Eds.), Agent-Mediated Electronic Commerce. VIII, 249 pages. 2007.

Vol. 4451: T.S. Huang, A. Nijholt, M. Pantic, A. Pentland (Eds.), Artifical Intelligence for Human Computing. XVI, 359 pages. 2007.

Vol. 4441: C. Müller (Ed.), Speaker Classification. X, 309 pages. 2007.

Vol. 4438: L. Maicher, A. Sigel, L.M. Garshol (Eds.), Leveraging the Semantics of Topic Maps. X, 257 pages. 2007.

Vol. 4434: G. Lakemeyer, E. Sklar, D.G. Sorrenti, T. Takahashi (Eds.), RoboCup 2006: Robot Soccer World Cup X. XIII, 566 pages. 2007.

Vol. 4429: R. Lu, J.H. Siekmann, C. Ullrich (Eds.), Cognitive Systems. X, 161 pages. 2007.

Vol. 4428: S. Edelkamp, A. Lomuscio (Eds.), Model Checking and Artificial Intelligence. IX, 185 pages. 2007.

Vol. 4426: Z.-H. Zhou, H. Li, Q. Yang (Eds.), Advances in Knowledge Discovery and Data Mining. XXV, 1161 pages. 2007.

Vol. 4411: R.H. Bordini, M. Dastani, J. Dix, A.E.F. Seghrouchni (Eds.), Programming Multi-Agent Systems. XIV, 249 pages. 2007.

Vol. 4410: A. Branco (Ed.), Anaphora: Analysis, Algorithms and Applications. X, 191 pages. 2007.

Vol. 4399: T. Kovacs, X. Llorà, K. Takadama, P.L. Lanzi, W. Stolzmann, S.W. Wilson (Eds.), Learning Classifier Systems. XII, 345 pages. 2007.

Vol. 4390: S.O. Kuznetsov, S. Schmidt (Eds.), Formal Concept Analysis. X, 329 pages. 2007.

Vol. 4389: D. Weyns, H. Van Dyke Parunak, F. Michel (Eds.), Environments for Multi-Agent Systems III. X, 273 pages. 2007.

Vol. 4386: P. Noriega, J. Vázquez-Salceda, G. Boella, O. Boissier, V. Dignum, N. Fornara, E. Matson (Eds.), Coordination, Organizations, Institutions, and Norms in Agent Systems II. XI, 373 pages. 2007.

Vol. 4384: T. Washio, K. Satoh, H. Takeda, A. Inokuchi (Eds.), New Frontiers in Artificial Intelligence. IX, 401 pages. 2007.

Vol. 4371: K. Inoue, K. Satoh, F. Toni (Eds.), Computational Logic in Multi-Agent Systems. X, 315 pages. 2007.

Vol. 4369: M. Umeda, A. Wolf, O. Bartenstein, U. Geske, D. Seipel, O. Takata (Eds.), Declarative Programming for Knowledge Management. X, 229 pages. 2006.

Vol. 4363: B.D. ten Cate, H.W. Zeevat (Eds.), Logic, Language, and Computation. XII, 281 pages. 2007.

Vol. 4343: C. Müller (Ed.), Speaker Classification I. X, 355 pages. 2007.

Vol. 4342: H. de Swart, E. Orłowska, G. Schmidt, M. Roubens (Eds.), Theory and Applications of Relational Structures as Knowledge Instruments II. X, 373 pages. 2006.

Vol. 4335: S.A. Brueckner, S. Hassas, M. Jelasity, D. Yamins (Eds.), Engineering Self-Organising Systems. XII, 212 pages. 2007.

Vol. 4334: B. Beckert, R. Hähnle, P.H. Schmitt (Eds.), Verification of Object-Oriented Software. XXIX, 658 pages. 2007.

Vol. 4333: U. Reimer, D. Karagiannis (Eds.), Practical Aspects of Knowledge Management. XII, 338 pages. 2006.

Vol. 4327: M. Baldoni, U. Endriss (Eds.), Declarative Agent Languages and Technologies IV. VIII, 257 pages. 2006.

Vol. 4314: C. Freksa, M. Kohlhase, K. Schill (Eds.), KI 2006: Advances in Artificial Intelligence. XII, 458 pages. 2007.

Vol. 4304: A. Sattar, B.-h. Kang (Eds.), AI 2006: Advances in Artificial Intelligence. XXVII, 1303 pages. 2006.